Water Markets for the 21st Century

GLOBAL ISSUES IN WATER POLICY

VOLUME 11

For further volumes:
http://www.springer.com/series/8877

K. William Easter • Qiuqiong Huang
Editors

Water Markets for the 21st Century

What Have We Learned?

Springer

Editors
K. William Easter
Department of Applied Economics
University of Minnesota
St. Paul, MN, USA

Qiuqiong Huang
Department of Agricultural Economics
and Agribusiness
University of Arkansas
Fayetteville, AR, USA

ISSN 2211-0631 ISSN 2211-0658 (electronic)
ISBN 978-94-017-9080-2 ISBN 978-94-017-9081-9 (eBook)
DOI 10.1007/978-94-017-9081-9
Springer Dordrecht Heidelberg New York London

Library of Congress Control Number: 2014945761

Printed on acid-free paper

Springer is part of Springer Science+Business Media (www.springer.com)

Preface

A growing demand for water combined with the impact of climate change on the timing and quantity of water availability have dramatically changed how policy makers view water resources. Pressure is building on water managers to do a better job of conserving and allocating water resources. Adding to this is the demand for more water to meet environmental needs, which makes it even more imperative that we substantially improve our water use. One mechanism for doing this is the increased use of water markets.

One of the objectives of this book is to provide the reader with a clear picture of what we have learned about water markets. Since we completed our first water markets book in the 1990s (Easter, Rosegrant, and Dinar, 1998, *Markets for Water: Potential and Performance*) a lot has happened around the world which has improved our understanding of the possible problems and opportunities for the future use of water markets. The book is a combination of results from new research and surveys of water markets in key parts of the world. Water markets in countries across five continents are examined. Australia has discovered the problem of sleeper water rights and has begun to use the water market to buy water for environmental flows. We have also become aware of local water markets that have been operating in Oman for centuries. These and many other experiences need to be spelled out in one place to help guide our future use and development of water markets.

The second objective of the book is to assess where we are with water markets and what suggestions we can make for their use in the twenty-first century to help us adapt to the impact of climate change and population growth on the availability of water resources. It is clear that with climate change water will need to be reallocated and used more effectively. We need to use less water and stop polluting it. The big question for us is how water markets can better help us address these tasks. Changes in water institutions will be a key part of the process as will the allocation of water rights or water use rights. Some of the biggest new insights come from the experience Australia has had with water markets over the last two decades in which they made major changes in their water institutions. Other countries such as Chile, Spain, and the USA have made modest changes, but more are needed. Yet countries such as South Africa may eliminate the use of water markets because politicians have raised concerns regarding the equity of the current ownership of water rights. Many seem to forget that water markets are just a tool to help manage and allocate water. The actual ownership and allocation of water rights or use rights is a separate issue. Water markets do not determine the initial allocation of rights. Markets can come into play once the water rights or use rights have been established. Since water rights are generally quite valuable their distribution can be highly political. The findings in this book suggest that a holistic approach should be taken to consider the physical environmental as well as institutions and politics in developing effective water markets.

St Paul, MN, USA K. William Easter
Fayetteville, AR, USA Qiuqiong Huang

Acknowledgements

The editors would like to thank all the authors who contributed chapters for this book. It is their knowledge of water markets that make this book a must to read for anyone interested in water allocation and management issues. The editors also want to thank Diane McAfee who has been a great help in organizing and processing the book for delivery to the publishers and Susan Pohlod for her advice and help especially with the figures. Finally, we want to thank George Norton, Jay Coggins, Ariel Dinar, Ruth Meinzen-Dick, K. Palanisami, and Jim Nickum who all reviewed chapters and made suggestions for changes and additions that strengthened each chapter they reviewed.

Contents

Contributors

Abdullah Al-Ghafri Aflaj Research Unit, The University of Nizwa, Nizwa, Sultanate of Oman

Henning Bjornlund School of Commerce|UniSA Business School, Center for Regulation and Market Analysis, University of South Australia, Adelaide, SA, Australia

Economics Department, University of Lethbridge, Lethbridge, AB, Canada

Nicholas Brozović Department of Agricultural and Consumer Economics, University of Illinois at Urbana-Champaign, Urbana, IL, USA

Javier Calatrava Escuela Técnica Superior de Ingeniería Agronómica, Universidad Politécnica de Cartagena, Cartagena, Spain

Bonnie Colby Department of Agricultural and Resource Economics, University of Arizona Tucson, Tucson, Arizona, USA

Guillermo Donoso Department of Agricultural Economics, Pontifica Universidad Católica de Chile, Santiago 22, Chile

K. William Easter Department of Applied Economics, University of Minnesota, St. Paul, MN, USA

Dustin Garrick Department of Political Science, McMaster University, Hamilton, ON, Canada

Walter G. Booth School of Engineering Practice, McMaster University, Hamilton, ON, Canada

Alberto Garrido Department of Agricultural Economics and Social Sciences, Research Centre for the Management of Agricultural and Environmental Risks, Centro de Estudios e Investigación para la Gestión de Riesgos Agrarios y Medioambientales, CEIGRAM, Universidad Politécnica de Madrid, Madrid, Spain

Christopher Goemans Department of Agricultural and Resource Economics, Colorado State University, Fort Collins, CO, USA

Robert Hearne Department of Agribusiness and Applied Economics, North Dakota State University, Fargo, ND, USA

Richard E. Howitt Department of Agricultural and Resource Economics, University of California at Davis, Davis, CA, USA

Jikun Huang Center for Chinese Agricultural Policy, Chinese Academy of Sciences, Beijjing, China

Qiuqiong Huang Department of Agricultural Economics and Agribusiness, University of Arkansas, Fayetteville, AR, USA

Lana Jones Economic Consultant, Corvallis, OR, USA

Adam Loch School of Commerce|UniSA Business School, Center for Regulation and Market Analysis, University of South Australia, Adelaide, SA, Australia

Laura McCann Department of Agricultural and Applied Economics, University of Missouri, Columbia, MO, USA

Michael O'Donnell Bureau of Business and Economic Research, MSC06 3510, 1 University of New Mexico, Albuquerque, NM, USA

Dennis Powers Aflaj Research Unit, The University of Nizwa, Nizwa, Sultanate of Oman

James Pritchett Department of Agricultural and Resource Economics, Colorado State University, Fort Collins, CO, USA

Dolores Rey CEIGRAM, Universidad Politécnica de Madrid, Madrid, Spain

Claudia Ringler Environment and Production Technology Division, International Food Policy Research Institute, Washington, DC, USA

Mark W. Rosegrant Environment and Production Technology Division, International Food Policy Research Institute, Washington, DC, USA

Scott Rozelle Freeman Spogli Institutional Studies, Stanford University, Stanford, CA, USA

R. Maria Saleth Madras Institute of Development Studies, Chennai, India

John Tisdell Tasmanian School of Business and Economics, University of Tasmania, Hobart, Tasmania, Australia

Jinxia Wang Center for Chinese Agricultural Policy, Chinese Academy of Sciences, Beijjing, China

Sarah Wheeler School of Commerce|UniSA Business School, Center for Regulation and Market Analysis, University of South Australia, Adelaide, SA, Australia

Wei Xu Department of Geography, University of Lethbridge, Alberta Water and Environmental Science Building, Lethbridge, AB, Canada

Mike Young Water and Environmental Policy, Global Food Studies Program, The University of Adelaide, Adelaide, SA, Australia

Gough Whitlam and Malcolm Fraser Chair in Australian Studies, Harvard University, Cambridge, MA, USA

Richael Young Department of Agricultural and Consumer Economics, University of Illinois at Urbana-Champaign, Urbana, IL, USA

Slim Zekri Department of Natural Resource Economics, College of Agricultural & Marine Sciences, Sultan Qaboos University, Muscat, Oman

Lijuang Zhang Center for Chinese Agricultural Policy, Chinese Academy of Sciences, Beijjing, China

Tingju Zhu Environment and Production Technology Division, International Food Policy Research Institute, Washington, DC, USA

Alec Zuo School of Commerce|UniSA Business School, Center for Regulation and Market Analysis, University of South Australia, Adelaide, SA, Australia

List of Figures

List of Tables

Chapter 1
Water Markets: How Do We Expand Their Use?

K. William Easter and Qiuqiong Huang

Abstract This chapter outlines how the book is organized and how it examines water markets in many different parts of the world. Water markets can be a very effective tool for many countries as they address their growing water scarcity problems. These markets can help reallocate water to higher valued users while improving both water use and allocation efficiency. With tools such as option contracts, water markets can also help in the environment of climate change and the accompanying increase in hydrologic variability. The constraints to water markets are also discussed along with suggestions regarding how such constraints can be eliminated. Effective institutional arrangements including a sound legal system with water rights separate from land are critical for reducing constraints particularly the potential of negative third party impacts. Markets in eight different countries and three different U.S. states are considered ranging from the village level water markets in Oman to basin wide formal water markets throughout the Murray-Darling River basin in Australia. Much has been learned about water markets in the past quarter century particularly that stakeholder participation is critical for the development and effectiveness of water markets with robust institutional arrangements supporting them.

Keywords Future demands • Water rights • Water markets • Constraints • Designing institutions

---

K.W. Easter (✉)
Department of Applied Economics, University of Minnesota, 1994 Buford Avenue, 231 Ruttan Hall, St. Paul, MN 55108-6040, USA
e-mail: kweaster@umn.edu

Q. Huang
Department of Agricultural Economics and Agribusiness, University of Arkansas, 221 Agriculture Building, Fayetteville, AR 72701, USA
e-mail: qqhuang@uark.edu

K.W. Easter and Q. Huang (eds.), *Water Markets for the 21st Century: What Have We Learned?*, Global Issues in Water Policy 11, DOI 10.1007/978-94-017-9081-9_1,

1.1 Introduction

This book was written to give individuals working on water resource problems a good understanding of what we have learned about water markets over the past two decades. During this period there has been growing demands for water and mounting concerns about our ability to meet these demands. As a result of these concerns more questions have been raised regarding how water might be reallocated to meet the growing demands. It is now more widely recognized that our current water allocations are far from optimum and users, in many cases are not given incentives to take into account the scarcity value of water and, therefore, use it accordingly. In general, water is and has been too cheap and used too lavishly. For example, groundwater is used to irrigate rice in Northwestern India where they get free electricity to pump the water from declining aquifers.

Because of this growing demand for water different allocation mechanisms such as water markets are being considered more seriously. For water markets to be used more widely to help reallocate water resources we need to learn what works and under what conditions they work. To do this we need to look at what Chile, the U.S. West, Australia, and other countries have learned as they have used water markets to reallocate water. We also need to consider the experience of other countries such as Spain and Oman who have used water markets at the community level for centuries (Maass and Anderson 1978).

Given the 2012 drought in the agricultural bread basket of the U.S, this may be an opportune time to see how water markets might help. It appears such droughts will be more frequent in the future and mechanisms such as water markets need to be adapted to help countries deal with such droughts (see Chaps. 3, 4 and 5). Can market mechanisms be used to encourage water users to store and save groundwater so that the groundwater can be tapped during drought periods as happened in 2012 particularly in Nebraska (Starita et al. 2012)? If so, what form might these markets take?

1.2 Different Market Types

Water markets can take a number of different forms. In the 1998 water markets book we made the distinction between formal and informal markets (Easter et al. 1998). Formal markets allow water trading over fairly large areas and are governed by a set of state and/or federal laws and rules. Informal markets are developed locally to allow the trade of water among neighboring farmers and are operated based on rules informally developed at the community level. In the formal markets you can have trading of water rights or just a trading of the right to use a set amount of water during a certain time period such as the growing season or just for next week or next month. In the informal markets the trade is primarily for water use in the next day, week or irrigation turn.

In the case of water right trades between the agricultural and urban sectors, the trades may be fairly abrupt or gradual. For example, a city may buy up water rights from farmers in the surrounding area but only use some of the water in the first several years. As the city grows and needs more water it will take more of its water allocation. This gradual increase in city water use allows the surrounding area more time to adjust to the loss of water and agriculture. A more abrupt transfer of all the water will likely have a more negative impact on the area selling the water. How much an area is impacted by either the sale of water or the purchase of water will depend on what percentage of the irrigation water is sold. The sale of a small percentage of the irrigation water supply, 5–10 %, will have a small impact on the exporting area since efficiency improvements can make up much of the difference in most cases. This is clearly a win-win exchange when efficiency improvements equal the amount traded. For large amounts of water the negative impact on the exporting region may be significant. A lot depends on labor mobility and the diversity of the exporting economy (Bourgeon et al. 2008).

Finally, the use of option markets offers an effective way for high risk water users to protect against water shortages during periods of drought (see Chaps. 4 and 5). For example, urban areas need to have a high level of assurance that they can deliver water to urban consumers. Consequently, they are willing to buy options to use farmers' water during periods of water shortages and drought. In this way farmers can use the water during normal periods and then let cities use the water when water is in short supply. The cities may hold an option contract that allows them to draw a certain quantity of water for three to five times in a 10 year period. The cities would buy the contract from the farmers, agreeing to pay farmers a set price for the water when the cities use it. Farmers with perennial crops may also enter into the options market to protect their trees and vines. Another alternative would be for cities or farmers to buy more senior water rights to assure their water supply during periods of drought.[1]

1.3 Impact of Water Markets

The key benefits from using water markets are the potential gains in both spatial and temporal allocational efficiencies and the significant reductions in rents earned by those who administer the water allocations. In most cases employees of the ministry or department of irrigation or water are the ones who make the major water allocation decisions and in the process extract economic rents from farmers. If the potential savings from reducing these rents are large, government officials have a strong incentive to resist the establishment of water markets. Furthermore, the actual size of these rents is difficult to determine because government officials do not want

[1]This works if water rights are based on appropriative rights with seniority established based on the time the water was first appropriated.

them revealed. In contrast, the gains from improving water allocation are easier to estimate. Improving the water allocation generally results in some fairly large gains from trade if the trade is between agriculture and industry. The gains will be much more modest if the trade is between farmers in the same area growing the same crops.

At a smaller scale, issues about rent seeking and high transaction cost can be minimized by using informal water markets. Informal markets among farmers in the same village or groups of villages can be used to allocate water without large investments in management or infrastructure. These markets can make groundwater available to small farmers who cannot afford to install their own wells. This increases their crop production and income as well as providing added income for well owners. These markets are for groundwater and the water is often sold based on the amount of time required to pump it. Usually the transaction costs are low because the enforcement of water rights is done informally at the local level. Yet these markets are very limited in scope and don't involve the exchange of water over a very large area or among different sectors. Still they keep costs low by not requiring extensive management and infrastructure. The major problem that may arise is an over drafting of the groundwater in areas with limited groundwater recharge and cheap electricity for pumping (see Chap. 13).

1.4 The Nature of Constraints

Given the growing scarcity of water why haven't markets expanded more rapidly in many arid or semi arid parts of the world? Part of the problem is the resource itself. It is costly to transfer water over long distances and it is not easy to quantify. In many cases large upfront costs for infrastructure, including canals and control structures, are required for an active formal water market. One area where infrastructure is not a problem is California with its extensive canal system (Chap. 5). Also, it is difficult to define water rights and determine what part can be traded and where and for what uses it can be traded (see Chap. 9). Does the water traded have to be used in the same sector and is the trade limited to consumptive uses? The issue is how you protect downstream users who may depend on upstream return flows that can be lost through upstream water trades. Another problem is that historically water has been quite cheap and thought to be abundant with no need for water markets to help reallocate water to its higher valued uses. However, in many areas water is no longer or never was abundant but water is still cheap and used with little regard for its economic value. In most cases there is no price for the water resource itself. Users only pay for part of the cost of supplying the water or gaining access to it. For example, in shallow groundwater aquifers it is inexpensive to install a well and the electricity for pumping may be highly subsidized or free of charge (India for example). This has resulted in the over use of the groundwater in areas such as the Punjab area of India where the groundwater has declined from 18 to 27 m in the past decade (Kaur et al. 2012).

Even when water is recognized as a scarce resource, people may resist the use of water markets for fear that they may not have access to water for their basic household needs. This is a particular concern for low income families living in developing countries that face water scarcity and have poor legal systems. Yet in some very dry areas in Oman, villages have been using community based water markets to allocate water among farmers for centuries (see Chap. 8).

Another drag on the development and expansion of water markets is the transaction costs of establishing and registering water rights. These costs include the cost of enacting laws that allow water users to establish water rights that are enforceable and tradable. These costs can be quite high because water rights must be separated from land rights and conditions set for dealing with potential third party impacts (return flows). If water rights can be designed to facilitate trading then the transaction costs for the actual exchanges can be minimized. In practice what tends to happen is that other goals besides the efficient allocation of water become important in determining under what conditions water can be traded. For example, in South Africa water trading has been seriously constrained by a requirement that the trade must benefit "small" farmers. In California the increasing requirements to allocate more water for environmental uses has had a dampening effect on water exchanges. Both are noble goals, but the way they are pursued may mean that water markets are ineffective or quite limited. The question is how can we overcome these constraints and establish effective water markets and still contribute to other important goals?

1.5 Meeting New and Changing Demands

As a number of authors have argued we are moving into an era with changing water supplies and rapidly growing demands for water including a whole set of new water uses ranging from extracting natural gas from shale to producing biofuels. So far water markets have not played a large role in providing for these expanding new uses. However, as water supplies become more limited particularly in irrigated areas of the U.S. West, water markets may be a good means for moving water from agriculture to the new energy uses. In areas where groundwater is the main source of water this would be a better strategy than gas companies buying the land to obtain access to the groundwater.

Growth in the urban sector particularly in developing countries is also putting pressure on water supplies. In many cases this involves taking water from agriculture. Again this is an area where water markets can play a larger role so that governments are not encouraged to take the water from farmers with little compensation and minimal concern regarding the most efficient mean of making such water transfers. In China, for example, agriculture is losing its priority in water allocation as more water moves to urban uses. This likely means more water will be taken from farmers with little or no compensation.

Another pressure on water supplies is the growing demands to maintain river levels for their environmental and recreational benefits. This is particularly important in developed countries where the demand for environment services has increased with the growth in income and population. For example, in the western U.S. where most of the senior water rights are owned and used by farmers, water rights are being purchased to main river flows. Also in countries such as India, not too long ago, rivers flowing into the ocean were considered a waste of water. They now realize that drying up rivers has adverse impacts on environmental services that flowing rivers provide, particularly healthy fisheries. Again water markets can play a role in helping meet these growing demands for the environmental services water provides.

1.6 Designing Institution

For water markets to foster the efficient exchange of water and water rights a number of institutions need to be in place. Clearly the basic laws or rules and means for their enforcement are critical for formal water markets to operate effectively. The laws and rules need to require the registration of water rights separate from land and specify under what conditions water and water rights can be traded. For example, can trades be made between users in different watersheds and for different water uses and under what conditions?

Institutional arrangements will be needed to resolve conflicts between water rights holders as well as between water right holders and third parties impacted by water trades. These impacts could be from changes in return flows or the pollution of return flows. Sometimes water user organizations can play this role, in other cases it may require a water court or a state official who has final review authority over water exchanges.[2]

The laws or rules governing the establishment of water rights and their exchange need to specify a number of characteristics. First, if water rights are not in perpetuity, what is their duration and under what conditions can they be renewed? In many cases the duration of the rights to use the water is from 10 to 99 years in length. The duration of the rights can have a significant impact on the value of the rights as can the ease of their renewal. For example, a use right of 10 years with no assurance that it can be renewed will likely limit the investments owners are willing to make for infrastructure to improve the efficiency of water delivery and use.

A second key feature of water rights is the conditions under which they can be traded. Who can they be traded to and do the buyers have to use the water in certain locations and for selected activities. A third closely related facet of water rights is their divisibility. In other words can you sell a 100 acre feet out of a right to use 10,000 acre feet? In addition, water rights need to specify whether a water allocation

[2]The water court in Colorado for example.

under an entitlement must be used. Does it need to be used during specified time periods and is it extinguished once the period has passed or can the allocation be used over a number of years, e.g., stored underground until needed? Also can the right be lost if it is not used annually? Both conditions limit the owners' ability to save water for future use.

Water rights may specify priorities among entitlement holders, e.g., appropriative rights (priority based on the time when right was granted) or priorities may be based on end use (agricultural vs. domestic uses). Rights may also set exchange rates particularly when return flows are important. For example, if water is traded for use in another sector or location, so that there is a reduction in return flows, then trades may be limited to only 50 % of the right with the other 50 % released as return flow.

In some countries such as Chile water rights don't have any within sector priorities. Thus during times of water storage all water allocations are reduced proportionately depending on the severity of the shortage and the sector. This means users with high risk will have a strong incentive to buy more water rights than they need for normal years to protect against droughts. Furthermore certain priority sectors such as urban water use may have smaller proportionate cuts in their allocation.

In some small water systems with community based water markets, these institutions develop and trading rules are established within the community. Water rights may be registered in the community and rules developed over the years regarding how water can be traded. Many times in small community markets, return flows are not a problem since trading is only among water users in the community. However, when disputes do arise they are, generally, handled by the village leader or village council (see Chap. 8).

1.7 Lessons for the Twenty-First Century

One of the key lessons emanating out of the past several decades of water market development is the importance of user support for their use and a well thought out institutional foundation for their operation. In a number of cases there has been a strong anti-privatization sentiment and resistance to water trading and the private ownership of water rights. Several countries have dealt with this problem by only giving rights to water use and for a specific time period such as 20 or 30 years. They also maintain the right to take back the use right under certain conditions. A problem with such use rights is the uncertainty that is created if the rights are for too short a period or are frequently abrogated by the government. Both conditions can cause water users not to invest in infrastructure or perennial crops because of the uncertainty created concerning the future of their use rights.

To help build support for water markets a sense of "fairness" should be established in the allocation of water rights or the use rights. This can be built into the institutional arrangements and conditions for water access. For example, the allocation of rights should take into account past water use in deciding who should receive water rights. The allocation rules should also consider how a particular user will use the water (irrigation vs. mining) and how much should be allocated to an individual user.

Another important lesson is that it is not a simple task to establish formal water markets that effectively allow water trading and allocation among different types of water users. One of the central concerns is the potential for third party effects. Will water trading impact downstream water users or environmental services? Both of these impacts have become central concerns in Australia as they make adjustments to their current water trading and allocation rules. Clearly their experience shows that institutional arrangements and rules must be in place to resolve third party impacts fairly and at a reasonable cost. The same is true for any negative impacts on environmental services. In Chap. 5 Howitt provides some good suggestions for resolving these issues upfront.

Finally, a transparent decision will need to be made concerning whether or not public ownership of water rights or use rights should trump private ownership and under what conditions (see Chaps. 7 and 9). If the public can take the private right then it is important to specify how the owner should be compensated and by what amount? Water rights that are subject to "public taking" without any compensation will have a much lower value and have a negative impact on user investment decisions. A country or community considering water markets may also need to provide water users assurances concerning a basic minimum quantity of fresh water for domestic uses including in some cases water for "small" gardens and domestic livestock (see Chap. 8). This will be particularly important for low income communities with expanding demand for agricultural or industrial water uses.

1.8 Organization of Book

The book includes 15 chapters, plus the introductory and concluding chapters. The chapters cover water markets in eight countries across five continents, including the three countries with the largest area of irrigation land; the U.S., India and China. Chapter 2 sets the framework for looking at transaction costs and the design of policy and institutions so that water markets can help allocate water and improve water use efficiency. Chapter 3 considers climate change and how water markets can be used as an effective adaptive response to growing water uncertainty and scarcity caused by climate changes and population growth. Chapter 4 also considers the impact of climate change and how, with advances in measuring and monitoring technologies, water acquisition programs using options contracts can be used to improve future water supply reliability under climate change. Chapter 5 focuses on water markets in California with an emphasis on option or spot markets and how environmental restrictions have been used to severely constrain their use and effectiveness.

Chapter 6 is the first of nine chapters that look at water markets in Chile, Spain, Oman, Australia, Canada, India, and China. In Chap. 6 the authors look at how water markets developed in Chile and how well they are meeting the country's need to reallocate water and use it efficiently. This is followed by Chap. 7 which describes how water markets have developed mostly in southeastern Spain, but are constrained by twentieth century regulations on water trading. An even older group of water markets in Oman are described in Chap. 8 along with an assessment of their sustainability, efficiency and equity.

Chapter 9 is the first of three chapters which consider water market development in Australia. It focuses on the institutions and organizations that have evolved in Australia to facilitate the operation of water markets and water management. Chapter 10 focuses primarily on how water markets have developed in the Murray-Darling River Basin of Australia. They find that water markets have had a net positive impact on water use in the basin. This is in contrast to Chap. 11 which emphasizes the negative environment impact that water trading has had in the Murray-Darling Basin. However, both chapters find that many of the negative impacts have now been corrected and the future outlook for their water markets is quite positive. Chapter 12 looks at why Alberta, Canada has been reluctant to use water markets even though it is facing increased water scarcity in its irrigated plains of Southern Alberta. Significant concerns about social equity and environmental issues appear to cause stakeholder to resist water trading.

The next set of three chapters focuses on groundwater markets in India, China and the U.S. Chapter 13 starts out with a review of how informal water markets have developed in India and then looks at why they seem to have stagnate. Chapter 14 paints a different picture for groundwater markets in China with the government taking a more positive role in supporting water markets than in the case of India. The authors find significant economic gains from village level water markets in China and India. Chapter 15 finds that water markets are developing for the trading of groundwater pumping rights in the Republican River Basin in Nebraska. This has been stimulated by the restrictions on groundwater use in the basin driven by interstate litigation over the impacts of pumping on surface water levels. Chapter 16 looks at how the institutional setting has impacted the performance of water markets in the western U.S. Finally, Chap. 17 concludes the book with a realistic picture of how water markets could change our future water use.

References

Bourgeon J-M, Easter KW, Smith RBW (2008) Water markets and third-party effects. Am J Agric Econ 90(4):902–917

Easter KW, Rosegrant MW, Dinar A (1998) Markets for water potential and performance. Kluwer Academic Publishers, Boston, 298 p

Kaur B, Singh S, Garg BR, Singh JM, Singh J (2012) Enhancing water productivity through on-farm resource conservation technology in Punjab agriculture. Agric Econ Res Rev 25(1):79–85

Maass A, Anderson RL (1978) . . . And the desert shall rejoice, conflict, growth and justice in arid environments. The MIT Press, Cambridge, MA

Starita J, Dahab M, Lenton R (2012) Drought hits the heartland of USA. Stockh Water Front 4:6–8

Chapter 2
Transaction Costs and Policy Design for Water Markets

Laura McCann and Dustin Garrick

Abstract This chapter synthesizes the growing empirical literature on transaction costs to identify pragmatic design recommendations for water markets and related institutions. The New Institutional Economics literature recognizes that appropriate policy choice and design will be a function of the specific characteristics of the problem. The physical and institutional determinants of both transaction costs and transformation costs should be considered in the design of water markets due to potential interactions between them. Analysts also need to incorporate the extent to which the technologies, institutional environment, governance structures, or policy designs can be changed; some factors can only be adjusted to or "designed around" while others can be designed differently. This framework highlights the importance of property rights, historic water use patterns, and path dependency since transaction costs will be incurred to obtain or retain property rights to water. The physical complexity associated with water resources increases transformation costs as well as transaction costs. Uncertainty and changing societal preferences highlight the importance of flexibility and conflict resolution mechanisms in institutional design. Sequencing of policy changes is also revealed as a key design consideration.

Portions of this chapter have previously appeared in McCann (2013) and are used with permission of Elsevier. Partial funding from USDA-NIFA Multistate Project W2190 is gratefully acknowledged.

L. McCann (✉)
Department of Agricultural and Applied Economics, University of Missouri,
212 Mumford Hall, Columbia, MO 65211-6200, USA
e-mail: McCannL@missouri.edu.

D. Garrick
Department of Political Science, McMaster University, 1280 Main Street West,
Kenneth Taylor Hall Room 534, Hamilton, ON L8S 4 M4, Canada

Walter G. Booth School of Engineering Practice, McMaster University, Hamilton,
ON L8S 4 M4, Canada
e-mail: dgarrick@mcmaster.ca

K.W. Easter and Q. Huang (eds.), *Water Markets for the 21st Century: What Have We Learned?*, Global Issues in Water Policy 11, DOI 10.1007/978-94-017-9081-9_2,

Keywords Institutions • Policy design • Property rights • Transaction costs • Water markets

2.1 Introduction

For some environmental and natural resource issues, it is difficult to model cause and effect, the problem definition may change over time, and there may not be consensus about the policy goal. Examples of these so-called "wicked" problems include climate change, nonpoint source pollution, water resource scarcity, and biodiversity conservation (Batie 2008). Water resource allocation can increasingly be viewed as a wicked problem. Population and economic growth have increased water demand across an expanding number of uses (e.g. agriculture, cities, energy and ecosystems) while supply is becoming more variable and uncertain due to climate change and deteriorating infrastructure.

Water trading and associated institutional reforms are a potentially attractive option to help manage these challenges. The potential benefits of water trading are two-fold. In terms of allocative efficiency, water trading maximizes welfare by allocating water to its highest and best use, often involving a shift from lower valued irrigation of annual crops to perennial crops, or a shift from irrigation to municipal or industrial uses. Water trading also contributes to productive efficiency by incentivizing water saving technologies since any conserved water can then be sold. However, benefits achieved depend on the design of the water markets and associated institutions. The design challenges for water markets relate principally to: (i) establishing diversion limits (the cap) and (ii) creating and/or modifying a tradable water rights system.

Design of policies and economic instruments is a relatively neglected area in applied economics according to King (2012a, b), and he has therefore encouraged applied economists to devote more attention to this task. There is relatively little literature on transaction costs[1] and design of environmental policies, but the role of transaction costs in the design of water markets has received increasing attention (Bennett 2005; Easter et al. 1998; Howitt 1994; McCann and Easter 2004; Garrido 2007; Griffin et al. 2012; Garrick et al. 2013.). Transaction costs should be a key consideration in policy design, especially for wicked problems, which are likely to entail higher transaction costs. Transaction costs ranged from 3 to 70 % of total costs in empirical studies of water markets with costs becoming a higher proportion of total costs for more complex transactions (e.g. for environmental flows) (Garrick et al. 2013).

[1]The definition of transaction costs used in this paper is that of Marshall (2013) "*Transaction costs are the costs of the resources used to (i) define, establish, maintain, use and change institutions and organizations, and (ii) define the problems that these institutions and organizations are intended to solve.*" This definition expands on the definition in McCann et al. (2005) and thus is broad enough to examine the institutional environment (North 1990).

Water trading activity in Western U.S. and Australian water markets has overcome initial impediments through market-enabling policy reforms to water rights, monitoring systems and trading rules. Strategies to reduce transaction costs are an important policy design challenge as water markets emerge, as illustrated by current experiences in China and South Africa where transaction costs remain a barrier to water trading (Grafton et al. 2011). However, transaction costs reduction remains a priority even in maturing water markets. For example, the 2004 National Water Initiative of Australia identified transaction costs reduction as a policy priority to expand water trading by improving information flows, coordinating licensing systems and water rights registries, and removing interstate barriers to trade (see e.g. clauses 25 and 58 of the National Water Initiative). Bjornlund (2004) identified several factors that drive transaction costs and impede water markets, including poorly defined property rights, jurisdictional barriers, and environmental uncertainty. Reducing these costs by improving policy design is especially important given government budget deficits and large potential gains from trade across sectors and users.

The gap between the theory and practice of water markets has been the focus of a well-developed literature. Saliba and Bush (1987) identified the sources of market failure tied to public goods provision, market power, externalities and third party effects. Policy responses to these market failures have yielded insights about design that draw from the institutional economics literature. Institutional and transaction costs analysis of water markets highlight the need to account for the development (and transition costs) of market-enabling policy reform in addition to the transaction costs of reallocation. Garrick and Aylward (2012) further emphasize the need for ongoing institutional change to address unintended consequences of prior reforms and adapt to shifting water use patterns and the associated social and environmental externalities.

The objective of this chapter is to synthesize the growing theoretical and empirical literature on transaction costs in order to identify recommendations for the design of water resource policies. A broad and pragmatic approach is taken by incorporating insights from neoclassical economics, new institutional economics, and classical institutional economics to examine factors affecting both transformation costs and transaction costs of environmental and natural resource policy. Examining both types of costs is important due to potential interactions between them. Minimizing, or at least reducing, the sum of these costs for a given level of water reallocation, both in a static and dynamic sense, is the evaluation criterion used in this chapter. Transformation costs include production and abatement costs. Water conservation – defined as less water per unit of output – is an example of economizing on production costs, while costs of mitigating water pollution and other externalities are an example of abatement costs.

The appropriate choice and design of a policy instrument will depend on the nature of the water allocation problem, both the physical and socio-economic context. Design of feasible policies requires consideration of the extent to which the technologies, institutional environment, governance structures, or policy designs can be changed. Some factors can only be adjusted to or "designed around" while others can be designed differently.

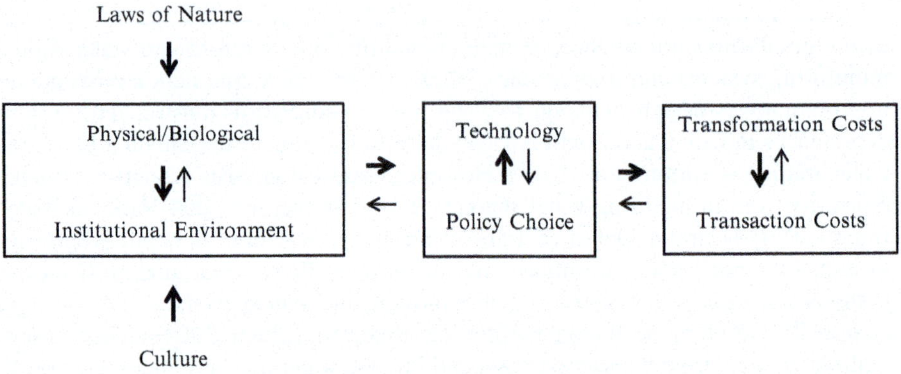

Fig. 2.1 Physical and institutional effects on transaction costs and transformation costs (Note: *Dark arrows* indicate a stronger effect and *arrows in both directions* indicate potential interactions or feedback effects)

The next section briefly summarizes the neoclassical, new institutional and classical institutional perspectives on transaction costs and their relevance for the design of water market institutions. In the third section, physical factors that affect transaction costs and transformation costs are examined, beginning with those that are least amenable to change. The fourth section examines the effect of institutions, beginning with deeper levels such as culture. Figure 2.1 presents the conceptual framework that is developed from the analysis of the physical and institutional issues, and which will be referred to throughout the chapter. It shows that some important factors are not amenable to change on the time scales addressed by policy design, i.e. laws of nature and culture. It also shows that, while discussed in separate sections, there are interactions between physical and institutional factors. The concluding section provides a synthesis of insights that is then used to develop design recommendations.

2.2 Alternative Perspectives on Transaction Costs and Water Resource Policy Design

The institutional economics literature recognizes that there are different levels of institutions and institutional analysis with more superficial levels being nested within deeper levels. Williamson's 2000 paper examines four different levels of institutional analysis: (1) informal institutions, (2) laws and policies (similar to North's institutional environment), (3) governance structures and/or policy instruments, and (4) price effects. The nested institutional framework of Williamson has been used to look at water management institutions and to inform transaction cost measurement (Easter and McCann 2010; McCann and Easter 2004) and will be used in this chapter. This section begins with the relatively superficial neoclassical treatment of transaction costs in environmental and natural resource policy analysis/design and proceeds to deeper levels of institutional analysis.

Transaction costs are increasingly being included in policy design and policy analysis, along with other costs and benefits of the policy (Krutilla and Krause 2011; McCann et al. 2005; Pujol et al. 2006; Stavins 1995). Typologies of transaction costs have been developed to facilitate measurement but they may also enable researchers to think about design more effectively (McCann et al. 2005). Garrick et al. (2013) have adapted generic transaction costs typologies across the major elements of policy design, implementation and adaptation in cap-and-trade water allocation systems.

The New Institutional Economics (NIE) literature consists of several branches including Williamson's transaction cost economics (TCE) (Williamson 1985), Coasian bargaining (Coase 1960), and collective action (Ostrom 1990). Some recent literature (e.g. Bougherara et al. 2009; Boutry 2011; Coggan et al. 2010) uses Williamson's concept of discriminating alignment to provide insights into environmental and natural resource issues. Coase's seminal paper explicitly examines the role of transaction costs in policy choice for addressing environmental impacts. Ostrom's work looks at the nature of the common pool resource and also the social context in order to develop design principles for natural resource management institutions. All three of these literatures recognize that appropriate choices, of governance structures or policies, will be a function of the specific characteristics of the problem.

Deeper levels of institutional analysis, such as studies of the institutional environment, are especially relevant for the design of solutions to wicked problems (Batie 2008). Institutional economics has examined environmental and natural resource issues (e.g. Bromley 1991; Schmid 2004; Vatn 2005). In this literature, property rights, including water rights, are an important concept affecting both the distribution and magnitudes of costs.

This brief overview and comparison of some of the literatures relating to transaction costs and institutions provides some background for readers who may not be familiar with these literatures, but a comprehensive review is beyond the scope of this chapter. The rest of the chapter incorporates useful concepts and insights from all of these literatures rather than being in the tradition of any single one of them. The next section examines a variety of physical factors that affect transaction costs as well as transformation costs.

2.3 Physical Factors Affecting Transaction Costs and Transformation Costs

Fundamental physical, biological, and technical factors will affect transformation costs and transaction costs and thus should affect the choice of policy instrument, and the design of policy. Batie (2008) indicates that wicked problems are typically interdisciplinary in nature while Schmid (2004) highlights the fundamental physical features underlying interdependencies between agents. The geographical area involved, time lags, amount of change needed, heterogeneity, internal versus external effects, measurability, economies of scale, uncertainty, asset specificity, and technology are other physical factors discussed in this section. These factors are presented in three subsections starting with those that are least amenable to change (the single dark arrow in the upper left of Fig. 2.1), followed by those that could change in a generation or two, and ending with those that could change in a few years based on changes in technology and/or institutions (the center box of Fig. 2.1). These cutoffs are somewhat arbitrary, as with the categories in Williamson (2000).

2.3.1 Physical Factors That Are Least Amenable to Change

Fundamental laws of physics and biology are examples of factors that are not amenable to change and which cannot fruitfully be the object of design. These general factors underlie many of the more specific issues addressed in this section.

2.3.1.1 Scale

The geographic scale of intervention that is needed to resolve the problem will affect policy design. Many water quality and quantity issues should be addressed on a watershed scale since both the transfer of the pollutant in space, as well as the quality of the receiving water body, matter for economic efficiency. This involves more coordination and thus higher transaction costs than would be necessary if location did not matter. This is particularly true if a water resource issue crosses political boundaries and since rivers often form boundaries of states or countries, this is quite common (Perry and Easter 2004). For example, the scale of water trading activity will affect policy design. Moving from informal, local spot markets to larger scale activity requires coordination across irrigation districts, state jurisdictions and river basin scales. It also increases the range of water uses involved. Nested governance of water institutions provides a strategy to coordinate local jurisdictions within larger and more diverse regional and national contexts (Challen 2000; Garrick et al. 2011). Grafton et al. (2011) identify the importance of river basin management for integrated assessment of water market performance. They use examples from the Murray-Darling Basin (which involves four states and one territory in Australia) where a comprehensive basin plan was used to establish sustainable diversion limits. Crase et al. (2013) discuss the transaction and transformation costs of policies to provide environmental flows in the basin. South Africa adopted catchment management authorities as part of its 1998 National Water Act, but implementation has lagged. The lack of transnational water trading is a sign of such policy design barriers, although recent reforms in the Colorado River Basin along the US-Mexico border include new provisions for international water banking (Makridis 2012).

2.3.1.2 Time Lags

In general, time lags create challenges for design of water markets. The time lag from groundwater pumping to noticeable impacts on surface water flows, and the lag between improved management (including environmental water recovery) and noticeable impacts will also have effects on transaction costs. In addition, the lag varies from region to region according to hydrogeological conditions (Skurray et al. 2012). In the Deschutes River of Central Oregon, the Oregon Water Resources Department, Deschutes River Conservancy, and irrigation districts created water banks that facilitate voluntary, compensated retirement and conversion of surface water rights into instream flow rights to offset the impact of new groundwater pumping in closed river basins. While generally perceived as successful, there are concerns with administrative capacity and enforcement provisions in part due to the time lags of groundwater pumping impacts and the associated biophysical complexity (Liberherr 2011).

2.3.2 Physical Factors That Are Somewhat Amenable to Change

2.3.2.1 Magnitude of Change

A very important physical factor related to both transformation and transaction costs, is the amount of change needed to address a problem.[2] The familiar upward sloping marginal abatement cost curve indicates that as more clean-up is required, the costs will increase (e.g. Roberts et al. 2012). There are empirical studies that seem to indicate that transaction costs and abatement costs both increase as the level of abatement increases (Garrick and Aylward 2012; Krutilla and Krause 2011; Laurenceau 2012; McCann and Hafdahl 2007; Rorstad et al. 2007). This is consistent with Krutilla and Krause (2011) who argue that the higher the potential losses to firms, the higher the levels of lobbying to prevent a new environmental policy, thus increasing transaction costs. Large changes in water use may necessitate the re-structuring of irrigation districts and associated infrastructure and also affect transaction costs. As a consequence, many irrigation districts have imposed restrictions or taxes (known as 'exit fees') on the volume of water exiting their service areas to ensure the district retains sufficient water and associated fees to operate and maintain the irrigation canals and distribution system (Libecap 2011). This became an issue in the Murray-Darling where recent basin planning efforts

[2]This chapter's framework takes the benefits of reallocation as given although the optimal amount of change will depend on both costs and benefits, both of which may change due to preferences and technical change. Nevertheless, in some cases the amount of change needed to solve a problem is a function of physical and biological factors.

yielded a controversial draft proposal to reduce cumulative basin-wide diversions by 27–37 % of the historic average (MDBA 2010). However, the proposed reductions were concentrated in specific sub-basins, with some regions facing even higher reductions that could bankrupt irrigation districts. The proposed step change reduction in diversion limits triggered political backlash that has substantially raised the transaction and transformation costs of policy reform (Crase et al. 2013).

2.3.2.2 Heterogeneity

An issue that is especially important for environmental and natural resource problems is heterogeneity in all its forms. Heterogeneity is a necessary condition for water markets, as there must be variation in the marginal productivity of water across different uses to allow for gains from trade. For example, Libecap (2005) notes the vast price differentials across different water uses in the Western U.S., citing the example of San Diego paying $225 per acre foot of water for which farmers paid $15.50. However, heterogeneity also poses problems in water markets when property rights are poorly specified and fail to account for different sources of water and their hydrologic interactions (Young and McColl 2009). Heterogeneity therefore becomes a problem for trading when water rights and their reliability are difficult to compare. The water markets of the Western U.S. are hamstrung by the high levels of heterogeneity in water rights. The prior appropriation doctrine, for example, requires all water rights in a region to be defined relative to each other rather than as proportional shares of the consumptive pool as is done in Australia (Ruml 2005).

2.3.2.3 External Effects

Technological externalities are transmitted through some physical medium, and enter the utility function or production function of another agent directly (physically), rather than indirectly through prices. There are often third-party effects resulting from water reallocation. Downstream water rights are often dependent on the (lagged) return flows from upstream water use. Return flows are a classic form of externality in water allocation and water markets. The buying and selling of water affects water use patterns and the ensuing return flows. Regulatory safeguards have limited the third party impacts of water trade to ensure that changes in return flows do not impair downstream users, particularly in the Western U.S. (e.g. Brown 2006); such restrictions can impose a significant barrier to trade by requiring case-by-case assessment of cropping intensity, irrigation efficiency and hydrology.

2.3.2.4 Excludability

A non-excludable good (or bad) is one for which it is not possible to exclude an additional user (or sufferer) at reasonable cost. Excluding people from such a good would involve either high costs to physically exclude potential users (e.g. irrigation headgates) or high costs of monitoring and enforcement (ditch riders and watermasters to measure and enforce water diversions). Both interventions would involve a variety of transaction costs for design and implementation. Technological changes may affect excludability, such as the construction of stream gauges and diversion weirs and associated measurement of water availability and use. Excludability is a challenge for water market design because costs of exclusion vary across multiple sources of water and are often higher for groundwater than surface water. In such cases, capping surface water use may have unintended consequences and increase pressure on groundwater reserves with costly excludability (Young and McColl 2009; Aguilera-Klink and Sanchez-Garcia 2005). Excludability also relates to the public goods characteristics of water. Environmental flows are a public good with indivisible benefits and insufficient incentives for private contributions to their provision and maintenance. As one example, overallocated basins encounter challenges due to the concentrated, private costs on irrigators (who must reduce or sell their water rights, often under duress) and the distributed, public costs and benefits of environmental restoration. The irrigation interests are therefore politically motivated to oppose reallocation, while free riding will make it comparably difficult to organize on behalf of the environment.

2.3.2.5 Measurability/Observability

Measurability and observability are somewhat related issues, and while often grouped with uncertainty, are distinct from that concept. They are also related to excludability in that activities/effects that are measurable would facilitate exclusion. In some sense, observability can be thought of as an extreme form of measurability-able to be measured with the senses. Measurability and observability have effects on transaction costs incurred by public agencies, particularly monitoring and enforcement costs, and thus affect what policies are feasible. For agri-environmental policy, the fact that measuring emissions would entail very high monitoring costs is what distinguishes nonpoint source pollution from point source pollution. Measurability may also affect the potential for new policy instruments. Bougherera et al. (2009) examined whether an environmental issue can be addressed using private property rights as a function of whether the good can be defined, defended and divested, each of which relates to measurability and excludability.

Similarly, water rights must be clearly bounded and measurable (Young and McColl 2009; Libecap 2005) for water markets to function appropriately. Monitoring requirements include water inflows, water diversions and use, and return flow impacts of water trading. Grafton et al. (2011) also identify the need to monitor public interest impacts, such as environmental flows. New regulatory and technological innovation has improved measurement and monitoring to encourage water markets. For example, Australia created its National Water Market System to streamline data collection and compatibility and ensure a consistent national registry of water rights and trading. Australia also committed 60 million AUD over 5 years as part of the 2008 Water for the Future initiative to improve data and information systems underpinning water trading.

2.3.2.6 Economies of Scale/Scope

Economies of scale that are possible are also important for policy design and to some extent this is a function of technology and industry structure as well as the magnitude of the change required. The high fixed costs of water trading have historically biased trading toward large volumes except where water banks (for an irrigation district) or infrastructure (reservoir project) exist. Reservoir projects allow the establishment of a consumptive pool of water rights that can defined as shares and traded more easily than water systems with several diversion rights.

Any policies that can exploit economies of scope would tend to be more efficient. Grolleau and McCann (2012) suggest that paying farmers near Munich to implement organic practices addressed many environmental issues at once and thus reduced transaction costs compared to a policy that addressed fertilizers, pesticides, etc., separately. In general policies should be designed to take advantage of this situation by taking advantage of the multi-purpose design of irrigation infrastructure to optimize irrigation, flood control and hydropower, as well as downstream urban water use.

2.3.2.7 Number of Agents

All else equal, total transaction costs will increase with the number of agents involved. If there are many similar entities, average transaction costs may be decreasing but this effect will be reduced if agents are heterogeneous. This is related to the frequency attribute of transactions (Williamson 1985). Higher frequency results in lower transaction costs per unit due to the ability to standardize procedures, but there may be fixed costs to set up these systems so total costs need to be compared. Along these lines, Coggan et al. (2013) in the case of offset schemes, and Cacho et al. (2013) in the case of greenhouse gas offsets, recommend standardizing policies and procedures to reduce transaction costs. Water banks and water trading registry systems achieve a similar function for large numbers of buyers and sellers. In general, policies need to be designed that involve smaller numbers of agents or that enable economies of scale in development of procedures but still have the flexibility to deal with heterogeneity.

2.3.2.8 Uncertainty

To some extent factors relating to uncertainty have been discussed. Time lags, natural variability in space and time, biological diversity, heterogeneity of agents, measurement difficulties, etc. all increase uncertainty and thus pose problems for the design of environmental and natural resource policy. Risk and uncertainty reduce utility for risk-averse agents, and reduce efficiency in general, but they also increase transaction costs. Due to uncertainty, complete contracts cannot be written, resulting in increased *ex-post* transaction costs (Williamson 1985). McCann (1998) indicates that uncertainty may not be immutable; as the state of knowledge improves some types of uncertainty may be reduced. However, improving information by new research is costly so decision-makers will typically have to act in a state of imperfect knowledge (e.g. Pannell et al. 2013) and changing conditions. The high level of uncertainty regarding ecological benefits of increased water flows in the Murray-Darling Basin, together with a poorly designed consultation process, increased transaction costs and resulted in inefficient outcomes (Crase et al. 2013). As Garrick et al. (2013) and Marshall (2013) point out, even if an ideal policy were implemented at one point in time, changes in technology, preferences, etc. would mean that institutions and policies would need to be revised over time. Moreover, uncertainty about future inflows, theft by opportunistic water users and legal and institutional fragmentation all pose risks for water markets, which require informed trading decisions regarding clearly defined water rights.

2.3.2.9 Asset Specificity

Asset specificity is another interesting insight from Williamson. If a resource (such as a pipeline to a city, or distribution system for an irrigation district) is unique to a specific transacting partner, and cannot be easily redeployed for transactions with other partners, transaction costs are increased since the owner will want to ensure a return on his or her investment. The design and scale of water distribution systems thus affects asset specificity.

The heterogeneity of water rights contributes to asset specificity. Physical interactions between water sources, socioeconomic interdependencies among water users, and legal protections of return flows for downstream users can vary considerably from location to location. In the Western U.S., the prior appropriation system of water rights is based on first-in-time, first-in-right principles which establish a seniority system that is highly location specific. Market power and bilateral monopolies become an issue when the supply and demand are spatially concentrated in a given source or destination area (e.g. Libecap 2008).

2.3.3 Physical Factors That Are Amenable to Change

2.3.3.1 Technical Change

As mentioned in the previous discussion, technology is an important factor affecting transformation and transaction costs. The current state of technology, for production, conservation and monitoring, but also the potential for technological change, needs to be taken into account by policy makers. While costs and benefits are affected by technology, technological change over time is affected by policy due to changes in relative prices, a phenomenon known as induced technical innovation (Hicks 1963; Hayami and Ruttan 1971).

In theory, tradable water rights create incentives for technological innovation by establishing an opportunity cost for water use; water users can sell or lease water saved through on-farm or distribution efficiency savings and therefore may

Table 2.1 Physical factors that affect transaction and transformation costs and thus design

Factors	Attributes	Water market policy design considerations
–Least amenable to change–	Increasing physical scale of problem Longer time lags	River basin planning to establish diversion limits and coordinate water licensing systems across multiple jurisdictions Account for groundwater – surface water interactions in water rights reform
–Somewhat amenable to change–	Magnitude of change needed Increasing heterogeneity Non-excludable External effects Private/public costs aligned Measurable/Observable Potential economies of scale/scope Number of agents involved Higher uncertainty Asset specificity	Establish water rights as shares of consumptive pool (instead of fixed allocation) Robust accounting of return flows in trading rules Low-cost measurement of inflows, diversions and return flows Link trading systems to storage and distribution infrastructure when possible, to decrease heterogeneity, and asset specificity
–Amenable to change–	Technical change	Provide extension services and allow trading to promote water conservation through private irrigation efficiency improvements

invest in efficient technologies that can maintain or maximize productivity with lower water use. Australian water markets suggest that such incentives have led to farmer and irrigation district investment in efficiency savings to reduce water used in distribution systems and on the farm (NWC 2006).

This section has outlined a range of physical factors that can affect the transaction and transformation costs of water resource policies (summarized in Table 2.1). The next section presents institutional factors that affect transaction costs and transformation costs. However, there are interactions between physical and institutional factors that make them difficult to separate (shown in Fig. 2.1 as arrows going in both directions in the boxes).

2.4 Cultural and Institutional Environment Factors Affecting Transaction Costs and Transformation Costs

As highlighted in Sect. 2.2 there are nested levels of institutional analysis (Williamson 2000). These different levels of institutions affect an agent's actions simultaneously, for example price incentives and thus agents' choices at more superficial levels are fundamentally affected by the deeper levels such as property rights. Williamson also indicates that changes in prices may happen immediately while changes in governance may happen at the end of a contract. The deepest level, culture, may take hundreds of years to change and thus affects other institutions, but is not itself amenable to change. Put another way, the transaction costs of effecting change at this level would typically be prohibitively high. The following section starts with deep institutional factors that are least amenable to change, and thus must be designed around, and ends with policy instruments, which are the most likely objects of design.

2.4.1 Institutional Factors That Are Least Amenable to Change

2.4.1.1 Culture

Culture may affect how people are socialized, what choices or actions they do not consider to be in their choice set, their fundamental values, the level of trust within the society, notions of fairness and their interest in the common good, etc. (Schmid 2004; Vatn 2005). Informal institutions such as custom, folklore and religion will also affect the formal institutional environment that each country has (indicated by the single dark arrow in the lower left of Fig. 2.1). In the case of water markets, Bauer (1997) identifies cultural and psychological factors associated with water's symbolic and livelihood significance in irrigation societies in Chile. Farmers have been reluctant to separate water rights from land rights out of concern that water will be traded out of irrigation. The concerns of irrigation communities about the long-term effects of water trade on their economic and cultural viability has impeded the emergence of spot markets in the Western U.S., Canada and Chile, as well as early stages of Australia's water market reforms (as discussed by Howitt (2014) in Chap. 5, this volume, Bjornlund et al. (2014) in Chap. 12, this volume, and Hearne and Donoso (2014) in Chap. 6, this volume). In active markets of Australia, trading activity and government acquisitions of water for the environment have led to cultural arguments grounded in rural values – both economic and cultural.

2.4.1.2 Institutional Environment

The formal institutional environment consists of constitutions, legal systems, laws and policies (Williamson 2000). One component of the institutional environment that is least amenable to change is a constitution which provides the rules for making rules. Constitutional provisions related to water are difficult to change. In Australia, for example, Section 100 of the constitution reserves powers to state governments to regulate water use for irrigation and navigation (see Connell 2007). This has created a fragmented institutional framework that has limited interstate trade until recently (Bjornlund 2004). In addition, given the common law tradition that the U.S. inherited from Britain, previous legal decisions provide precedent for, or preclude, some policy instruments (Richards 2000; Kubasek and Silverman 1997). Common law affects water allocation by imposing norms of 'no harm' and associated regulatory safeguards to support agrarian values and protect the public trust and third parties (Schorr 2005). This highlights the issue of path dependency which is discussed at greater length below.

The system of government is typically not amenable to change. The policy-making in democracies can be quite messy (and thus involve high transaction costs). Friedman, in his book "Hot, Flat and Crowded" (2008) has a chapter entitled "China for a day" in which he suggests that, if China wanted to, it could make the hard environmental policy choices that the U.S. has been unable to make. China's recent interest in water markets and water trading illustrates this point. The government authorized and funded pilot trading activity in the Jiao River Basin, accelerating the reform process that may take decades in other regions (Grafton et al. 2011). This brings up the general concept of the capability of governments which Birner and Wittner (2004) recognize as an important constraint to environmental improvement in developing countries.

The legal system and the courts also affect the transaction costs associated with alternative policy instruments. A legal system that effectively enforces contracts enables contractual relationships that may improve economic efficiency. The governance literature based on Williamson assumes that there is a well-functioning legal system that can enforce the contracts that agents make. This is not the case everywhere (Birner and Wittner 2004). Ostrom (1990) indicates that rapid, low-cost conflict resolution mechanisms are important for successful collective action institutions. Such institutions prove important in many emerging water markets by allowing local and informal conflict management to avoid more costly and cumbersome administrative hearings and court cases. The next section includes several institutional environment issues that are more amenable to change than the legal system.

2.4.2 Institutional Factors That Are Somewhat Amenable to Change

2.4.2.1 Physical Versus Administrative Boundaries

The location of political and administrative boundaries can affect transaction costs, particularly for water management. This demonstrates the importance of considering both physical and institutional factors. Administrative boundaries that do not coincide with environmental areas of interest (e.g. counties, states or countries versus watersheds) make cooperation more difficult and increase transaction costs, particularly if small administrative units are the ones that have authority for environmental and natural resource issues (Perry and Easter 2004). Multiple agencies with responsibilities for solving a problem will also increase coordination costs (Laurenceau 2012). In some cases new umbrella organizations that can facilitate coordination across agencies or political boundaries may be helpful (e.g. the Murray-Darling Basin Authority in Australia, or the Columbia Basin Water Transactions Program in the Northwest USA). Related to this, Batie (2008) indicates that creating boundary organizations that mediate between scientists, resource managers, and stakeholders may be useful for wicked problems. While entailing transaction costs to create and operate, they may ultimately reduce transaction costs, especially in situations of conflict. Schlager and Blomquist (2008) note the difficulties of integrated river basin management because institutions are costly to develop, but that some water allocation challenges can be addressed without river basin level alignment of hydrological and political borders.

2.4.2.2 Lobbying

Krutilla and Krause (2011) argue that the transaction costs at the enactment stage, such as lobbying over a policy at both the legislative and agency (bureaucratic) levels, may be higher than the transaction costs to implement a policy. Typically these costs are ignored by economists and only the transaction costs of implementing and operating a new program are evaluated. Crase et al. (2013) argue that the consultation process in the Murray-Darling enabled lobbying by irrigators which ultimately resulted in poor policy decisions.

2.4.2.3 Property Rights

The general issue of property rights is fundamentally important both for distributional impacts but also for efficiency. Demand for changes to the bundle of property rights, which entails transaction costs, may arise due to changes in technology (Demsetz 1967) or preferences. Garrick et al. (2013) and Crase et al. (2013) provide the example of preferences for environmental flows leading to changes in water rights. Young and McColl (2009) note the importance of separating land and water rights, and also aligning water rights with hydrological interactions of groundwater, surface water and farm dams. A multi-phase legislative process has established a strong tradable permit system in Australia. The process has involved over a century of reform with strong state control, followed by state and national legislation to address environmental needs and coordinate basin-wide trade in the Murray-Darling (Tisdell (2014), Chap. 9, this volume). Bromley (1992) and Stavins (1995) point out that those who do not have the property rights (e.g. the rights to be free from pollution) are those that will incur costs to change the property rights structure. Also, when governments create brand new rights, transaction costs are incurred to obtain those rights (Krutilla and Krause 2011).

Schmid (1995, 2004) argues that because the efficient outcome assumes a particular system of property rights (e.g. you have to pay for mineral resources but not for the right to pollute) one cannot determine an efficient outcome independent of the property rights assignment. However, some parties may be able to make changes at lower cost than others and the transaction costs of regulating some groups may be lower than regulating others, as discussed earlier in the section on physical factors. Assignment of property rights thus may affect the magnitude as well as the distribution of transformation or transaction costs.

The question of where rights and responsibilities should be assigned should also consider which party has better information or is better able to use information. Furthermore, there has been an increasing need for vertical integration within nested water institutions to coordinate at the river basin level (Easter and McCann 2010; Schlager and Blomquist 2008). Nested property rights, or institutional hierarchies (Challen 2000) have developed to manage the externalities of water use and adjust private and irrigation district water rights to match the broader public interests.

2.4.2.4 Market Structure

While also discussed earlier in the section on physical factors, market structure may affect transaction costs in another way; a monopsony structure may facilitate bargaining, while bilateral monopoly can impede it. Schmid (1995) highlights the fact that when property rights were with a large cement plant in Florida, the local citrus growers did not organize to bargain with the cement plant to reduce their dust emissions. When a legal change transferred the property rights to the citrus growers (to not have their harvests diminished), the cement plant then bought property near the plant. Grolleau and McCann (2012) indicate that water utilities in Munich and New York were able to negotiate with farmers more easily than if all the individual water customers had had to do so. Irrigation districts have had a profound impact on market structure by facilitating trades within districts and impeding transfers out of districts (Carey et al. 2002; Libecap 2011).

2.4.2.5 Existing Laws and Policies

Specific legislation, such as the National Water Act in Australia, affects what policy instruments can be used, how they can be implemented, and the transaction costs of making changes. It is recognized that there are interactions between water quantity and environmental quality but the existing legislation made it very difficult to coordinate policy instruments to address both issues until the 2007 Water Act. The 2007 Act and the National Water Initiative that preceded it in 2004 have attempted to consolidate market-enabling reforms and streamline regulatory changes for water rights (Young and McColl 2009). Existing laws may also preclude consideration of some environmental effects. In the western U.S., water laws precluded consideration of instream environmental effects, although this is changing (Easter and McCann 2010; Garrick et al. 2013).

It is thus necessary to recognize that previous policy decisions can either enable or constrain the design of efficient and effective policies. Challen (2000) points out that once water rights are vested at lower levels of decision-making it is difficult (i.e. it would incur high transaction costs) to move private use rights back up the institutional hierarchy. Garrick et al. (2013) and Marshall (2013) also emphasize the importance of path dependency and lock-in in determining the costs of switching to new water resource management regimes. More generally, possible interactions among existing policies, or between existing policies and new policies, need to be considered.

2.4.3 Institutional Factors That Are Amenable to Change

Choice of governance for market transactions and choice of policy instrument for addressing environmental or natural resource problems represent a less deep level of analysis than changing the institutional environment (Williamson 2000). At this latter level the objective is typically to design new institutions (center box of Fig. 2.1) in contrast to "designing around" immutable factors. Typically the literature examines the choice of one "best" policy for the situation, e.g. water pricing or water markets. However, policies have feedback effects so choices at this level should take account of not only static effects, but also dynamic effects, especially the incentive for technological change (indicated by arrows going both ways in the center box of Fig. 2.1).

2.4.3.1 Sequencing and Timing

Sequencing of policy matters. While there is very little literature on sequencing, one would expect to have higher transaction costs to implement a draconian policy, if less restrictive, more popular policies, such as education efforts, have not been tried previously. Ervin and Graffy (1996) suggest picking the low hanging fruit first, i.e. implementing policies that have low total costs (transformation plus transaction costs). Batie (2008) indicates that adaptive management may be helpful; a policy is implemented, the results are observed and then adjustments are made. History does show that expecting companies to make immediate adjustments to regulatory changes often does not work well (e.g. lower volume toilets (Fernandez 2001), Clean Air and Water Acts (Tietenberg 2005)). Therefore having some lead time, or a gradual ratcheting up of policies, may be helpful. On the other hand, transaction costs of multiple policies are incurred if the policies subsequently need to be changed so designing policies to allow for sequencing is desirable. Crase et al. (2013) also suggest that initially requiring changes that are too small to result in observable environmental impacts may be problematic as far as support for further change is concerned. Garrick et al. (2013) note the importance of a multiphase sequencing of institutional transitions to support water trading, identifying at least three broad phases, market emergence, market strengthening, and adjustment. Sequencing matters, particularly to allow for experimentation and learning through informal trading as well as balance between security (water rights reforms and diversion limits) and flexibility (adjustment of rules to address externalities).

2.4.3.2 Intermediaries

Use of intermediaries (e.g. brokers) may reduce transaction costs, especially for infrequent transactions that require specialized knowledge (Coggan et al. 2013). This is related to the discussion of economies of scale and scope in the previous section. In the case of water markets, water banks provide a clearinghouse function to pool buyers and sellers and decrease the transaction costs of administrative review, price discovery and enforcement due to economies of scale associated with streamlined procedures for a large block of water (rather than an individual transaction) (Clifford et al. 2004). Water districts may also provide many of the same functions as water banks, or even create formal water banks, but with lower transaction costs because they are locally managed. These institutional innovations are frequently linked with economies of scale in infrastructure, such as reservoirs.

The various cultural and institutional factors affecting transaction costs are summarized in Table 2.2. Cultural factors that are least amenable to change (and thus with negligible feedback effects) are shown by a single dark arrow in the bottom left of Fig. 2.1. Those that are somewhat amenable to change are shown as the institutional environment in the leftmost box. Policies and policy instruments, the primary objects of design, are shown in the center box.

2.5 Conclusions and Design Recommendations

Water market design involves the establishment of diversion limits and tradable water rights, as well as periodic adjustments to address unintended consequences of prior reforms and changing natural conditions. Our analysis focuses on situations where property rights systems and governance allow the development of formal water markets. In these cases, where water markets are the focus of policymaking and planning efforts in water management, transaction cost analysis offers some insights about the types of physical and institutional factors that can be changed and the strategies to work around other factors to reduce transaction costs.

One of the benefits of incorporating transaction costs, as well as transformation costs, into the design of institutions and policy instruments is that it enables the analyst to bring in practical issues that are normally ignored. Transaction cost analysis also allows one to examine factors such as biophysical complexity (and associated exclusion, heterogeneity and scale issues), as well as cultural values, conflict and lobbying that are often seen as beyond the scope of economics but which are crucial to making progress on wicked problems.

Table 2.2 Cultural and institutional factors that affect transaction and transformation costs and thus design

Factors	Attributes	Water market policy design considerations
–Least amenable to change–	Culture with trust, social capital Institutional environment: Democracy Effective legal system High level of proof	Establish trust and social capital with local stakeholders through effective planning when developing diversion limits Develop information systems and water rights registries to ease burden of proof Provide extension services to navigate complex administrative procedures
–Somewhat amenable to change–	Mismatch of physical and administrative boundaries Institutions that increase lobbying Property rights assigned to those who cannot easily make changes or are hard to regulate Market structures that foster economies of scale and scope Well-designed previous legislation	Establish river basin organizations to coordinate multiple local, state and federal agencies and sectors Low cost conflict management and resolution mechanisms to limit transfer protests Identify needs of irrigation districts system-wide operations and maintenance to reduce barriers to trade out of irrigation districts Enable periodic review of diversion limits and minor adjustments in water rights as information and preferences change
–Amenable to change–	Appropriate sequencing and timing of policy interventions Use of behavioral economics concepts such as choice architecture, especially defaults Intermediaries	Use effective pilots and spot market trading before engaging in comprehensive reform Invest in extension services to inform irrigators of incentives for voluntary reallocation and private investment in conservation technology Encourage water banking and brokerages to assist in trading procedures

More generally, including transaction costs in the analysis and design of policy highlights the importance of the institutional environment, i.e. the political and legal system, as well as the specific existing policies that both enable and constrain our choices. Property rights, and conflict over property rights, which results in high costs of enactment, are revealed as fundamental determinants of transaction costs.

In addition, it helps us think about unintended consequences of policies. Previous decisions affect not only environmental quality and natural resource use, but also norms and the institutional environment, e.g. the issue of path dependence or lock-in that is raised by Challen (2000); Crase et al. (2013); Garrick et al. (2013); Libecap (2011) and Marshall (2013) to understand the difficulty of adjusting historic water use patterns as preferences and availability change. Path dependence, and the interaction between transformation costs and transaction costs implies that examination of the sequencing of policies, rather than just choice of policies may be useful.

Many physical factors affecting water market performance are difficult to change because of complex connections between different users and infrastructure systems. Hydrological interactions and time lags across different phases of the water cycle are difficult to change without inter-basin transfers and capital intensive infrastructure. Policymaking efforts can work around these constraints by establishing a nested set of diversion limits that accounts for hydrological interactions across scales and sources, e.g. groundwater and surface water. The flexibility to adjust these constraints periodically is paramount given uncertainty and changing social preferences. This has been illustrated by the recent basin planning experience in the Murray-Darling Basin of Australia.

A range of policy design considerations can address other physical factors associated with water's biophysical complexity: heterogeneity, externalities, asset specificity and economies of scale and scope. These policy design strategies can take advantage of water rights reforms that support low-cost monitoring and conflict resolution, such as water entitlements as shares of available supplies, instead of fixed volumes.

Like physical factors, institutions and culture often prove difficult and slow to change, raising challenges for policymaking. Recognizing the factors that can be changed versus those which must be worked around can be useful in identifying design strategies and sequencing of water market reforms. Social capital, democratic institutions, the rule of law and burden of proof are characteristics of the wider society that are difficult to change. Water markets in developing regions will often struggle to move beyond informal trading because of their weak legal systems and limited social capital. In these contexts, the ability to cultivate trust among users is critical; water users associations should be included in planning and rulemaking to build on social capital. Information systems and brokerage or extension services can be useful in more formal settings where bureaucratic challenges impede progress.

Several other institutional and cultural factors are more amenable to policy changes that will reduce transaction costs, including the mismatch of physical and administrative boundaries, lobbying by affected third parties and the impacts of prior legislation. For example, river basin or catchment level organizations can harmonize diversion limits across administrative jurisdictions. Such organizations are also well positioned to work with local users to anticipate concerns and prevent lobbying or protests of transactions. One recommendation is to make use of existing institutions, policies and forms, where applicable, to reduce transaction costs. If possible, build on existing policies (water rights reforms in Australia), and/or prevent conflicts with existing policies (through national frameworks to coordinate state water allocation policy). For example, irrigation extension services can be used to expand access to incentive-based programs and to help irrigators navigate complex water trading regulations. Research on water saving technologies can create win-win options for irrigators.

In conclusion, while water market institutions that are more efficient may arise spontaneously (e.g. Demsetz 1967), in general they should be the focus of design, especially in the case of water policy issues. Applied economists have typically focused on the design of policy instruments, and to some extent technical change, but including transaction costs in the analysis means that we also should think about design in the context of the institutional environment. The effect of physical factors on transaction costs, and their interaction with institutional factors, also needs to be considered. This type of analysis implies economists and policy makers ought to consider the dynamic effects of policy choices on both technological change and institutions. Creating general policies and procedures that can be adaptable to heterogeneous and changing situations would be useful.

Ultimately, policy choice and policy design need to be matched to the specific physical and institutional characteristics of the problem. Some specific policy recommendations flow from incorporating transaction costs in water policy design. It is helpful to think of this process as a hierarchy, evaluating and trying easier solutions (e.g. the use of brokerage, licensing registries and extension services) first and then making more fundamental changes in policy, technology, or even the institutional environment (creating river basin organizations and adjusting diversion limits) if needed or when the amount of change required is large.

References

Aguilera-Klink F, Sanchez-Garcia J (2005) Water markets in Tenerife: the conflict between instrumental and ceremonial functions of the institutions. Int J Water 3(2):166–185

Batie SS (2008) Wicked problems and applied economics. Am J Agric Econ 90(5):1176–1191

Bauer CJ (1997) Bringing markets down to earth: the political economy of water rights in Chile 1975–1996. World Dev 25(5):639–659

Bennett J (2005) The evolution of markets for water. Edward Elgar Publishing, Northampton

Birner R, Wittner H (2004) On the efficient boundaries of the state: the contribution of transaction-costs economics to the analysis of decentralization and devolution in natural resource management. Environ Plann C Gov Policy 22(5):667–685

Bjornlund H (2004) What impedes water markets? Water 31(7):47–51

Bjornlund H, Zuo A, Wheeler S, Xu W (2014) Exploring the reluctance to embrace water markets in Alberta, Canada. In: Easter KW, Huang QQ (eds) Water markets for the 21st century: what have we learned? Springer, Dordrecht

Bougherara D, Grolleau G, Mzoughi N (2009) The 'make or buy' decision in private environmental transactions. Eur J Law Econ 27(1):79–99

Boutry O (2011) Agriculture et environnement: une analyse néo-institutionnelle de l'évolution des pratiques agricoles. Le cas de la gestion quantitative de la ressource en eau en Charente-Maritime. Ph.D. Dissertation, University of Poitiers, France

Bromley DW (1991) Environment and economy: property rights and public policy. Blackwell Publishers, Cambridge
Bromley DW (1992) Entitlements and public policy in environmental risks. In: Bromley DW, Kathleen S (eds) The social response to environmental risk: policy formulation in an age of uncertainty. Kluwer Academic Publishers, Boston
Brown T (2006) Trends in water market activity and price in the western United States. Water Resour Res 42(9):1–14
Cacho OJ, Lipper L, Moss J (2013) Transaction costs of carbon offset projects: a comparative study. Ecol Econ 88:232–243
Carey J, Sunding DL, Zilberman D (2002) Transaction costs and trading behavior in an immature water market. Environ Dev Econ 7(4):733–750
Challen R (2000) Institutions, transaction costs, and environmental policy: institutional reform for water resources. Edward Elgar, Cheltenham/England
Clifford P, Landry C, Larsen-Hayden A (2004) Analysis of water banks in the Western States. Washington Department of Ecology. Publication number 04-11-011
Coase RH (1960) The problem of social cost. J Law Econ 3:1–44
Coggan A, Whitten SM, Bennett J (2010) Influences of transaction costs in environmental policy. Ecol Econ 69(9):1777–1784
Coggan A, Buitelaar E, Whitten SM, Bennett J (2013) Factors that influence transaction costs in development offsets: who bears what and why? Ecol Econ 88:222–231
Connell D (2007) Water politics in the Murray-Darling Basin. Federation Press, Sydney
Crase L, O'Keefe S, Dollery B (2013) Talk is cheap, or is it? The cost of consulting about uncertain reallocation of water in the Murray-Darling Basin, Australia. Ecol Econ 88:206–213
Demsetz H (1967) Toward a theory of property rights. Am Econ Rev 57(2):347–359
Easter W, McCann L (2010) Nested institutions and the need to improve international water institutions. Water Policy 12:500–516
Easter KW, Dinar A, Rosegrant WM (1998) Chap. 1, Water markets: transaction costs and institutional options. In: Easter KW, Rosegrant WM, Ariel D (eds) Water markets: potential and performance. Kluwer Academic Publishers, Boston
Ervin DE, Graffy EA (1996) Leaner environmental policies for agriculture. Choices Fourth Q 11:27–33
Fernandez F (2001) Cutting edge fixture technology: manufacturers perspective. PM Eng 7(7): 38–42
Friedman T (2008) Hot, flat and crowded: why we need a green revolution and how it can change America. Farrar, Straus & Giroux, New York
Garrick DE, Aylward B (2012) Transaction costs and institutional performance. Land Econ 88(3):536–560
Garrick D, Lane-Miller C, McCoy A (2011) Institutional innovations to govern environmental water in the western United States: lessons for Australia's Murray-Darling Basin. Econ Pap 30(2):167–184
Garrick D, Whitten S, Coggan A (2013) Understanding the evolution and performance of market-based water allocation reforms: a transaction costs analysis framework. Ecol Econ 88:195–205
Garrido A (2007) Water markets design and evidence from experimental economics. Environ Resour Econ 38:311–330
Grafton RQ, Libecap G, McGlennon S, Landry C, O'Brien B (2011) An integrated assessment of water markets: a cross-country comparison. Rev Environ Econ Policy 5(2):219–239
Griffin RC, Peck DE, Maestu J (2012) Myths, principles and issues in water trading. In: Maestu J (ed) Water trading and global water scarcity: international experiences. RFF Press, New York
Grolleau G, McCann L (2012) Designing watershed programs to pay farmers for water quality services: case studies of Munich and New York City. Ecol Econ 76:87–94
Hayami Y, Ruttan VW (1971) Agricultural development: an international perspective. John Hopkins University Press, Baltimore
Hearne R, Donoso G (2014) Water markets in Chile: are they meeting needs? In: Easter KW, Huang QQ (eds) Water markets for the 21st century: what have we learned? Springer, Dordrecht

Hicks JR (1963) The theory of wages, 2nd edn. St. Martin's Press, New York
Howitt RE (1994) Empirical analysis of water market institutions: the 1991 California water market. Resour Energy Econ 16(4):357–371
Howitt RE (2014) Are lease water markets still emerging in California? In: Easter KW, Huang QQ (eds) Water markets for the 21st century: what have we learned? Springer, Dordrecht
King RP (2012a) The science of design. Am J Agric Econ 94(2):275–284
King RP (2012b) Climate change policy and the science of design. Distinguished fellows address presented at the Australian agricultural and resource economics society annual conference, Fremantle, Western Australia, 10 February, 2012
Krutilla K, Krause R (2011) Transaction costs and environmental policy: an assessment framework and literature review. Int Rev Environ Resour Econ 4:262–354
Kubasek NK, Silverman GS (1997) Environmental law, 2nd edn. Prentice Hall, Upper Saddle River
Laurenceau M (2012) A transaction cost approach for environmental policy analysis: the case of the water framework directive in the Scheldt International river basin district. Ph.D. thesis, ENGEES, Strasbourg, France
Libecap G (2005) The problem of water. National Bureau of Economic Research working paper. Available at: http://www.aeaweb.org/annual_mtg_papers/2006/0108_1300_0702.pdf
Libecap G (2008) Chinatown revisited: Owens Valley and Los Angeles—bargaining costs and fairness perceptions of the first major water rights exchange. J Law Econ Organ 24(2):1–28
Libecap G (2011) Institutional path dependence in climate adaptation: Coman's "some unsettled problems of irrigation". Am Econ Rev 101: 1–19
Liberherr E (2011) Acceptability of the Deschutes groundwater mitigation program. Ecol Law Curr 38:25–35
Makridis C (2012) Multilateral water governance: prospects for transboundary water banking. Global Water Forum Discussion Paper
Marshall G (2013) Transaction costs, collective action and adaptation in managing complex social-ecological systems. Ecol Econ 88:185–194
McCann L (1998) Implications of a deterministic coin toss for economic modeling. Assoc Environ Resour Econ Newsl 19(1):18–19
McCann L (2013) Transaction costs and environmental policy design. Ecol Econ 88:253–262
McCann L, Easter KW (2004) A framework for estimating the transaction costs of alternative mechanisms for water exchange and allocation. Water Resour Res 40(9) July 2004. Reprinted in The economics of water quality. K. William Easter, Naomi Zeitouni (eds) Ashgate Publishing, Williston 2006
McCann L, Hafdahl A (2007) Agency perceptions of alternative salinity policies: the role of fairness. Land Econ 83(3):331–352
McCann L, Colby B, Easter KW, Kasterine A, Kuperan KV (2005) Transaction cost measurement for evaluating environmental policies. Ecol Econ 52(4):527–542
Murray-Darling Basin Authority (MDBA) (2010) Guide to the basin plan. Murray Darling Basin Authority, Canberra
National Water Commission (NWC) (2006) Investing in irrigation: achieving efficiency and sustainability. Available online at: http://archive.nwc.gov.au/data/assets/pdffile/0020/18173/Investing-in-Irrigation-PUB-0306.pdf
North DC (1990) Institutions, institutional change and economic performance. Cambridge University Press, Cambridge
Ostrom E (1990) Governing the commons: the evolution of institutions for collective action. Cambridge University Press, Cambridge
Pannell DJ, Roberts AM, Park G, Alexander J (2013) Improving environmental decisions: a transaction cost story. Ecol Econ 88:244–252
Perry J, Easter KW (2004) Resolving the scale incompatibility dilemma in river basin management. Water Resour Res 40 (2004), W08S06, doi: 10.1029/2003WR002882
Pujol J, Raggi M, Viaggi D (2006) The potential impact of markets for irrigation water in Italy and Spain: a comparison of two study areas. Aust J Agric Resour Econ 50:361–380

Richards K (2000) Framing environmental policy instrument choice. Duke Environ Law Policy Forum X(2):221–285
Roberts AM, Pannell DJ, Doole G, Vigiak O (2012) Agricultural land management strategies to reduce phosphorus loads in the Gippsland Lakes, Australia. Agr Syst 106(1):11–22
Rorstad PK, Vatn A, Kvakkestad V (2007) Why do transaction costs of agricultural policies vary? Agric Econ 36:1–11
Ruml CC (2005) The Coase theorem and western U.S. appropriative water rights. Nat Resour J 45(1):169–200
Saliba BC, Bush D (1987) Water markets in theory and practice: market transfers, water values and public policy. Westview Press, Boulder
Schlager E, Blomquist W (2008) Embracing watershed politics. University Press of Colorado, Boulder
Schmid AA (1995) The environment and property rights issues. In: Daniel WB (ed) Handbook of environmental economics. Basil Blackwell Ltd., Cambridge, MA
Schmid AA (2004) Conflict and cooperation: institutional and behavioral economics. Blackwell Publishing, Hoboken
Schorr D (2005) Appropriation as agrarianism: distributive justice in the creation of property rights. Ecol Law Q 32:3–71
Skurray JH, Roberts EJ, Pannell DJ (2012) Hydrological challenges to groundwater trading: lessons from south-west Western Australia. J Hydrol 412–413:256–268
Stavins RN (1995) Transaction costs and tradeable permits. J Environ Econ Manage 29(2):133–148
Tietenberg T (2005) Environmental and natural resource economics, 7th edn. Addison Wesley, Lebanon
Tisdell J (2014) The evolution of water Legislation in Australia. In: Easter KW, Huang QQ (eds) Water markets for the 21st century: what have we learned? Springer, Dordrecht
Vatn A (2005) Institutions and the environment. Edward Elgar Publishing, Cheltenham
Williamson OE (1985) The economic institutions of capitalism. Free Press, New York
Williamson OE (2000) The new institutional economics: taking stock, looking ahead. J Econ Lit 38(3):595–613
Young MD, McColl JC (2009) Double trouble: the importance of accounting for and defining water entitlements consistent with hydrological realities. Aust J Agric Resour Econ 53(1):19–35

Chapter 3
Water Markets as an Adaptive Response to Climate Change

Mark W. Rosegrant, Claudia Ringler, and Tingju Zhu

Abstract The chapter reviews the status of and potential for water markets as an adaptive response to climate change, with implications for developing countries. An analysis of several climate change scenarios suggests increasing pressure on water supplies relative to demand, and points to regions of the world where water markets could help efficiently move water among users. Growing water scarcity and variability due to climate change might well propel development of new water markets given the significant potential of water markets to alleviate growing water stress both in the long run and in response to short term water variability.

Keywords Water market • Climate variability • Climate change • Water scarcity • Adaptation

3.1 Introduction

Vulnerability of water supplies to periodic and long-term shortages, with adverse implications for global food production, has been a concern for decades. The prospective effect of climate change on water is heightening this concern. In response to growing pressures on water resources, several developing countries have begun to incorporate elements supporting water markets into their water legislation, and informal water markets continue to expand. In addition, countries such as China have experimented with market-like innovations in water allocation (Calow et al. 2009; Ringler et al. 2010). Nevertheless, no new formal water markets have been created in developing countries over the past decade.

M.W. Rosegrant (✉) • C. Ringler • T. Zhu
Environment and Production Technology Division, International Food Policy Research Institute, 2033 K Street NW, Washington, DC 20006-1002, USA
e-mail: m.rosegrant@cgiar.org; c.ringler@cgiar.org; t.zhu@cgiar.org

K.W. Easter and Q. Huang (eds.), *Water Markets for the 21st Century: What Have We Learned?*, Global Issues in Water Policy 11, DOI 10.1007/978-94-017-9081-9_3,

Given the significant interactions between water and climate, water-based and water-related climate change adaptation mechanisms will be critical for successfully adapting to climate change. Water-based climate change adaptation strategies tend to focus on improving availability and reliability of water supplies. The two most common adaptation options found in the literature are development of more efficient irrigation and development of water storage (World Bank 2009).

Water markets, on the other hand, are not usually proposed as a mechanism for climate change adaptation. However, growing water scarcity and variability due to climate change might well propel development of new water markets given the significant potential of water markets to alleviate growing water stress both in the long run and in response to short-term water variability. At the very least, climate change is increasing the value of successful implementation of water markets because of the increased variability, risk, and uncertainty in water availability and therefore the need for flexibility in water allocation that is provided by water markets.

This chapter first describes the key water-climate interactions that water markets and other water-based climate change adaptation options would need to address. Next, an analysis of four climate change scenarios estimates potential impacts on water supply and demand and identifies regions and countries where pressures on water resources might increase from a supply perspective. This analysis is followed by a critical assessment of the literature regarding the potential role of water markets as an effective adaptation response to climate change. Finally, the chapter explores alternatives to water markets that might be helpful if development of water markets is not practical or possible.

3.2 Uncertain Effects of Climate Change on Water Availability and Variability

Available evidence suggests that climate change may lead to substantial changes in mean annual streamflows, the seasonal distributions of flows, melting of snowpack, and increased probability of extreme high- or low-flow conditions. Specifically, climate change impacts on water resources include changes in the timing of water availability due to changes in glaciers, snow and rainfall; changes in water demands due to increased temperatures; changes in surface water availability and groundwater storage; an increased number and intensity of extreme climatic events (droughts and floods); changes in water quality; and sea-level rise. For water markets, the most relevant among these will be changes in the timing of water availability, changes in water demands, changes in surface water availability, and extreme events.

The ultimate outcome of climate change and its effects on water availability are not known with precision. Unknowns include geographic location, direction of change (less/more precipitation), degree of change in precipitation (low/high), change in precipitation intensity (low/high), and timing (within the next 5 years or over multiple decades). Shifting precipitation patterns and warming temperatures could increase water scarcity in some regions while other areas may experience increased soil-moisture availability, which could increase opportunities for agricultural production (Malcolm et al. 2012).

Climate change may considerably alter hydrological regimes, affecting both the demand and supply sides of a water management system. In a study for the Maipo Basin in Chile, Meza et al. (2012) analyzed impacts of climate change on irrigation demand and streamflow, by statistically downscaling the PRECIS-HadCM3 regional climate projections forced by the SRES A2 and B2 emission scenarios to the Maipo Basin. They found that summer stream flow declines while irrigation water demands increase by 60–80 % during the traditional peak of irrigation under climate change. Allowing proportional reductions with stream flow below demand, they found that the probability of permanent water use rights being unmet increases from 15 % under the baseline climate to as much as 50 % under climate change. This represents a serious impediment to the viability of irrigated agriculture at its current level of intensity, cropping patterns, and water use efficiency and favors more flexible water allocation methods.

The World Bank points out that climate change will make it harder to manage the world's water because it will affect the entire water cycle (World Bank 2009, p. 137). Warming will speed up the hydrological cycle, increasing precipitation in some areas and for the world as a whole. At the same time, increased evaporation will make droughts more prevalent across wide swaths of the world by the middle of the twenty-first century.

For basins whose surface water supply is provided by a mix of rainfall, glacial melt, and snowmelt (such as in Pakistan, Mongolia, or California in the United States), glacier melt alone will affect both seasonal and long-term water availability. Bates et al. (2008) expect several changes in climate change-surface water interactions, including increased precipitation in higher latitudes and parts of the tropics combined with declines in rainfall in lower/mid-latitude regions and parts of the subtropics. As a result, surface runoff is predicted to increase at high latitudes and in some wet, tropical areas. On the other hand, runoff is expected to decline in dry regions at mid-latitudes and in the dry tropics. In most areas, impacts remain somewhat to highly uncertain. with intensity of precipitation expected to rise, regardless of whether total precipitation is decreasing or increasing. Increased intensity will likely pose challenges for agricultural and other users of water when trying to capture and manage available water supplies.

Many of these regions are already net food importers, such as much of Africa, the Middle East, and Central America. Other regions where average annual runoff is expected to decline sharply include major agricultural producers, such as much of Europe, and parts of South America, North America, and Australia. The average number of consecutive dry days could increase by up to 20 days in many of these regions.

Even less certain is how reduced water availability from climate change might affect local food production and global food demand. Such uncertainty calls for methods of allocating water and other resources that are sufficiently flexible to accommodate various potential outcomes, particularly those where water becomes increasingly scarce and local food production more tenuous.

Uncertainty affects decision-making related to policies and investments for coping with potential declines in water availability due to climate change. This concern parallels the more general issue of increased pressure on global water available as a result of a growing, wealthier global population and the concomitant increased demand for food. How should countries and communities decide to allocate their water resources? These policies will affect water availability for food production as well as for other uses, including urban, recreation, and environmental water uses.

Allocation of global water use is currently based on a variety of methods. Economic value is used to allocate water to its highest-valued use in well-developed water markets within Australia, the United States (California), Chile, and Mexico (World Bank 2009). Informal water markets flourish inside formal markets and in many other water-scarce environments, generally at a smaller scale. Well-studied examples include India (Saleth 1998) and Pakistan (Meinzen-Dick 1996). Elsewhere, water is allocated administratively or through civil societies/water user organizations, or some combination of market, administrative, and user-based methods. In many countries, water allocation remains rather ad-hoc and uncoordinated, with water allocations influenced by new investments with objectives that sometimes conflict with each other.[1] Thus, more generally, the current mix of methods for allocating water reflects historical precedence, ad-hoc allocation through development, and the relatively slow movement in some parts of the world to market-based solutions.

3.3 Climate Change Impacts on Water Resources: Insights from Four Climate Change Scenarios

Climate change has implications for water availability by affecting runoff and altering local supply and demand for water. Understanding the magnitude of these potential changes, particularly where and how irrigation water supply might change relative to demand, would help policymakers and researchers target efforts to adapt to these changes using water markets or other adaptation strategies.

To analyze prospective shifts in water supply and demand, four climate change scenarios, representing global climatic conditions around 2050, are assessed. These scenarios are constructed for the global food and water projection model, IMPACT

[1] For example, a new hydropower station might be built by a county's Ministry of Energy at the same time that a new water supply or sewage treatment plant might be built just downstream. Meanwhile, the local irrigation water district might be adding more pumps or widening canal intakes, whereas elsewhere in the irrigation system, irrigated land is removed from production due to peri-urban build-up.

(Rosegrant et al. 2012), to estimate impacts on annual runoff and changes to irrigation water supply relative to demand. The scenarios are based on climate projections[2] of two Global Circulation Models (GCMs), CSIRO-Mk3.5 and MIROC3.2. The projections for MIROC are generally hotter and wetter whereas the CSIRO generates projections that are somewhat drier with lower temperature increases.

Each GCM is driven by the same two emissions scenarios, SRES A1b and B1 (Jones and Thornton 2013). The A1b scenario assumes rapid economic growth and has medium-high greenhouse gas emissions growth. The B1 scenario also has rapid economic growth but with more effort on ecological sustainability and a faster shift from an industrial to a service and information economy, and therefore somewhat slower growth in greenhouse gas emissions.

The naming of the four climate scenarios here is based on the combinations of GCMs with emissions scenarios: CSIRO-a1, CSIRO-b1, MIROC-a1, and MIROC-b1. A fifth simulation serves as the baseline. It uses historical climate data, and population projections are the medium variant population growth rate projections from the population statistics division of the UN. Income projections are estimated by the authors, drawing upon the Millennium Ecosystem Assessment (2005) and the World Bank EACC study (Margulis 2010).

The IMPACT model simulates water demand and supply at the sub-national level for the domestic, industrial, livestock and irrigation sectors. First, total water supply is optimized, driven by the objectives of minimizing both annual water shortages and maximum monthly shortages within a year, as well as maximizing the end-of-year storage to ensure that a certain amount of water is carried over to the next year. The purpose of minimizing maximum monthly shortages is to avoid severe shortage concentration in one or a few months. The optimization is done annually, while the water supply and demand balances are conducted on a monthly basis. Second, non-irrigation sectors, including household and industrial use, receive priority access to water, with irrigation as residual user (Rosegrant et al. 2012). This allocation mechanism assumes automatic transfers of water from irrigation to non-irrigation sectors when there is scarcity or when scarcity increases as a result of increasing demand or declining supply.

3.3.1 Results: Annual Mean Runoff

The percent changes in annual mean runoff in 2050 as compared to the historical climate (1971–2000) are estimated for the four climate scenarios (Fig. 3.1). All four scenarios show significant annual declines in runoff for the Middle East and North Africa region. Similarly, all four scenarios show much lower runoff in southern

[2]A statistical downscaling process was used to convert coarse-scale GCM projections to higher resolution suitable for local impact assessment by statistically relating large-scale climate variables to local-scale variables.

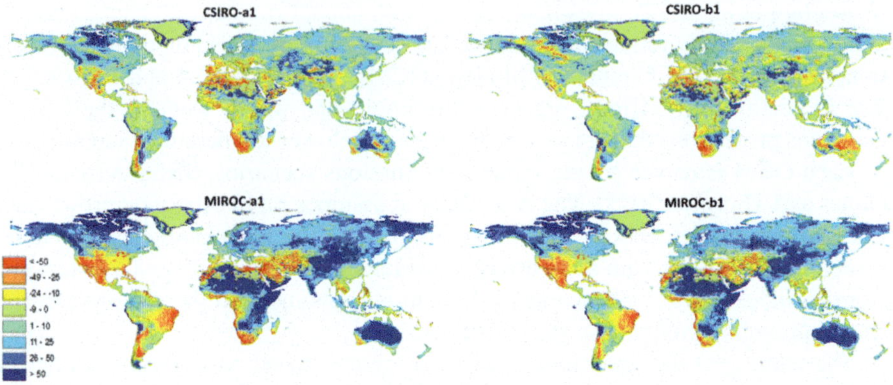

Fig. 3.1 Percent changes in annual mean runoff in 2050 compared to the 1971–2000 period for four climate change scenarios (Source: Authors)

Africa and southern Spain. Two regions with active water markets—Chile and the southwestern United States—are also expected to experience runoff declines as a result of climate change under the four scenarios examined. Given their active water markets, these two regions are expected to cope with reduced runoff better than other regions. Results for runoff are inconclusive for much of Africa south of the Sahara and much of Asia, with the two global climate models diverging on changes in runoff over time. However, further declines in runoff are likely for parts of China and Southeast Asia.

3.3.2 Results: Irrigation Water Supply Reliability

Another key measure of climate change impacts on water resources is how those impacts will affect the supply of and demand for irrigation water. Irrigation water supply reliability (IWSR), defined as the annual average ratio of irrigation water supply to irrigation water demand for 2000, 2030 and 2050, for the baseline and four climate change scenarios is shown in Fig. 3.2. Global average irrigation water supply reliability shows a declining trend, from approximately 0.77 in 2000 to 0.69 in 2050 as growth in total water demand outpaces growth of supply and efficiency improvements. This means that pressure on moving water out of irrigation increases considerably as a result of more rapidly growing non-irrigation demands.

Two of the scenarios, MIROC-a1 and MIROC-b1, lead to nearly identical IWSR outcomes at the global level, with IWSR in both scenarios declining more slowly than the baseline. Under CSIRO-a1 and CSIRO-b1, on the other hand, IWSR declines faster compared to the baseline. The MIROC-a1 and MIROC-b1 scenarios suggest a future with fewer irrigation water shortages while the CSIRO-a1 and CSIRO-b1 scenarios suggest that climate change may worsen irrigation water

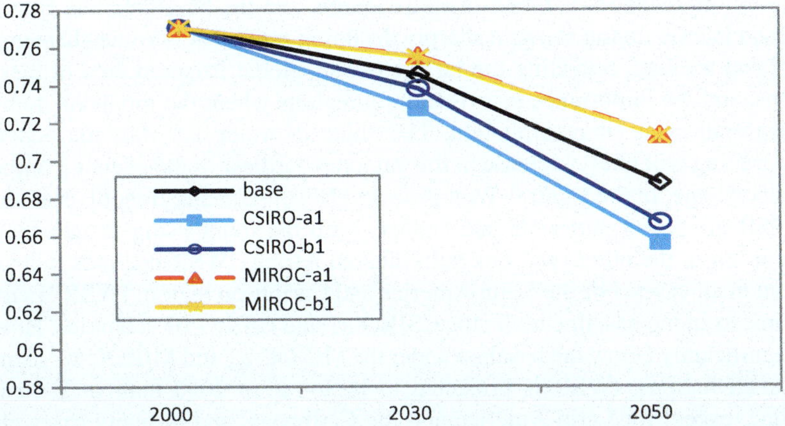

Fig. 3.2 Projections of global average irrigation water supply reliability (Source: Authors)

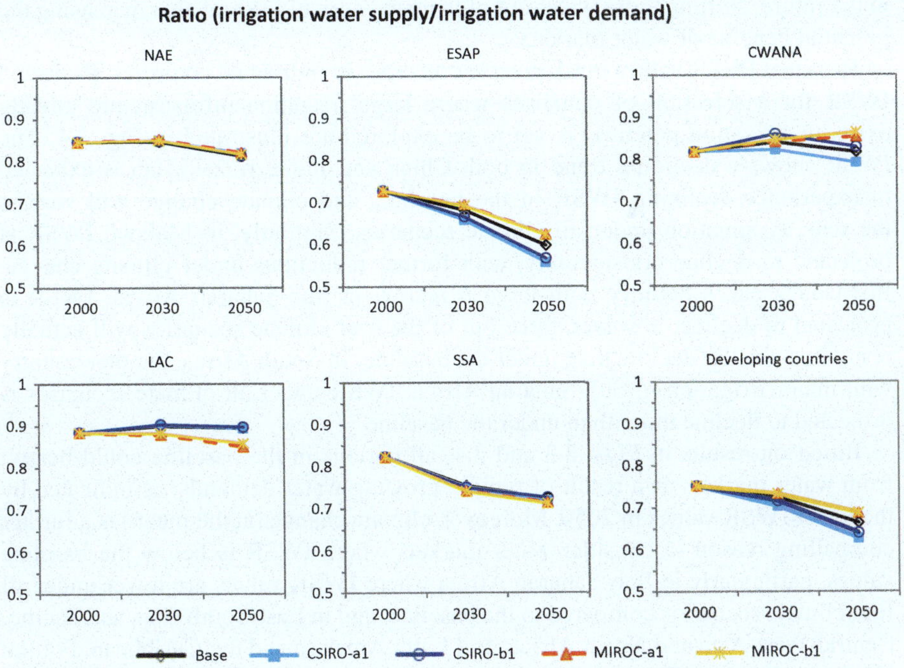

Fig. 3.3 Estimated irrigation water supply reliability by region in 2000, 2030 and 2050 (Note: *NAE* North America and Europe, *ESAP* East-South Asia and Pacific, *CWANA* Central-West Asia and North Africa, *LAC* Latin America and Caribbean, *SSA* Sub-Saharan Africa. Source: Authors)

shortages in the future with potentially more water transferred out of the irrigated agriculture sector to meet domestic and industrial demands.

Regionally, diverse trajectories of IWSR are found, as illustrated in Fig. 3.3. The North America and Europe region shows slightly declining IWSR out to 2050, with essentially no changes across climate scenarios and the baseline. The East-South Asia and Pacific region shows a sharply declining trend that is somewhat similar to the global average, reflecting the large share of global irrigated area in these two regions, and the importance of irrigation in regions where the monsoon dominates the seasonal cycle of precipitation. The value of water and thus the benefits of water trading will likely increase in this region regardless of the eventual impacts of climate change. In the Central-West Asia and North Africa region, diverse changes in IWSR are found under climate change, with the trend rising in two scenarios and uneven in the other two. For Sub-Saharan Africa, IWSR declines in a similar fashion in all cases. For the Latin America and Caribbean region, IWSR is virtually the same as in the baseline under the CSIRO-a1 and CSIRO-b1 scenarios. However, it is significantly below the baseline under the MIROC-a1 and MIROC-b1 scenarios.

For all regions, IWSR is projected to be lower in 2050 than in 2000 in the baseline, except for Latin America and the Caribbean, and in many cases climate change reduces IWSR. Declining irrigation water supply relative to demand over time due to climate change suggests that water demand and supply gaps will widen in the future, shifting more water out of irrigated agriculture and also increasing the potential benefits of water markets.

Regional IWSR values mask diverse outcomes in individual countries. Projected IWSR for a selection of countries where large irrigation infrastructure already exists or irrigation potential is yet to be explored are illustrated in Fig. 3.4. The IWSR shows a declining trend in both China and India. Kazakhstan is expected to experience declining IWSR in the baseline, and climate change will worsen the scarcity situation under all climate scenarios. Similarly, in Malawi, IWSR is projected to decline considerably, with further reductions under climate change. In Madagascar, a country with large irrigation in Sub-Saharan Africa, IWSR is projected to decline; however, three out of the four climate scenarios will actually result in a smaller decline than under the baseline. In South Africa, another country with major irrigation in Sub-Saharan Africa, IWSR under all climate scenarios is projected to decline more than under the baseline.

Based on results in Figs. 3.3 and 3.4, all regions in the baseline could benefit from water markets as a result of rapidly growing water demands, as indicated by the lower IWSR values in 2050. Moreover, climate change (at the mean) is a further compelling reason to consider water markets where IWSR is below the baseline values, particularly in Sub-Saharan Africa where IWSR values are lower under all four climate scenarios compared to the baseline, and in East-South Asia and Pacific, Central-West Asia and North Africa, and Latin America and the Caribbean. For the selected countries, Kazakhstan, Malawi and South Africa exhibit growing scarcity and thus a need for enhanced water allocation efficiency under all four climate scenarios.

This assessment does not consider the potentially large benefits of water markets in reducing uncertainty concerns associated with growing climate variability. The potential for water markets to address uncertainty are discussed below.

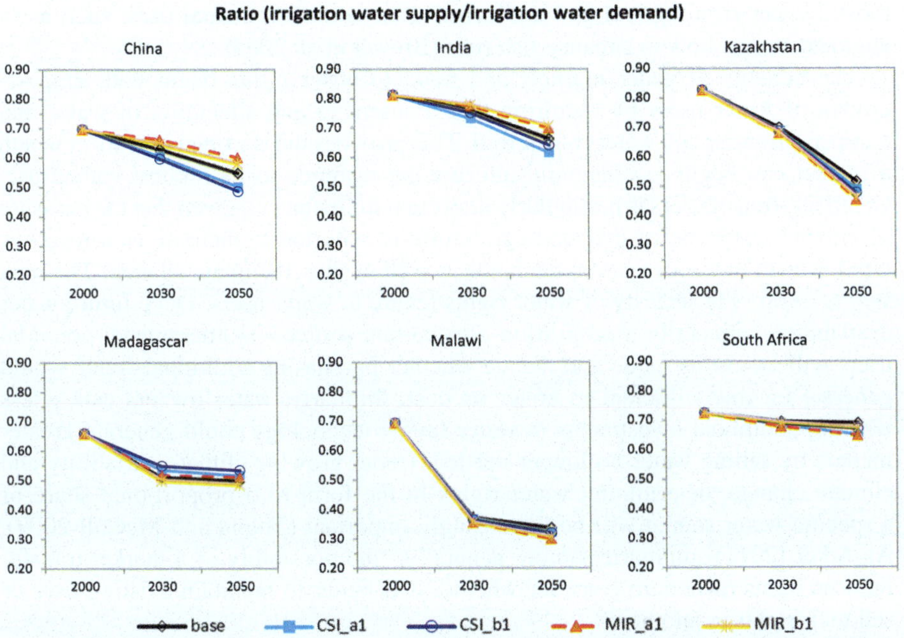

Fig. 3.4 Estimated irrigation water supply reliability in 2000, 2030 and 2050 for selected countries (Source: Authors)

3.4 Water Markets and Tradable Water Rights: Advantages and Challenges[3]

Numerous researchers have pointed out that the most fundamental requirement for a workable water market is well-defined property rights for water. These rights have economic value that reflects the underlying asset (use of water), and if adequately protected, will retain their value indefinitely. As in any market, these rights must be subject to transfer, including purchase or rent for a given amount of time. The physical conveyance structure must also be available, and relevant transaction costs must be low enough to encourage trading.

Private property rights empower individuals to pursue profit through either accumulating more rights or selling them. They can also empower private groups to pursue either profit or non-monetary objectives such as environmental goals. For example, environmental organizations might purchase water rights to increase stream flow for wildlife. In the Western US, for example, 9 % of all water trades

[3]See other chapters for a more extensive discussion of pros and cons of water markets. Much of this section is summarized from Adler (2011) and Rosegrant and Binswanger (1994).

moved water from agricultural and urban uses to environmental uses, such as to augment stream flow to enhance fish runs (Brewer et al. 2006).

The benefits of water markets and tradable water rights begin with empowerment of water users by requiring their consent to any allocation of water and compensation for any water transferred. This provides the starting point from which a market can begin to efficiently allocate the resource to its highest-valued use. Compensation of users is particularly important for irrigators, given that the pressure of moving water out of irrigated agriculture is expected to increase as a result of rapid growth in non-irrigation demands as well as due to climate change. The next step is to provide security of water rights tenure to water users. Once future water availability—generally a proportion of a certain source—is guaranteed, potential users will assess its value and decide whether purchasing additional water would generate income sufficient to cover its cost. Similarly, water owners can assess whether additional investments in water-saving technology could generate greater income by selling water no longer needed. Under growing climate variability and climate change, determining water rights in the form of a proportional share of a specific water source will be increasingly important (Young and McColl 2009). As Adler (2011) summarizes more generally, "the possibility of a market transfer induces rights owners to consider whether it is better to maintain existing uses or sell their rights to another."

Water markets also have benefits relative to other means of allocating water. Grafton et al. (2010) state, "Reliance upon voluntary exchanges of water to meet environmental [and other] objectives is valuable because when parties reach mutually-beneficial agreements they avoid the divisive, time-consuming rancor that has characterized arbitrary judicial and administrative water reallocation." Moreover, they add, "mandated regulation of water use and rationing fail to internalize the opportunity cost of water and do not encourage conservation or efficient reallocation of water."

Societal gains from water markets or a system of marketable rights to water are created primarily by forcing the entire range of water users (water owners, sellers, buyers) to consider its value in alternative uses. If water is more valuable to a specific entity at a particular point in time, then others will be inclined to find other, less costly means to reach their objective. This is again an especially valuable characteristic under climate variability because the resulting changes in supply and demand of water will alter its value, perhaps with increasing variability. This growing fluctuation can best be accommodated by a market with individual water rights holders and buyers deciding how much another unit of water is worth to their operation, assuming that they have sufficient information on changes in water availability.

Moreover, tradable water rights can provide incentives for water users to take account of external costs imposed by their water use, reducing the pressure to degrade resources through excess withdrawal and runoff that increases the levels of agricultural chemicals and salts in the water. The potential efficiency and welfare gains from the transfer of water rights can be very high, with some estimates indicating net welfare gains perhaps greater than the water rights themselves (Adler 2011).

Establishment of transferable water rights helps formalize existing rights to users, which facilitates acceptance by current water users. A common alternative is simply imposing certain pricing mechanisms. An example is volumetric pricing, whereby water is priced administratively at higher levels for additional increments of use. Although pricing mechanisms can help establish more appropriate water values, they can be viewed by water owners as an expropriation of traditional water rights and can face stiff opposition from current users.

The ultimate advantage of tradable water rights as the basis for a water market is that it provides maximum flexibility in responding to changes in agricultural output prices (crops or livestock products), which affects water demand by agricultural users, changes in water demand for urban, industrial, or environmental purposes, and changes in water availability, driven by climate change or other factors. A market-based system can be more responsive than centralized allocation of water, which depends on manual calculations and analytic judgments to determine where benefits would be maximized. Political considerations are also more likely to come into play in non-market water allocation.

In sum, Adler (2011) states that "markets facilitate the aggregation of individual choices and preferences so as to encourage the deployment of resources to their greatest and highest valued uses . . . the more robust water markets become, the more powerful [incentives to consider alternative uses] will be—and the more pressure there will be for more efficient water use."

While the economic benefit of tradable water rights and water markets is generally acknowledged, the challenges and constraints to water markets have turned out to be persistent and strong. As a result, movement toward water markets has been slow, especially in developing countries, although there has been progress in China and elsewhere in market-like water allocation.

Challenges include high transaction costs, which can stem from the initial need for improved conveyance and other features for on-demand delivery of water, including metering and enforcement of contracts. This can be particularly troublesome where water is utilized to irrigate low-value crops such as rice and wheat. Separately, a challenge for tradable rights occurs when there is incomplete definition of property rights to surface and groundwater.

Another concern for water markets in developing countries is the development of market power and its potential adverse consequences for income and wealth distribution, particularly on the poor. For example, protecting property rights [for water] can reinforce pre-existing economic and social inequities (Adler 2011). Moreover, governments in developing countries are often worried about the potential transfer of water out of agriculture, which may not be politically acceptable. At the same time, agriculture-to-urban transfers continue apace—through investment decisions that tend to favor urban areas—and farmers would thus benefit from markets as they would allow for compensation of these water transfers.

Policymakers must also acknowledge the "special" nature of water in some communities. In this case, water may not be viewed simply as a commodity but as an asset holding additional social value that should be shared equitably and not sold when scarce. These concerns in some instances can supersede economic arguments for water markets to allocate water, which could slow the potential adoption of market solutions or render them a less viable option.

A final significant concern is 'third-party' effects from return flows that benefit downstream users. For example, the downstream user may receive less water when an upstream irrigator transfers water to a user that does not contribute return flows to the downstream user. Other third-party effects may include reduced employment opportunities or general negative effects for the local economy (and thus government tax income) when water is transferred out of the region, or long-term sustainability issues if groundwater is transferred out. Water markets tend to work best in those locations where mitigation of impacts on third parties is not required or not desired.

3.5 Empirical Evidence of Successful Water Markets Under Climate Variability

Australia is often cited as a success story for water markets. Researchers have pointed out how water markets have been used by irrigators since the mid-1990s to manage water scarcity and that water markets could continue to serve this purpose as scarcity increases due to climate change. In general, success lies in the adjustment process afforded by the development and implementation of water markets. The characteristics of market performance in Australia, and in the United States, indicate that the value of water markets will increase due to climate change-induced increases in drought and variability in precipitation and runoff.

Fargher (undated) concludes that water trading in Australia has been a major success in supporting agricultural productivity, and the country's experience can be instructive for other countries facing water scarcity. The basic concept is "cap and trade" whereby the cap is the total water available for consumption (accounting for sustainable amounts of extraction), and water users are provided with entitlements to a share of the total. The entitlements are tradable and priced by the market through trades by buyers and sellers.

Fargher attributes the success in water markets in part to the underlying characteristics of the water resources and industry mix. Key factors include large interconnected water systems, extremely variable water flows, and diverse agriculture enterprises that lead to varying demand for water. Illustrated with volumetric data for temporary allocations and permanent entitlements, water trading in the Murray-Darling Basin began in the early 1980s at very low levels but increased substantially over the subsequent years. Growth was punctuated with spikes in years when seasonal water availability dropped sharply, such as in 1994–1995, and during droughts such as in 2002–2003. The immediate response by water buyers and sellers to the weather conditions and seasonal water availability demonstrates the market's ability to supply water to the highest-valued use. Increases in water trade occurred again in 2006–2007 and 2007–2008 during very dry seasonal spells, indicating that the returns to water markets will increase under climate change, which is projected to increase the frequency of droughts.

Fargher also cites pricing data showing how the market adjusts to seasonal water conditions, with an inverse relationship between availability of water and prices for allocation trades. Prices peaked in 2007–2008 but then declined substantially in 2010–2011 upon improved water availability.

Similarly, Bjornlund and Rossini (2010) reported that between 1993 and 2009, water prices in the Goulburn Murray Irrigation District in Victoria increased substantially, with average annual growth nearly 27 % for annual allocations and 13 % for the underlying ('permanent') entitlements. Interestingly, for allocations, prices were cyclical and driven by allocation levels as well as levels of evaporation. Entitlement prices followed a steady trend, reflecting a steady growth in the value of the underlying asset. The researchers determined that the two prices were linked, with allocation prices driving entitlement prices. As a result, the water market has developed to the point where potential water users can purchase allocations, entitlements or even long-term options to water allocations, which provide additional flexibility in prices and help secure future water supplies. Such an arrangement may be of interest to certain buyers who want to pay only for water that is actually delivered in future years rather than owning the underlying asset. Or some farmers might retain ownership of the entitlement (to secure water for high-value use) and temporarily sell a portion of the allocation that is not needed to buyers who use it for irrigation or environmental purposes, for example.

Bjornlund and Rossini (2010) report that water markets have allowed irrigators to achieve the highest possible return from their declining seasonal water allocation while reducing the economic hardships associated with the decline. In South Australia, for example, 90 % of the additional water purchases were used to expand horticultural production. In contrast, farmers in Victoria purchased water to maintain (not expand) crop production because there had been a decline in seasonal allocations. As water demand increased over time, irrigators were reportedly reluctant to sell their entitlements (the underlying asset) because of increasing values and instead sold their water allocations periodically if necessary. In subsequent years water scarcity worsened and the allocation market became more mature, with farmers opting to either sell their water allocation or use it to grow a crop. However, in some cases, the economic health of communities was threatened when irrigators sold their annual water allocations (and/or long-term entitlements) to other users, reducing economic activity in the local area.

Bjornlund and Rossini (2010) conclude that water markets have played an important role in allocating water in Australia and allowed farmers to manage drought. The authors expect the need for this function to increase as water scarcity worsens under climate change.

As a result of the water market, water moved from lower valued rice crops with flexible irrigation demands to higher-valued horticultural crops with inflexible demands due the long-lived asset (that is, fruit trees). According to Fargher, the trading is indicative of significant economic efficiency gains while providing cash flow to individual farm businesses who sell their water allocations. In the future, to the extent that prices of rice increase relative to alternative crops, water markets in Australia would be expected to shift some water back to the production of rice.

Fargher also points out challenges in developing water markets, specifically technical issues such as establishing workable registers of water entitlements and accounting for water use. Parochial interest of individual states can also adversely affect market development, as can community concerns regarding possible emergence of "water barons" who might exert control over water supplies. Despite these challenges, Fargher concludes that the water markets have had a significant and positive impact on agricultural productivity in Australia, with benefits for the wider economy as the country adapts to fluctuating water conditions and competing demands for water supplies.

Other researchers have also reviewed the Australian case, drawing similar conclusions about the benefits of reallocating water to higher-valued uses, including various agricultural and urban uses, and pointing out that higher water prices have encouraged additional investments to improve on-farm water efficiency (Grafton et al. 2010). They conclude that water markets have clearly helped allocate increasingly scarce water supplies among competing users and facilitated economic development while supporting agriculture and assisting with environmental objectives and that the Australian drought significantly increased water trading, thus mitigating adverse impacts on individual water users (Grafton et al. 2010).

Experiences have been similar in the western United States. Grafton et al. (2010) summarize water trading data from 1987 to 2008 for 12 western states, where most trade is from the agriculture sector (77 % of all water transactions and 62 % of total volume traded). The number of trades increased from 91 in 1987 to 287 in 2008, and the most numerous trades were from agricultural to urban users. The median price of water leased in these cases was four times higher than the lease price paid within the agricultural sector, indicating substantial value of gains from cross-sector trading.

The authors conclude that, "Growing demand for fresh water in the presence of climate change and the likelihood of greater supply variability underscore the importance of smooth, low-cost reallocation made possible by water markets." They also point out how the market structure illustrates current misallocation of water (and the need for further water market reform) because farmers, unlike urban households, typically pay only for pumping and conveyance costs for water and do not yet pay for the value associated with water scarcity. Another issue is relatively high transaction costs that appear to impede water trade in California, with users complaining that administrative and facility charges are too high.

Elsewhere in the world, Bekchanov et al. (2013) find higher gains from water trading under growing water scarcity for the Aral Sea Basin using an integrated economic-hydrologic model of the basin. And when considering water management, Luo et al. (2003), in a mathematical programming analysis of water allocation under uncertainty with increased probability of drought conditions due to climate change, find that water resource management with water trading is more efficient than without water trading. Water trading is also found to be a viable method for adaptation to climate change impacts under water scarcity.

The flexibility of markets in addressing uncertainty in water availability can be further improved through the use of options markets, which are being developed in Australia (see above) and the United States. In options markets, irrigators or water agencies purchase water at some point in the future, if water conditions turn out to be dry. The premium represents the value of the flexibility gained by the buyer from postponing the decision whether or not to purchase water (Hansen et al. 2006). Research indicates that the value of options markets could be high. According to an experimental analysis by Hansen et al. (2008), the use of tradable options in markets simulating California conditions results in significant gains from trade, with average returns increasing by 46 % in competitive markets and by 63 % in monopsonistic markets. Gains from trade under options markets are also much more evenly distributed between the single buyer and the many sellers in the case of monopsony. Options markets for temporary transfers can also facilitate acceptance of water trading and reduce transaction costs, enhancing the prospects for water markets more broadly (Tomkins and Weber 2010).

As analyses and success stories in some of the countries show, water markets "thrive" under the water scarcity and variability conditions that are expected to prevail under climate change.

3.6 Climate Change Outcomes and Potential Responses: Markets and Alternatives

As described earlier, climate change may affect water availability in a variety of ways. The potential for water markets to assist in allocating scarce water resources shows promise for dealing with both lower and more uncertain water supplies in the future. Conversely, the constraints mentioned above imply that other types of climate change adaptation need to be considered in conjunction with the development of markets, primarily because markets may not be practical or sufficient in some areas. Furthermore, many such non-market alternatives would likely complement activities of water markets and possibly enhance their impact; and advanced technologies can partly substitute for markets through improved efficiency of water use.

A summary of adaptation strategies is shown in Table 3.1. Two precipitation scenarios are shown in the columns: (1) drier/no change, and (2) wetter and/or greater rainfall frequency or intensity. The two scenarios are separated into two categories of agricultural practices: (a) irrigated, and (b) combined irrigated/rainfed as some strategies could apply to both.

Irrigation development is generally considered to be a (or the) key water-based adaptation mechanism. Irrigation is obviously a pre-condition for subsequent water market development. For both precipitation scenarios, there appears to be a role for increased water storage, both above and below ground, as a complement to market-based strategies to deal with increased water supply variability. According to the

Table 3.1 Response strategies and adaptation options for precipitation scenarios under climate change

Type of agriculture	Change in precipitation: Drier/no change	Change in precipitation: Wetter and/or greater rainfall frequency/intensity
Irrigated	Agricultural intensification through irrigation Increase irrigation efficiency (e.g., drip/sprinkler irrigation, laser land levelers) Improve water use (use brackish water where possible, line canals) **Water rotations/scheduling (concentrate irrigation in periods of peak growth)** **Change irrigation water pricing** **New reservoirs** **Watershed management**	**New reservoirs** Wetland restoration Conjunctive use of ground and surface water (including artificial recharge) Joint man-made and natural storage management
Irrigated or Rainfed	Change tillage and farming practices to conserve soil moisture and nutrients, reduce run-off (harvest rainwater), and control soil erosion (reduced tillage; ridge and furrow, mulch, stubble and straw; rotate crops; avoid monocropping; use lower planting densities; use of cover crops) Change timing of farming practices (advance sowing dates to offset moisture stress during warm periods) Change crops/varieties (research on drought-tolerant/water-efficient varieties) Change land topography to improve water uptake such as subdivide large fields, maintain grass waterways or strips, roughen land surface, install flat-channel or other types of terraces, build bunds (dikes) to increase soil moisture **Irrigation**	Investments in drainage infrastructure (e.g., diversion ditches, tile-drains) Water management to prevent waterlogging, erosion, and nutrient leaching Promotion of varieties adapted to waterlogging

Source: Various sources including Bryan et al. (2011) and Malcolm et al. (2012)
Note: Strategies that could include elements of water markets in **bold font**

World Bank (2009), storage is a key aspect of any strategy to manage variability, both during droughts when water is needed to produce crops and during floods when excess flows are harmful or are needed for future use. Building reservoirs and enhancing groundwater recharge to increase storage capacity would make water available for additional crop production or other uses, and dams would need to be built to accommodate future changes in rainfall and runoff. However, market-based solutions, as described above, should also be considered as a viable alternative to more efficiently allocate the resource. Dams are also important infrastructure components for water markets, as releases can be more easily managed for water trades than natural flows and release volumes can be more easily communicated to water traders.

Extending the water storage concept to the soil is another option. The use of cover crops and other production techniques such as conservation tillage and mulching can help the soil retain water that would otherwise be lost to evaporation, thereby increasing crop yields. Changing the timing of farming practices, such as advancing sowing dates, can be used to compensate moisture stress during warm periods. These improved farming techniques make more water available to plants and can help prevent catastrophic crop losses while widening the cropping (and varietal) choices where adopted.

Switching to crops and/or varieties that are more suitable for reduced moisture environment can also stretch available water supplies, as would additional research and dissemination of drought-tolerant/water-efficient varieties. Developing varieties resistant to cold could increase water productivity if their growing season occurs during the winter when heat and moisture stress is reduced (World Bank 2009).

Structural improvements to landscapes can improve water availability as well. Small-scale rainwater storage or "water harvesting" can be used where land is available. Other changes to topography that improve water uptake include subdividing large fields, maintaining grass waterways or strips, roughening land surface, installing flat-channel or other types of terraces, and building bunds (dikes). To address the growing risk of flooding under climate change, combining man-made with natural storage management, such as wetlands or groundwater, is increasingly being considered and practiced.

Total water supplies can also be increased by desalinating brackish water and reusing it. This alternative is expected to become more attractive as energy-efficient filtering systems become less costly and more widely available. It is already widely practiced in the Middle East and North Africa and to a lesser extent in southern Spain (see Chap. 7).

For irrigated areas, more efficient irrigation techniques include drip irrigation and simple upgrades of canal ditches (e.g., lining) to prevent water loss. Also, in irrigated areas where appropriate, conjunctive (combined) use of surface and ground waters can improve operation flexibility and water supply reliability, thus increasing the chances of water transfers and often improving the benefits of water management. More active conjunctive management of surface and groundwater can divert this resource from wet years to dry years and can result in enhancing the availability of this limited water supply. Artificial groundwater recharge is often a means for enhancing conjunctive use. Artificial recharge (AR) is the enhancement of natural groundwater supplies using man-made conveyances such as infiltration pits and basins, or injection wells. Storage of excess water in an aquifer during the wet season using AR and pumping it for irrigation in the dry season or dry years is a way to increase the reliability and resilience of water management systems and to augment natural water sources as climate change and development pressures increase demand on finite groundwater supplies (EPA 2013).

Groundwater banking is a form of AR that has shown huge potential in the state of California. With groundwater banking, groundwater is stored underground during wet years and extracted through pumping during dry years. In a groundwater banking arrangement, procurement of water is performed from willing sellers during wet years; then stored and extracted during dry years. The economic value of this type of storage project is that water is scarcer, and hence more valuable, in dry years than in wet years. Such projects also have relatively modest costs and can be privately financed and operated.

Groundwater banking is already active, with the three major water banks (Arvin–Edison, Kern, and Semitropic) having a combined storage capacity of about 3 million acre-feet, more than five times the amount of water in Millerton Lake, one of the larger reservoirs feeding the Central Valley surface-water system (Christian-Smith 2013). In addition, several smaller banking programs have been launched by the Buena Vista Water Storage District, Rosedale-Rio Bravo Water Storage District, and Kern Delta Water District. Altogether, groundwater banks in Kern County can currently store over 800,000 acre-feet a year and return 700,000 acre-feet annually (Christian-Smith 2013).

Together with the development of water markets, irrigation water pricing can be altered to reflect the actual resource cost of water. Water rotations and scheduling could also be incorporated into the water market price to encourage more efficient use of water as well as providing financial incentives for water to be allocated to its best economic use while encouraging conservation across a wide range of users.

3.7 Conclusion

The expansion of water markets appears to be an effective adaptation response to climate change. Based on successes in existing water markets described in the literature and research results presented in this chapter on prospective declines in irrigation water supply reliability under climate change, many regions of the world would likely benefit from the development of water markets, particularly in the Middle East and North Africa region, southern Africa, southern Spain, parts of the United States and South America, parts of China and parts of Southeast Asia.

According to Fargher, water trading would be most beneficial and would likely see the most success in situations where: water resources are fully developed for consumptive use; seasonal water availability is variable within and among connected water systems; and demand for water is increasing for multiple uses, including agricultural, environmental, urban, and industrial uses. Also required is a large number of connected water users who have varying demands, different marginal benefits and degrees of flexibility to respond to water shortages and are exposed to fluctuations in global agricultural commodity markets. For effective functioning of the market, though, the author cites the necessity of setting a cap on total sustainable water extractions and establishing entitlements that are specified, monitored and enforced so all parties know what can be bought and sold in a regulated trading environment with proper accounting of water use. Another key to success of water markets is ensuring that there are no third-party effects (Rosegrant and Binswanger 1994).

Grafton et al. (2010) also note that successful transfers of water from low to higher valued use depend on a separation of water and land rights, an appropriate institutional and legal framework for trading, and an acceptable allocation of water rights based on past usage. Ideally, water rights should be assessed as flow shares to facilitate allocation under growing climate variability and climate change. Analysis by these and other researchers show empirically an increase in trading and value of water trading with increasing water scarcity in Australia and the United States. This supports the argument that increasing water scarcity under climate change will be conducive to development of water markets and a valuable complement to the typically proposed irrigation and storage development options.

We thus conclude that water markets are an excellent climate change adaptation tool, possibly together with other, complementary developments, such as expansion of storage, lining of canals and enhanced on-farm water use efficiency. Despite this apparent potential, we are not aware of water markets yet being proposed in any of the national climate change adaptation plans developed under the United Nations Framework Convention on Climate Change or elsewhere, largely due to the same factors that hold back market development in general. Climate change and growing variability in water supplies will, however, likely require policymakers to consider developing water markets as the value of water increases and the need for flexibility in allocating it among users continues to grow.

References

Adler J (2011) Water rights, markets, and changing ecological conditions. Paper for presentation at the environmental law symposium on twenty-first century water law. Lewis & Clark Law School, Portland, OR, US. http://law.lclark.edu/live/files/11168-421adlerpdf

Bates BC, Kundzewicz ZW, Wu S, Palutikof JP (eds) (2008) Climate change and water. Technical paper of the intergovernmental panel on climate change. IPCC Secretariat, Geneva, 210 pp

Bekchanov M, Bhaduri A, Ringler C (2013) How market-based water allocation can improve water use efficiency in the Aral Sea basin?, ZEF- Discussion papers on development policy no. 177, Center for Development Research, Bonn, April 2013, p 47

Bjornlund H, Rossini P (2010) Climate change, water scarcity and water market—implications for farmers' wealth and farm succession. 16th Pacific Rim Real Estate Society Conference. Wellington, New Zealand. http://ura.unisa.edu.au/R/AKBV2ACVJ1P9E8PR3P2MT4HM7YBGA5A9Y3DKK2EMGFFACBUT7Q-00342?func=results-brief

Brewer J, Glennon R, Ker A, Libecap G (2006) Water markets: western water transfers from agriculture to urban uses, 1987–2005. Paper prepared for ISNIE 2006: Institutions: Economic, Political and Social Behavior. Boulder, Colorado, September 21–24, 2006. http://www.isnie.org/ISNIE06/Papers06/02.5%20(no%20discussant)/kerrlibecap.pdf

Bryan E, Ringler C, Okoba B, Koo J, Herrero M, Silvestri S (2011) Agricultural management for climate change adaptation, greenhouse gas mitigation, and agricultural productivity—insights from Kenya. IFPRI discussion paper 01098. International Food Policy Research Institute, Washington, DC. http://www.ifpri.org/sites/default/files/publications/ifpridp01098.pdf

Calow RC, Howarth SE, Wang J (2009) Irrigation development and water rights reform in China. Int J Water Resour Dev 25(2):227–248

Christian-Smith J (2013) Improving water management through groundwater banking: Kern County and the Rosedale-Rio Bravo water storage district. Pacific institute farm water success stories: groundwater banking. 8 pp. Available at: http://www.pacinst.org/wp-content/uploads/2013/02/groundwater_banking3.pdf

EPA (2013) Aquifer Recharge (AR) and Aquifer Storage & Recovery (ASR). http://water.epa.gov/type/groundwater/uic/aquiferrecharge.cfm

Fargher W (Undated) Responding to scarcity: lessons from Australian water markets in supporting agricultural productivity during drought. Published paper on website of Organisation for Economic Co-operation and Development. http://www.oecd.org/tad/sustainable-agriculture/49192129.pdf

Grafton RQ, Landry C, Libecap GD, O'Brien RJ (2010) Water markets: Australia's Murray-Darling Basin and the US Southwest. NBER working paper no. 15797, Cambridge. Available at: http://www.nber.org/papers/w15797.pdf

Hansen K, Howitt R, Williams J (2006) Implementing options markets in California to manage water supply uncertainty. Selected paper prepared for presentation at the American Agricultural Economics Association Annual Meeting Long Beach, California, July 23–26, 2006

Hansen K, Kaplan J, Kroll S (2008) Valuing options in water markets: a laboratory investigation. DARE CSU working paper 001–08

Jones PG, Thornton PK (2013) Generating downscaled weather data from a suite of climate models for agricultural modelling applications. Agric Syst 114:1–5. Available at: http://linkinghub.elsevier.com/retrieve/pii/S0308521X12001242. Accessed 4 Mar 2013

Luo B, Maqsood I, Yin YY, Huang GH, Cohen SJ (2003) Adaption to climate change through water trading under uncertainty – an inexact two-stage nonlinear programming approach. J Environ Inform 2(2):58–68

Malcolm S, Marshall E, Aillery M, Heisey P, Livingston M, Day-Rubenstein K (2012) Agricultural adaptation to a changing climate economic and environmental implications vary by U.S. Region. Economic Research Report Number 136. Economic Research Service, U.S. Department of Agriculture, Washington, DC. http://www.ers.usda.gov/media/848748/err136.pdf

Margulis S (2010) Economics of adaptation to climate change: synthesis report. World Bank, Washington, DC

Meinzen-Dick RS (1996) Groundwater markets in Pakistan: participatin and productivity. Research report 105. IFPRI, Washington, DC

Meza F, Wilks D (2012) Impacts of climate change on irrigated agriculture in the Maipo Basin, Chile: reliability of water rights and changes in the demand for irrigation. J Water Resour Plann Manage 138(5):421–430. doi:10.1061/(ASCE)WR.1943-5452

Millennium Ecosystem Assessment (2005) Ecosystems and human well-being: general synthesis, Millennium Ecosystem Assessment series. Island Press, Washington, DC

Ringler C, Cai X, Wang J, Akhter A, Xue Y, Xu Z, Yang E, Zhao J, Zhu T, Cheng L, Fu Y, Fu X, Gu X, You L (2010) Yellow river basin: living with scarcity. Water Int 35(5):681–701

Rosegrant MW (2012) International Model for Policy Analysis of Agricultural Commodities and Trade (IMPACT): model description, Washington, DC. Available at: http://www.ifpri.org/publication/international-model-policy-analysis-agricultural-commodities-and-trade-impact-1

Rosegrant M, Binswanger H (1994) Markets in tradable water rights: potential for efficiency gains in developing country resource allocation. World Dev 22(11):1613–1625

Saleth Maria R (1998) Water markets in India: economic and institutional aspects. In: Easter KW, Rosegrant MW, Dinar A (eds) Markets for water: potential and performance. Kluwer Academic Publishers, Boston

Tomkins CD, Weber TA (2010) Option contracting in the California water market. J Regul Econ 37:107–141

Wang Y (2012) A simulation of water markets with transaction costs. Agric Water Manage 103:54–61. Available at: http://linkinghub.elsevier.com/retrieve/pii/S0378377411002897. Accessed 4 Mar 2013

World Bank (2009) World development report 2010: development and climate change. "Chapter 3: Managing land and water to feed nine billion people and protect natural systems." Washington, DC. http://siteresources.worldbank.org/INTWDR2010/Resources/5287678-1226014527953/WDR10-Full-Text.pdf

Young MD, McColl JC (2009) Double trouble: the importance of accounting for and defining water entitlements consistent with hydrological realities. Aust J Agric Resour Econ 53(1):19–35

Chapter 4
Supply Reliability Under Climate Change: Forbearance Agreements and Measurement of Water Conserved

Bonnie Colby, Lana Jones, and Michael O'Donnell

Abstract Climate change brings about hydrologic changes that include higher seasonal inter-annual and decadal variability in precipitation, streamflow and water in storage. This increased variability poses challenges for those responsible for providing reliable water supplies for urban and environmental needs. A new generation of risk sharing arrangements are being negotiated between owners of high reliability water supplies, often agricultural districts, and urban and environmental water providers. This chapter outlines the features of such arrangements, provides examples of where they are being implemented, discusses the challenges of measuring and monitoring reduced consumptive use on farms and implemented, and elucidates policy recommendations to help support more cost effective arrangements responsive to climate change needs.

Keywords Water supply • Climate change • Water conservation • Water transfers • Agricultural consumptive use

B. Colby (✉)
Department of Agricultural and Resource Economics, University of Arizona Tucson, 1110 E. James Rogers Way, Chavez Building, Tucson, AZ 85721, USA
e-mail: bcolby@email.arizona.edu

L. Jones
Economic Consultant, Corvallis, OR 97339, USA
e-mail: lanaesthetic@gmail.com

M. O'Donnell
Bureau of Business and Economic Research, MSC06 3510, 1 University of New Mexico, Albuquerque, NM 87131-0001, USA
e-mail: mo8684@unm.edu

K.W. Easter and Q. Huang (eds.), *Water Markets for the 21st Century: What Have We Learned?*, Global Issues in Water Policy 11, DOI 10.1007/978-94-017-9081-9_4,

Acronyms

AZMET	Arizona Meteorological Network
BOR	U.S. Bureau of Reclamation
BPA	Bonneville Power Administration
CDWR	California Department of Water Resources
CIMIS	California Irrigation Management Information System
CU	Consumptive Use
EAA	Edwards Aquifer Authority
ERS	Economic Research Service
ET(AW)	Evapotranspiration (of applied water)
EWRI	Environmental and Water Resources Institute
FAO	Food and Agriculture Organization of the United Nations
GAO	U.S. Government Accountability Office
GIS	Geographic Information System
GPS	Global Positioning System
IAG	Independent Monitoring Group
IDWR	Idaho Department of Water Resources
IID	Imperial Irrigation District
KWAPA	Klamath Water and Power Authority
LCRAS	Lower Colorado River Accounting System
LDCM	Landsat Data Continuity Mission
METRIC	Mapping Evapotranspiration at high Resolution with Internalized Calibration
MDBA	Murray-Darling Basin Authority
MODIS	Moderate Resolution Imaging Spectroradiometer
NASS	National Agricultural Statistics Service
NCAT	National Center for Appropriate Technology
NDVI	Normalized Difference Vegetation Index
ReSET	Remote Sensing of ET
SEB	Surface Energy Balance
SEBAL	Surface Energy Balance Algorithm for Land
SLC	Scan Line Corrector
TM	Thematic Mapper
UDWR	Utah Division of Water Rights
USGS	U.S. Geological Survey
USDA	U.S. Department of Agriculture
VRI	Variable Rate Irrigation
WRP	Wetlands Reserve Program
WTP	Water Transaction Program

4.1 Introduction

Public and private organizations worldwide are experimenting with innovative water sharing arrangements to match water supply and demand under climate change. Water markets can help communities balance competing needs like environmental use, food and fiber production and energy generation. This chapter examines design and implementation of water acquisition programs to improve water supply reliability in the face of the effects of climate change on water supply and demand. We explore the use of voluntary, negotiated water supply reliability contracts (reliability contracts), with a focus upon measuring and monitoring changes in agricultural consumptive use under such contracts. While reliability contracts have been used in the USA and elsewhere; they are not yet common and involve unique considerations to accomplish the goal of improving supply reliability given challenges posed by climate change.

Voluntary arrangements that pay farmers and irrigation districts to temporarily reduce or cease irrigation in order to provide water for use in other sectors and locations are common in many parts of the world (Grafton et al. 2010). In many regions, the largest and most reliable water entitlements are held in agriculture. In dry periods, environmental and urban users may face cutbacks leading to economic losses, degraded habitat and loss of endangered species. When cities, conservation groups and environmental agencies are willing and able to temporarily utilize water normally used to irrigate crops, there is potential for all water users to benefit—for example, when a city offers to pay more for water than what farmers would earn irrigating some of their crops for a season.

Water transfers from agriculture have been controversial. Impacts to local businesses, microenvironments, and the character and culture of rural areas must be considered. Although water transfer agreements have had problems, they rival other options for improving supply reliability in terms of environmental impacts and financial cost. In many regions, infrastructure to convey water from one purpose and location of use to another is already in place. Adequate legal frameworks for transferring the use of water also exist in many jurisdictions, though these frameworks can be improved (Grafton et al. 2010).

While many water acquisition programs involve similar features, terms vary. In this chapter the terms "forbearance" and "fallowing" refer to agreements to conserve water by suspending irrigation on cropped land.

In a later section, we discuss three approaches to measuring and monitoring conserved water; water delivery based methods, traditional evapotranspiration (ET) estimation, and satellite-assisted ET estimation. Although it is challenging to accurately measure changes in water consumption, such measurement is essential for reliability agreements that seek to free up irrigation water for other uses. Some past forbearance programs have been hindered in their credibility and effectiveness by inadequate attention to measuring and monitoring changes in agricultural water use (Young 2010; Keplinger and McCarl 1998).

The ability to carefully and cost-effectively measure and monitor reductions in agricultural consumptive use is particularly important for reliability agreements, as these involve intermittent (rather than permanent) reductions in agricultural water consumption. Due to this intermittency, net benefits are smaller than that of permanent water acquisitions and such agreements are sensitive to being derailed by high measurement and monitoring costs. Also, reliability agreements often must provide a specific amount of water at specific time periods, such as for seasonal habitat requirements. This timing specificity means that the effects of changes in irrigation use of water must be measurable to the satisfaction of the party contracting for the water. As we discuss later in the chapter, improved measurement technology can make time-specific and intermittent water transfers economically viable where they previously would have been impractical due to measurement costs and uncertainties.

Water supplies can be variable, and are becoming more difficult to predict as climate change progresses. Projected effects of climate change on water supplies vary by region. For much of the western U.S., climate change has the effect of increasing variability in precipitation and exacerbating extreme events such as drought and flooding. Moreover, the warmer temperatures increase evapotranspiration from cropland and also increase urban water demand for outdoor use and for electricity used to cool indoor spaces (Garrick and Jacobs 2006; Williams 2007; Garfin 2012). This combination of effects on water supply and demand poses challenges for maintaining reliable supplies for cities, farms and ecosystems.

There are many approaches to address supply variability. These include reducing water demand through pricing or regulations and increasing regional storage capacity to carry water over from wet periods to dry periods. Voluntary agreements to temporarily transfer water, triggered by pre-specified low water supply conditions, are another method for mitigating variability and sharing risk of shortage in regional water supplies (Colby et al. 2011a, b). Desirable characteristics of such agreements include: (1) matching willingness-to-pay for improved reliability with a water supply arrangement that provides the desired level of reliability, (2) low negotiation, implementation and conflict resolution costs (Michelsen and Young 1993; Colby 1990), and (3) reduced net regional economic losses during periods of low supply (Colby et al. 2011b).

We use the term reliability contract to refer to contractual arrangements made in advance of need under which a change in water use is triggered by low supply conditions. The potential usefulness of such arrangements have been recognized in the literature for some decades. Michelsen and Young (1993) evaluate option contracts for temporary use of irrigation water rights and find them a cost effective form of drought insurance for urban water agencies. They provide a framework for evaluating the economic benefits of water supply options and demonstrate that dry year water option are economically viable over a considerable range of economic and hydrological conditions. Howitt (1998) notes that option markets for longer-term contracts can have significant advantages over permanent purchases of water, especially where transaction costs of acquiring permanent water rights are prohibitive.

As with contracting devices in general, a reliability contract specifies payment and risk sharing between the contractor (the party seeking to procure more reliable water) and the contractee (the party providing water). Typically, a reliability agreement includes a negotiated upfront payment and identifies a trigger event. If the trigger occurs, then the contractor is entitled to use the volume of water specified in the contract under the terms specified. In the western United States, contractees are generally irrigators and irrigation districts, since irrigation withdrawals account for approximately 80 % of the freshwater withdrawals and because a large portion of the most senior and drought-proof water entitlements belong to agricultural interests (USGS 2009).

A reliability contract may extend for a single season or year, or for any number of contracted years (Mays et al. 2002). The time horizon of the contract will depend in part on the type of water supply variability that the contractor would like to mitigate (Mays et al. 2002). Reliability contracts allow the contractor to guard against several types of risk, described below, while the contractee receives consideration. Consideration may take a variety of forms: a monetary payment, debt forgiveness, favorable pricing for services, livestock feed to substitute for crops not grown, water management benefits or other services. For farmer contractees, consideration received can be treated as analogous to producing another "crop" (forbearance) in the farm's financial risk management portfolio (Jones and Colby 2010).

The contractor acquires protection against water shortage, as the option may be exercised if the trigger event occurs (Williams 2007). Also, by not purchasing or leasing the water outright, the contractor guards against the costs of holding too much water in wet years; including unnecessary purchase and storage costs (Williams 2007). The contractor guards against the risks associated with conflicts over permanent water transfers, which may limit the quantity and duration of a transfer agreement (Howe 1996). Finally, the contractor guards against the risk of volatility over time in the cost of acquiring water (Woo et al. 2001) by locking in a contractual price (which may vary with pre-specified indicators, such as regional inflation factors), for the life of the contract.

4.1.1 Types of Reliability Contracts

If an option contract framework is used, the agricultural water user selling the option receives a negotiated payment per volume of water and refrains from consuming that volume for the contractually specified period. The efficacy of this arrangement rests on the probability that water is actually available in the system for the entitlement holder. Under severe dry conditions, even very senior entitlements may not yield water. When the option is exercised, the contractor pays the contractee a specified additional amount of consideration (exercise payment) per volume of water obtained; the irrigator then fallows a portion of his land in order to transfer water that would have been used for irrigation to the contractor in accordance with the terms of the contract (Hass 2006). The upfront consideration (option premiums) and exercise payments can help to smooth out the typical variability in agricultural revenues by diversifying an irrigator's agricultural portfolio to include water leasing revenues (Mays et al. 2002; Jones and Colby 2010).

A key distinction between the typical option contract framework and other reliability contracts is the payment structure and/or the type of consideration exchanged. For instance, instead of paying an option premium, a contractor may elect to purchase an irrigator's land and then lease that land back to the irrigator and allow the irrigator to continue irrigation. If an agreed upon trigger occurs, the irrigator relinquishes the right to irrigate. This type of arrangement is sometime referred to as a contingent lease-back. In this case, neither an option premium nor an exercise payment is paid to the irrigator; however, this arrangement is considered a reliability contract because it is an arrangement made in advance of need that is triggered by a specific event. Examples of varying supply reliability agreements are provided below.

4.2 Designing and Implementing Reliability Contracts

In designing any reliability contract that utilizes irrigation fallowing, it is essential to clearly articulate the overall goal of the contract. Generally this will be to secure water at a reasonable price in order to make water available for another use when the supplies typically available for that use are short. Other goals of a contract could include securing water in a manner perceived as "fair", with minimal disruption to local economies; and instilling confidence in contractors while maintaining the goodwill of contractees.

A key consideration when engaging in a reliability contract is to determine what volume of water is needed to achieve the desired levels of supply reliability. Because it costs the contractor more money to keep a larger volume of water in option, the contractor's goal is to keep the minimum volume of water in option to achieve adequate insurance against supply shortfall. This balancing should incorporate available climate and hydrological models used for predicting supply variability, where practical (Hartmann 2005; Troch et al. 2008; Lyon et al. 2008; Tueling et al. 2007; Hirsch et al. 1993; Salas 1993; Stedinger et al. 1993). Streamflow and weather forecast models can assist in determining if a year will be relatively wetter or dryer. In dryer years, or in those years when reservoir storage is low, it may be appropriate to place more water in option.

Negotiating a price for a supply reliability arrangement can be complex. From the perspective of the contractor, it is optimal for them to use the minimum amount of money required to keep an option open. On the other hand, the contractee likely wants to receive the largest payment possible for enrolling a portion of their acreage. Likewise, if the option is called, the contractor would like to spend the minimum amount of money where the contractee would like to receive the maximum payment. Negotiations must be successfully concluded between the contractor and the contractee to determine a mutually acceptable trade.

Sophisticated approaches have been proposed in the literature to value option contracts and determine appropriate prices. Villinski (2003a, b) draws upon financial derivative pricing theory to value option contracts. She notes that option contracts for water can be more complex than those for financial instruments and that no simple pricing formula to calculate their value exists. Villinski builds on the Black-Scholes option-pricing framework and uses dynamic programming techniques to value option contracts for water.

Hansen et al. (2013) utilize experimental economics to analyze the effect of annual dry-year options on water markets, in the absence of adequate data for econometric analysis. They consider how market structure and option contract availability affect water price and allocation within a market. Their research suggests that gains from trade are on average higher when options can be traded. Further, once an options market is available, gains from trade are more evenly distributed between the single buyer and the many sellers in the case of monopsony. Hansen et al. (2008) use a hydro-economic simulation model to examine the value of transferring risk from one party to another through an options market, exploring the optimal price for options and gains from trade under varying conditions.

Determining how much to pay for the option premium and exercise payment can be a difficult task, particularly in regions with relatively rare transactions. The prices paid should reflect current market conditions for water rights and the level of risk associated with supply shortfalls. Put differently, the offer amount should be gauged against the benefits foregone by using the water in the proposed environmental or urban use and foregoing the usual irrigation use of the water (Jaeger and Mikesell 2002; Hansen et al. 2008). This may be difficult to accomplish, particularly when the contract extends for multiple years. Additionally, "the value of water varies enormously, depending on the supply source's reliability, quantity of water, access and cost of conveyance, duration and firmness of contractual commitments, and the buyer's type of use and alternative sources of comparable water supplies" (Water Strategist 1997). Economic conditions, federal farm programs, political climate and many other variables are also important factor in determining the value of water. In order for a deal between the two parties to be realized, the sum of money paid to the irrigator must equal or exceed the net income she would have received had his land not been fallowed (Hass 2006).

An important concern when structuring the option contract is to determine the date range within which the option may be called. In this determination, two main issues are important: first, the window to call the option must be timed such that the contractor is able to take delivery of the water when it is most likely to be needed. For instance, if the optioned water is needed in summer the contractor will want the call window to be in spring, not in fall. If the options are called too early, the contractor faces the risk that the optioned water will no longer be needed if late spring rains ease the drought.

Second, the contractor's optimal timing windows must be balanced against financial considerations for irrigators in their seasonal farm planning and operations cycle. If the call window is negotiated near to or after the planting cycle, then irrigators will demand a higher option premium in consideration of crop production costs already incurred. The closer to the planting cycle that the option window is open, the more costly it is for the irrigator to cease irrigation on short notice. The Metropolitan Water District of Southern California (MWD) encountered this timing issue in 2003 and increased the premium that it paid irrigators for keeping the option window open an additional month (Jenkins 2007).

4.3 Trigger Mechanisms

There is no clear cut method for determining what event to select that will activate the option. The trigger should be pre-specified, objective, not influenced by actions of parties to the agreement and observable to the participants so that they have a reasonable expectation of the outcome. The trigger should also be related to the ultimate purpose of the optioned water (Mayes et al. 2002). For instance, if securing adequate water flows for fish is a goal, then the option would be called as the stream approaches a predetermined critical level (Willis et al. 1998; CDWR 2000). Stream flow was proposed as a trigger for calling an option in the Snake River Basin to ensure adequate water levels for the salmon population (Willis et al. 1998). In areas where winter runoff provides an important water supply, winter runoff volume may be used as a trigger mechanism. Bonneville Power Administration (BPA) has used winter runoff volume as an indicator of when to activate forbearance agreements with irrigators to ensure water availability (Bonneville Power Administration 2002).

Reservoir elevation is a potential trigger for calling optioned water in regions where water elevation levels in specific reservoirs are good indicators of impending supply shortfalls (CDWR 2000). In areas where groundwater is used to supplement surface water, marked changes in groundwater levels or in groundwater pumping may be used as a trigger to call optioned water (CDWR 2000). A fall in groundwater levels or an increase in groundwater pumping may indicate drought conditions, as these can result from surface water resources being limited. In order for groundwater triggers to be effective, groundwater levels and/or pumping must be measured and a threshold for calling the optioned water must be developed. Where groundwater is used to supplement surface water supplies, then it may be valuable to create a trigger index based on some combination of reservoir levels and groundwater conditions.

Climate and streamflow forecast information may be useful in structuring dry year water use arrangements (Hartmann 2005). Climate and water supply forecasts may be useful in predicting how often a trigger condition would occur in a decade. This information can be valuable in structuring the contract as the contractor likely will want more frequent opportunities to exercise options and irrigators may wish to have higher option premiums to compensate them for more frequent disruption of farm operations. More general climate information, such as whether a particular year has a strong El Niño with snowpack likely to be above average, can be valuable to both contractor and irrigators in their planning. In addition, climate forecasts may also be useful to determine when to call optioned water within a specific year.

4.4 Dry-Year Reliability Contract Examples

Dry-year option contracts have been used intermittently in California beginning in the early 1990s (Jercich 1997). In 1995, the state of California's Water Bank negotiated contracts with local irrigation districts for the option to purchase 29,000 acre-feet of water, paying an option premium of $3.50 per acre-foot (Jercich 1997). The Bank was permitted to call the option by May 1995 and if the option was not called, the irrigators kept their option premiums (Jercich 1997). If the options were called, the price paid to the irrigators would have been a pre-negotiated price of $35.50 to $41.50 per acre-foot, in 1995 dollars. In this instance, the options were not called because the winter months were wetter than anticipated, so additional water was unnecessary (Jercich 1997).

In the winter of 2002, MWD negotiated with the Sacramento Valley irrigation districts for 1-year option contracts for 146,000 acre-feet of water (CDWR 2002; Jenkins 2007). Under the contract terms, MWD had until March 2003 to call the option and if the option was not called, the irrigators kept their option premium of $10 per acre-foot (CDWR 2002; Jenkins 2007). If the option was called, MWD was obligated to pay an additional $90 per acre-foot for the option water (Jenkins 2007). Because the end of 2002 and beginning of 2003 were dry, MWD called all of the options (MWD 2003a; Jenkins 2007). In April, after the options were called, it began to rain – making the called water unnecessary (MWD 2003b; Jenkins 2007). As a result, MWD had more water than could be stored and much of the option water flowed out to the ocean (Jenkins 2007).

In an effort to minimize the likelihood of repeating the 2003 experience, MWD negotiated with the irrigation districts for an additional year of option contracts. However, it negotiated with the irrigators to extend the deadline to call the optioned water from March to April in exchange for a higher option premium of $20 per acre-foot (MWD 2004; Jenkins 2007). In 2005, this contract modification was validated as a relatively heavy rain hit in April, making calling the optioned water unnecessary (Jenkins 2007). Although MWD paid a total of $1.25 million in option premiums in 2005, it would have had to pay $16 million if it had purchased the water outright.

In addition to relatively short-term option contracts, MWD has also entered into a long-term (35-year) fallowing contract with the Palo Verde Irrigation District (PVID) beginning in 2005. Under the terms of the agreement, a base load area of approximately 6,000 acres will be fallowed for each of 35 years and MWD may exercise its option on a maximum of 115 thousand acre-feet in ten of those 35 years (Bowles 2008). Participants are not allowed to switch to groundwater if options are called, and the agreement requires participating irrigators to participate in land management measures including weed control and erosion control (PVID 2004b). MWD determines the acreage for fallowing and that is based upon forecast demand, supply and storage conditions. MWD agreed to pay $3,170 per acre for the landowner's maximum fallowing commitment, where a maximum of 35 % of a particular landowner's land is eligible for the sign up payment (Trends 2004). If an option is called, MWD will pay an additional $602 per acre fallowed that year, to be adjusted over time based on the consumer price index (PVID 2004a).

In addition to the typical option contract framework, there are other interesting examples of agreements that may be made in advance of need and are triggered by a particular event. The first is the Bonneville Power Administration's (BPA) load reduction program. BPA is a federal agency headquartered in Portland, Oregon which markets hydro-generated power to the Pacific Northwest (BPA 2008). Because electricity generation is tied to water availability, a reduction in water volume can limit generation capacity. In dry years, BPA uses load buy-downs to achieve electricity load reductions in an effort to limit their own water demand (BPA 2002). This ensures that minimum stream flows for fish passage are met (BPA 2001). Years are considered 'dry' when winter runoff is below a predetermined volume (BPA 2006). The program works by BPA paying either irrigators or industries not to use electricity.

An important component to this arrangement is that BPA's dry-year buy-downs may only be used during specified times during the calendar year for some purposes. For example, a load buy-down is available whenever a large industrial electricity user is operating at high capacity and is willing to participate, whereas an irrigation buy-down is available only between April and September and must be implemented prior to planting (BPA/KC 2001).

Another example of an innovative supply reliability agreement occurred in Utah when a city paid a farmer $25,000 for a 25-year dry year option and agreed to provide $1,000 and 300 t of hay in any year that the option was exercised (Clyde 1965). Because of this agreement, the city was able to acquire the volume of water that it desired and the farmer was able to continue farming operations. A similar model was used by the Oregon Water Trust when it paid a farmer $6,600 to compensate him for not growing hay to feed his livestock (Anderson and Leal 1998).

A similar supply reliability contract is a contingent lease-back, as discussed above. A contingent lease-back is an agreement in which land and water are purchased by the entity desiring long-term control of the water and are leased back to the irrigator so that irrigation can continue except when water is needed to replace drought shortfalls (Colby 2003). This is similar to an option contract in the sense that the water may be called periodically, requiring irrigation to be suspended. In order for this arrangement to be attractive to farmers, the up-front payment by the water seeking entity to purchase the farm and water rights must be attractive, along with the timing of notice to cease irrigation and other terms of the lease. Such arrangements have been used in Arizona and Colorado, where cities purchased agricultural land and water entitlements anticipating future growth. Since the acquisitions occurred in advance of needing to transfer the water acquired out of agriculture to urban use, the cities leased the land and water back to farmers for continued agricultural use (Saliba and Bush 1986).

An ongoing and innovative region-wide arrangement exhibits some of the characteristics of a supply reliability contract and is discussed here to illustrate that arrangements of this type can cover large geographic areas. In 2007, after years of complex negotiations, the seven states of the Colorado River Basin agreed to a new approach to sharing shortages among themselves. Provisions to adjust how much water the states can take from the shared river are triggered when reservoir elevations in Lake Mead hit specified low levels. Lake Mead is the key storage reservoir in the lower portion of the basin (DOI 2007).

In 2012, Mexico entered into the Colorado River shortage sharing agreement by joining with the U.S. to amend the bi-national treaty governing allocation and water quality on the Colorado River (Minute 319). The Treaty amendment, Minute 319, allows Mexico, which lacks storage capacity, to store some of its Colorado River allotment in Lake Mead. This significantly enhances Mexico's storage capacity and supply reliability from the river and helps maintain more water in Lake Mead which assists water users in the U.S which rely on the reservoir. Minute 319 also allows the U.S. to send less water to Mexico in drought years, sharing more broadly dry year shortages otherwise borne solely by water users in the U.S. These new arrangements provide added clarity on how shortages will be managed and allow better planning for low water supply conditions.

4.5 Measuring and Monitoring Reduced CU

This section considers in some detail measuring and monitoring in supply reliability arrangements. Cost effective measuring and monitoring is particularly crucial for temporary and intermittent transfers because the net economic benefits of temporary transfers are small compared to outright purchase of agricultural water rights. Consequently, such transfers can readily be derailed by high measurement and monitoring costs. The aim of this section is to identify successful strategies to measure and monitor changes in agricultural consumptive use (CU). Failure to consider and address these issues can make supply reliability agreements infeasible, burdened by the weight of excessive measuring and monitoring costs.

The task of ensuring that the water being purchased is actually conserved can be broken down into two main components: (a) establishing a measurement protocol to assess reductions in consumptive use by the contractee, and (b) monitoring ongoing compliance by contractees. Defining and measuring conserved water is the more technically difficult part of the process. The expenses a program can reasonably incur in measuring conserved water will depend on the nature of the program and what the water will be used for. Technical measurement and monitoring choices should be based on a balancing of costs and benefits. What are the benefits of having a more precise measure of conserved water and what are the costs of that precision? With new remote-sensing based methods of measuring ET, it may be possible to get accurate measurements at a lower cost than what was the norm in the past.

A few definitions are helpful. Evapotranspiration, ET, is the amount of water used by plants plus the water that evaporates from soil and plant surfaces. Water that is applied but not consumed (such as drainage or recharge) is not part of ET. After water is diverted from its source and applied to fields, some of the water evaporates immediately, and some water is taken up and transpired by crops (together, these uses of water are *evapotranspiration*). Some of the water may run off the field and collect into side streams, while some water may percolate through the soil to reach an aquifer. The water which has returned to a water source, above the surface or below, is termed *return flow*. The remainder of the water, which was used by the crops or immediately evaporated, has been deemed to be consumed by the crops. This use of water is said to have been *consumptive use*, because the water did not return to a water source. The amount of water that is consumed for irrigation may differ from the amount that is available for transfer. The transferable quantity will vary based on applicable laws as well as on irrigation technologies used and the method by which water is being conserved.

Opportunities to reduce agricultural consumptive use fall into four general categories: (a) leaving cropland fallow, unplanted and not irrigated, (b) planting and harvesting the customary crop mix but applying and consuming less water than customary through deliberate deficit irrigation, (c) planting and harvesting the customary crop mix but applying and consuming less water than customary through changes in irrigation and water conveyance technologies (replacing flood irrigation with sprinkler irrigation or lining a dirt canal to reduce conveyance losses), and (d) changing to a crop mix with lower consumptive use than customary (Merchant 2005). Some of these changes in the crop production and irrigation process can minimally decrease (or even increase) net income from crop production while using less water. Various jurisdictions have differing laws that affect whether a farmer can become more efficient in the use of water and retain the right to use or sell that conserved water.

Advances in measuring and monitoring can make innovative supply reliability arrangements feasible. Deficit irrigation is one option. With deficit irrigation, crops are supplied water below their full transpiration potential. This is an attractive water transfer option because it allows farmers to grow their customary crops and conserve water at the same time. Yields decrease with deficit irrigation but profit losses can be offset by water conservation payments. Water savings per field might be small compared to fallowing, but deficit irrigation over large areas can add up to significant conservation. See Lindenmayer et al. (2011) or Geerts and Raes (2009) for some advantages and disadvantages of deficit irrigation.

We describe measurement and monitoring methodologies and practices based on a review of programs that seek to reduce consumptive use in crop production throughout arid regions of the world. Water conservation and transfer programs from California, Texas, Idaho, Oregon, the U.S. Department of Agriculture (USDA) and Australia are considered here. Fallowing programs in California, Oregon, and Idaho conserve water historically used to irrigate crops for transfer to other purposes. Recent iterations of these programs are examined. The Australian program discussed here is not a fallowing program, but rather an attempt to cap diversions in the Murray-Darling Basin. All the programs described originated in the 1990s or early 2000s. Many of the programs are still operating and have been continually refined with passing years.

As demands on water intensify, the precise amount of water being consumed in competing uses is important both for conservation management and for potential water transfers. Transfer of water conserved by cropland fallowing gives rise to a complex measurement problem. If a specific area of cropland is not irrigated, how do contractors know how much water is "saved" and potentially available to transfer? Answering the question of how much water would have been used on a field had it been planted, irrigated and harvested depends upon commingling influences. These include irrigation management practices; temperature and rainfall patterns; irrigation practices on neighboring farms; soil characteristics; economic incentives to plant competing crops; and return flow.

Return flow is an issue because decreases in return flow can damage other water users and habitats. Return flow is water that is applied to crops but not consumed. It includes drainage, recharge, and runoff. For example, suppose a farmer applies 50 acre-feet of water to his fields each year, his crops consume 30 acre-feet, and the remaining 20 acre-feet is return flow. Downstream farmers rely on 20 AF as return flow for use on their fields; it may also support habitat along drainage canals and fields. If the farmer sold all 50 acre-feet, the downstream farmers and habitat would be injured by the sale. Measuring fallowing savings in terms of applied water potentially exaggerate water savings and lead to insufficient water supply downstream. If the farmer is allowed to sell only his consumptive use (i.e. 30 acre-feet), the water that would have been return flow is left instream for other users. Return flow accounting is an important issue that should be evaluated. Return flow that cannot be used for irrigation could still be valuable. If it drains into a saline sink, for example, it might have important recreation or habitat value.

Another issue requiring careful consideration is the availability of reliable records of historic irrigation water use. Without these, the ability to measure consumptive water use savings is limited. Some early transfer programs, like the Edwards Aquifer Authority (EAA) Irrigation Suspension Program, had no concrete way to gauge past water consumption in order to calculate conserved water. Water meters at the farm and field level were uncommon, so the program aimed to reduce the amount of land in agricultural production instead of aiming to reduce water use by a specific amount. EAA selected land for the program based on the expected effects its temporary retirement would have on court-ordered flow levels for environmentally important springs. The land's location, past crops, and irrigation technology were used to rank its effect on spring flow. Keplinger et al. (1998) later estimated the program's water savings based on weather patterns and irrigation estimates. However, as water becomes scarcer and transfers more expensive, buyers demand more solid evidence of how much water their funds are conserving.

The technical methodology for estimating how much water will be saved through fallowing and made available for transfer needs to be agreed upon in advance if reliability agreements are to succeed. As the Water Transfer Workgroup (2002) reported, based on California's experience, "The inability of interested parties to agree on the volume of transferable water associated with the short-term fallowing of agricultural lands has caused substantial controversy and delays in approving water transfer proposals." Forbearance agreements must be based on accepted methods for estimating water consumption, consider local complexities such as return flow accounting, and be consistent with applicable laws.

4.6 Monitoring Compliance in Water Transfer Agreements

While accurate measuring is an important goal, careful monitoring is also important to ensure that actual conservation aligns with the amount agreed upon for reduction of consumptive use. Site visits are the most common form of monitoring in the programs surveyed.

Water buyers may be able to decrease the burden of monitoring with careful program structuring. For example, farmers enrolled in IID's fallowing program agree to have their field head gates locked, blocking water flow to enrolled fields. Water transfer contracts need to stipulate that farmers must not use water saved from an enrolled field to increase water use on a different field.

Significant resources have been spent in the U.S. and elsewhere to refine the methods for measuring water savings in voluntary programs to reduce agricultural use. Less attention has been paid to the monitoring and enforcement of programs. Monitoring may be comparatively straightforward in fallowing programs with the use of intermittent site visits and satellite imagery, as it is relatively simple to verify that crops are not being planted and irrigated. A proposal to reduce consumptive use through deficit irrigation, or through changing the mix of crops grown, requires more sophisticated monitoring techniques. Due to recent advances in remote sensing for crop and ET mapping, monitoring these may be easier than it has been in the past.

4.7 Innovations in Measuring and Monitoring

In this section we summarize ideas for improving measurement and monitoring protocols through use of GIS tools, remote sensing imagery, and other innovative ideas that have been either proposed or utilized in fallowing programs. We also look at strategies that may be adapted from other programs seeking to alter agricultural practices and to measure or monitor those changes.

Improving current measuring, monitoring and enforcement of temporary water transfers will benefit both contractors and contractees. If employing innovative techniques reduces the costs of monitoring water transfers, those savings can be transferred to participating farmers in the form of larger payments. The cost savings from monitoring improvements can also be used for additional water acquisitions. Water conservation programs lacking careful measuring, monitoring, and enforcement put themselves at risk of wasting money and water. In his 2010 report on the lessons learned from the Australian Water Reform Program, Young points out that although the Murray-Darling Basin Cap prevented states from issuing new water licenses, water use expanded considerably and water dependent ecosystems declined. Young (2010) attributes the over-allocation problem to the lack of on-farm monitoring of unmetered uses like tree plantings, and reduced return flows from irrigation efficiency improvements. He concludes, "it will be necessary to manage all forms of water use with greater precision."

Measuring and monitoring innovations discussed below range from those gaining acceptance in agricultural uses to those not yet tried or tested. Innovative techniques and technologies could increase measuring or monitoring accuracy and may not be expensive; however, the innovations need to be carefully matched to local needs and to appropriate uses. A technology useful for monitoring whether an enrolled field is being fallowed may be impractical for measuring the amount of water conserved by fallowing. Likewise, a technology suited to measuring water savings might be too complicated or expensive if all that's needed is verification that no irrigation has taken place on enrolled fields. New technology installation is required in some cases but existing technology or policy adjustment can also be used as a low-cost means to increase measuring, monitoring, and enforcing power.

4.8 Remote Sensing

Free remote sensing image archives, provided by NASA and the USGS, now span over three decades. This collection is an enormous resource for water managers. Although remote sensing is beginning to be used in official capacities, its full potential is not yet realized. The Bureau of Reclamation uses remote sensing data to map crops and riparian vegetation in the Lower Colorado River Basin (BOR 2009) and has also used remote images to monitor fallowing programs (DOI 2009). Researchers in the U.S. are estimating field-specific evapotranspiration (ET) using satellite images (Allen et al. 2007b; Elhaddad et al. 2011; Glenn et al. 2011; Piñón-Villarreal et al. 2010).

More and more conservation programs, including the USDA's Wetlands Reserve Program (USDA 2006) and the Idaho Department of Water Resources Water Transfer Program (IDWR 2006), are using satellite images to monitor participants. Measuring water savings (based on changes in ET) with satellite images is a promising, though more complex, innovation. Samani et al. (2007) recommend the use of remote sensing to calculate ET, "over a large scale for various crops, riparian vegetation and soil conditions without the complications associated with traditional methods of assessing ET."

Four uses of remote sensing to monitor and measure water use are examined here. The first two cases are primarily suited for monitoring participant compliance; the second two are suited for measuring water savings. In the first case, the Bureau of Reclamation's Lower Colorado River Accounting System (LCRAS) uses satellite imagery together with a GIS database to map vegetation annually. Once crops are mapped, the BOR estimates ET using classic crop-coefficient methods. In the second case, the National Agricultural Statistics Service's CropScape uses satellite images to produce crop cover maps for major crops all over the U.S. In the third case, surface energy balance methods combine thermal satellite images with ground-based weather data to estimate evapotranspiration. In the fourth case, vegetation index based methods combine indexes like the Normalized Difference Vegetation Index (NDVI) and traditional ET methods to estimate a relationship between local vegetative cover and ET.

4.9 Monitoring Reduced Consumptive Use

Programs wishing to adopt remote sensing for monitoring could start by monitoring fields using satellite images. With 30 by 30 m pixels (.22 acres), Landsat images are detailed enough to show fallowing on fields as small as a few acres. MODIS images, with a 250 by 250 m pixel size (15.44 acres), might be used to monitor large fields. The two examples below describe monitoring using remote sensing.

4.9.1 Lower Colorado River Accounting System

The Bureau of Reclamation Lower Colorado Regional Office (BOR) reports diversions, return flows, and consumptive use of water diverted from the Colorado River below Lee Ferry in their annual Decree accounting reports.[1] The LCRAS was created to improve estimates of consumptive use from Hoover Dam to the U.S.-Mexico border.[2] LCRAS reports provide estimates of evapotranspiration (ET) for three ground covers: irrigated agriculture, riparian vegetation, and open water. The BOR uses remote sensing data from satellite images and GIS mapping to identify the different ground cover areas along the Colorado River, which are then stored in a GIS database. The satellite images come from Landsat 5, Landsat 7, and Terra satellites (BOR 2009). Ground surveys are also collected on about 12 % of irrigated fields and these are used to classify crop group images and to ground-check the accuracy of the classifications (BOR 2009).

Using remotely-sensed spectral data and the spatial database, the BOR identifies the crop groups grown on each field each year (BOR 2009). In 2008, the BOR identified 28 crop groups, including alfalfa, small grains, lettuce, and fallow. The BOR calculates ET based on crop groups but does not estimate ET for fallow fields; however, this does not mean that fallow fields have zero water use. As noted by the California Department of Water Resources, water seeping from nearby fields and weed growth may allow significant amounts of water to evapotranspire from fallowed fields (CDWR 2009).

A fallowing monitoring policy could be adapted from the BOR's method for identifying fields and ground cover. Their crop group mapping accuracy in 2008 was over 90 % accurate (BOR 2009). Although accuracy in the detection of fallow versus crop groups was not reported separately, the importance of distinguishing between the two is most likely minimal as the difference between crops and fallow fields is easier to detect than differences between types of crops. Distinguishing between crop types would become particularly difficult late in the growing season when crop vegetation is mature.

---

[1] "Lower Colorado River Water Accounting," Bureau of Reclamation, modified January 2012, http://www.usbr.gov/lc/region/g4000/wtracct.html

[2] *Ibid.*

Accuracy depends in part on field size. Landsat TM pixels are 30 by 30 m.[3] As a rough rule of thumb, monitoring fields larger than one hectare by remote sensing should be feasible depending on field shape. In programs that wish to use Landsat as the primary monitoring method, minimum field size should be taken into account during enrollment.

Remote sensing data might also offer future potential for automated monitoring. Once enrolled fields were identified in a GIS database, satellite image retrieval and spectral comparison might be automated. If the spectral data for an enrolled field fell outside expected fallow parameters, the monitoring agency could verify on the ground.

4.9.2 CropScape and the Cropland Data Layer

CropScape[4] is the online geospatial data and visualization service of the Cropland Data Layer produced by the National Agricultural Statistics Service (NASS). The Cropland Data Layer (CDL) contains information about the type and location of over 100 major crops grown in the U.S. The CDL uses images from Deimos-1, UK-2, Waifs, and Landsat 5 satellites.[5] CropScape is the online portal for viewing the CDL. It doesn't require specialized software to view or download the spatial data. CropScape's uses include land cover monitoring and biodiversity assessment (NASS 2011).

CropScape imagery is produced at a sufficient scale to monitor individual fields—30 m. However, at this time it is only available for historic, not real time, monitoring. NASS does not release CDL data during the growing season because of its possible detrimental effect on the market and investors; and because it is unverified until the growing season is over (Mueller 2011). Early in the growing season it can also be difficult to distinguish fallowed fields from other fields, including rangeland (Mueller 2011). CropScape's release schedule limits its use as a real-time fallowing monitoring tool. However, it provides a detailed example of how GIS and satellite images can be combined for agricultural monitoring. Agencies might be able to develop a local real-time monitoring tool using CropScape as a guide.

For measuring actual reductions in water consumption, CropScape could be used to establish past cropping patterns and estimate fallowed water savings. Its suitability will depend on the program location, whether the crops grown are among the over 100 crop types that CropScape supports, and the complexity of cropping patterns.

---

[3]"Land Cover Analysis," NOAA, last accessed April 20, 2012, http://www.csc.noaa.gov/crs/lca/faq_gen.html

[4]"CropScape – Cropland Data Cover," USDA NASS, last accessed April 20, 2012, http://nassgeodata.gmu.edu/CropScape/

[5]"About CDL," USDA NASS, last accessed April 25, 2012, http://www.nass.usda.gov/research/Cropland/SARS1a.htm

4.10 Measuring Conservation

A more advanced application of remote sensing involves using images to estimate changes in actual farm field evapotranspiration. Remote sensing models to estimate ET have developed along two main lines: surface energy balance or vegetation index (Bastiaanssen et al. 2005). Researchers in the western U.S. have been working to tailor energy-balance models, such as SEBAL, to local climates and needs (Allen et al. (2007b) in Idaho; Piñón-Villarreal et al. (2010) in New Mexico; Taghvaeian (2011) in the Lower Colorado River Basin). ET models based on vegetation indexes have developed as an alternative and a complement to energy-balance models (Glenn et al. 2011; Hunsaker et al. 2007; Nagler et al. 2009).

The availability of historic as well as current remote sensing imagery gives users more options for measuring water conservation. Fallowing programs that estimate water savings using cropping pattern and crop coefficients could replace coefficients with remotely sensed ET. Alternately, fallowing programs could estimate water savings by comparing remotely sensed ET on fallow fields and cropped fields in the same time period.

Remote-sensing ET models are complex and the investment required to implement them may not be justified in all water acquisition programs.

4.10.1 Surface Energy Balance

Surface energy balance (SEB) models use thermal satellite images to estimate ET. They are based on the principle that the energy arriving at and leaving a field surface must balance. As water evaporates and transpires it uses energy, so ET is a factor that can be estimated in the surface energy balance. Researchers have developed many SEB models. An example model, METRIC, is described here. Others include REEM (Samani et al. 2007), and ReSET (Elhaddad et al. 2011).

METRIC is an energy-balance method for mapping evapotranspiration at a moderate scale (limited by thermal image spatial and temporal resolution). The maps can be used for water rights conflicts, groundwater management, and determining actual ET at many scales (Allen 2005; Allen et al. 2007a). METRIC uses thermal images from Landsat satellites. METRIC offers a way to measure field water use at a seasonal scale. As farmers irrigate, their fields become cooler because evapotranspiring water absorbs energy. This temperature difference is what the satellite images highlight. METRIC combines data from Landsat with ground-based reference ET to boost accuracy (Allen 2005).

METRIC offers a potentially cost effective, accurate way to measure water conservation using Landsat data. To date, six Landsat satellites have captured data going back over three decades,[6] so ET can be estimated for prior as well as current years.

[6]"From the Beginning," NASA, modified April 19, 2012, http://landsat.gsfc.nasa.gov/about/history.html

No major capital investments are required to implement METRIC although local weather stations are required. Allen (2005) lists three necessities for estimating ET using METRIC: high-resolution thermal satellite images, good quality weather data, and an "experienced, thinking human at the controls." Given that Landsat images remain free and that local weather station data is not hard to come by, the limiting factor for implementing these kinds of models is probably staff training.

4.10.2 Vegetation Index

Vegetation index (VI) methods are a less computationally intense way to estimate ET with remote sensing. Simplified versions of these methods might also be used to monitor conservation. Most vegetation index (VI) methods estimate the transpiration part of ET by comparing the amounts of red and near infrared light in an image. The Normalized Difference Vegetation Index (NDVI) is a commonly used VI. Its values range from -1 to 1. NDVI for barren areas is close to zero.

Vegetation indexes are used in conjunction with ground-based reference ET (ET_0) to estimate ET. As part of their implementation, VIs must be calibrated by being regressed against ground ET measurements in the area where ET will be estimated. Empirical relationships are established between the VI, ET, and weather data to find the local correlation between vegetation cover and ET (Senay et al. 2011). VI methods cannot estimate evaporation like an SEB can, but plant transpiration is the larger part of ET and often the major unknown (Nagler et al. 2009). Once calibrated, VI methods can predict actual ET.

The main advantage of VI methods compared to SEB is ease of calculation. Rafn et al. (2008) say their NDVI method, "is relatively fast, easy, and requires little knowledge of evaporation physics and aerodynamics." The primary disadvantage of vegetation index ET estimation is the focus on transpiration only.

4.10.3 Precision Irrigation

In the Flint River Basin, Georgia, a collaboration that began in 2004 to conserve water, has since saved billions of gallons of water (TNC 2010). The Nature Conservancy, Flint River Soil and Water Conservation District and the U.S. Department of Agriculture formed the Flint River Basin Partnership and have worked with farmers in the region to install the variable rate irrigation (VRI) systems responsible for the water savings. VRI systems are limited to use with sprinkler irrigation. They are also expensive, requiring regional and farm level infrastructure investments. Although the expense could be hard to justify solely for improved measuring or monitoring, VRI and other precision irrigation technologies could benefit farmers by improving yields, and lowering water use and associated input costs. This potential, combined with improved measuring and monitoring benefits for water transfer programs, could make them feasible in areas with multifaceted water conservation programs.

In a VRI system, a GPS is mounted on a central pivot or lateral move sprinkler irrigation system. A controller on the sprinkler uses soil moisture data collected from wireless monitors to manage water application rates. As the pivot moves around the field its position is tracked via GPS and valves on each sprinkler nozzle automatically open and close to provide precise irrigation (Watson and Scarborough 2010). The soil moisture monitors collect real-time data and send it via wireless broadband network so farmers can adjust irrigation schedules to apply water only as needed (Watson and Scarborough 2010).

A new system, called SWIIM (Sustainable Water & Innovative Irrigation Management), feeds field-based water and crop data instantly to computers and is being field-tested in Colorado. The system allows for irrigation to be automatically turned on and off when stream gauges or soil sensors indicate the need to do so. The system may be useful in support fallowing and forbearance programs as precise water use measurements taken and stored by the system can provide data for farmers' transfer proposals, making water transfers easier to arrange.

Precision irrigation systems can support fallowing programs in two ways: measuring conserved water and monitoring contract compliance. A few seasons after such a system was installed, applied water records could be used in crop shifting or deficit irrigation programs. Water application records could be used to validate crop water savings when irrigators shift to lower water using crops. They could also provide evidence of reduced water application in deficit irrigation agreements.

The soil moisture monitors used in precision irrigation could be used to monitor fallowing programs. Remote soil moisture monitors could be placed in fallowed fields with GPS units transmitting their position and soil moisture conditions to a central location. If the system detected unexpected moisture, a field visit would confirm if there was a sensor malfunction or if the field was being watered. Remote soil moisture monitors can save acquisition program managers and farmers significant time compared to driving from field to field to check soil (Watson and Scarborough 2010; Morris 2006). Monitors reduce the need to visit every farm engaged in fallowing.

4.11 Balancing Costs and Accuracy

The desired level of measuring and monitoring accuracy is an important determinant of the cost. Very accurate measurements of applied water could be accomplished with, for example, variable rate irrigation systems that send hourly readings to centralized computers via wireless Internet. However, with installation costs estimated at $138 per acre,[7] this accuracy cannot be achieved without significant infrastructure investments.

Some level of measurement and monitoring costs can be justified by the benefits of increased accuracy, increased contractor confidence that they are getting what they paid for, improved confidence in the water transfer program and increased participation by farmers and irrigation districts. Federal funding may also be available to reduce the upfront cost of these kinds of investments, particularly from USDA programs such as EQIP.[8] In the southeastern U.S., average on-farm costs of variable rate irrigation systems dropped over $100 dollars to around $34 per acre after federal cost-share assistance.[9] The investment likely would only be worthwhile in areas where variable rate irrigation furthers other goals - beyond improved water measuring or monitoring accuracy.

Acquisition programs need the goodwill of the irrigation community, including growers and district managers. Definitions of conserved water and measurement and monitoring protocols need to be developed with this goodwill component in mind. It is possible that a mutually agreed upon definition and set of protocols may represent a compromise between irrigation community acceptance and cutting edge technical understanding. Remote-soil moisture sensors, for example, would require significant involvement from irrigators to implement. The farming community may be willing to participate though, if the technology has been proven effective at both lowering costs and raising yields. Many of the innovative technologies discussed here have benefits to offer farmers, which should be emphasized when planning measurement and monitoring strategies. From a program point of view remotely sensed ET can provide accurate, reasonable-cost estimates of water savings. From a farmer point of view, it can provide real-time access to crop-water interactions and contribute to better irrigation scheduling.

Monitoring is costly, but a program that pays for water and fails to receive the amount of water agreed upon is also costly. Programs with good monitoring practices will be more credible and more likely to continue to provide viable means to improve water supply reliability. The water-buying agency must decide how much to monitor participating farmers. There is a trade-off between the amount of monitoring and the cost effectiveness of a program. At one extreme, complete monitoring could ensure total contract compliance by farmers at a very high cost. The efficient level of monitoring will vary by program. Generally, efficient monitoring should encourage enough farmer compliance so that the benefits of the program are assured while monitoring costs are reasonable.

---

[7]"Introduction to Precision Agriculture," Clemson Cooperative Extension, last accessed April 20, 2012, http://www.clemson.edu/extension/rowcrops/precisionagriculture.html

[8]"Environmental Quality Incentives Program," USDA, modified June 5, 2012, http://www.nrcs.usda.gov/wps/portal/nrcs/main/national/programs/financial/eqip

[9]"Introduction to Precision Agriculture," Clemson Cooperative Extension, last accessed April 20, 2012, http://www.clemson.edu/extension/rowcrops/precisionagriculture.html

4.12 Summary and Recommendations

Climate change promises to make providing reliable water supplies more challenging in many regions. Water supply reliability contracts are a promising tool to manage risks associated with decreased supply reliability. Under this type of contract, a temporary water transfer is activated only when pre-specified trigger conditions are met. Regional climate and streamflow models can help water managers to consider probabilities of trigger events. Those seeking to improve their water supply reliability must compare their likely costs under this type of arrangement with alternative approaches. Negotiating and implementing reliability contracts are time and labor intensive, due to the relative complexity of such contracts. Alternative methods may include expanding storage in a reservoir or underground, or obtaining water as it is needed through auctions, leases and purchases. Water providers can also choose to refrain from taking action and bear the costs of an insufficient water supply. The costs of reliability contracts must be balanced against the water shortage costs they are intended to circumvent.

There are various technical approaches discussed for measuring reduced consumptive water use and monitoring compliance with fallowing program contracts. We have discussed three basic methods to measure water savings: water delivery based methods, traditional ET, and satellite-assisted ET. The water delivery method is familiar and cost effective but likely to be the least accurate of the methods. The traditional ET method is accurate and cost effective if ET crop coefficients used to calculate water savings are up-to-date for the location needed.

The optimal approach for measuring conserved water in a fallowing program depends on the crops, region, and availability of relevant ET crop coefficients. If crop types and practices have not changed significantly since crop coefficients were developed in a region, then their continued use is likely to produce an acceptable level of accuracy. If crop coefficients are outdated, updating ET values based on emerging satellite-assisted estimators can be both more accurate and less costly than traditional ET estimators. The appropriate balance in tradeoffs between measurement costs and accuracy will differ across regions and time periods. In regions where water supplies have high economic and ecological value and there is intense competition for water, higher investment in accurate measurement may well be warranted.

Water acquisition programs that rely upon cropland fallowing should consider in which specific arenas of measurement and monitoring the greatest gains can be made to achieve program goals. Satellite-based technology, in particular, is becoming available at the temporal and spatial scales necessary to be useful in fallowing programs. High-resolution imagery may be freely available, and greater numbers of water sector professionals are receiving training in working with remotely sensed imagery. Adoption of satellite-based monitoring could be the first step. If it proves worthwhile, satellite-assisted ET estimation might be explored.

To summarize, reliability contracts are a potentially valuable tool for addressing the increased supply uncertainty and longer, more severe droughts that accompany climate change. The ability to create option contracts provides flexibility in response to a changing climate and provides more certainty in future acquisition costs. Purchasers of dry-year options prepare against the possibility of extreme drought, and also allow for the possibility that the purchase of extra water may not be needed after all.

Better measurement and monitoring brings greater confidence in water conservation and transfer programs, and can lead to more participants entering the temporary water transfer market. As temporary water transfer agreements and reliability contracts become more commonplace, the mechanisms for negotiating, approving, implementing and monitoring them become more refined and implementation costs can decrease. An efficient, flexible and mature transfer process can assist agricultural, urban, environmental and recreational water users to adapt more cost effectively to the water supply challenges posed by climate change.

Acknowledgments The authors appreciate skillful editing provided by Priya Sundareshan. Support for this work came from multiple sources: the Climate Assessment for the Southwest (CLIMAS) program at the University of Arizona, through a National Oceanic and Atmospheric Administration's Climate Program Office grant; the U.S. Bureau of Reclamation through the University of Arizona "Enhancing Water Supply Reliability"; a National Oceanic and Atmospheric Administration Sector Applications Research Program (SARP) grant; the US Department of Agriculture's Multistate Research Project, W2190, "Water Policy and Management Challenges in the West" and the Walton Family Foundation.

References

Allen RG (2005) METRIC: high resolution satellite quantification of evapotranspiration overview. http://www.kimberly.uidaho.edu/water/metric/FtCollins/Allen_Allen_METRIC_Overview_Moscow_2.pdf. University of Idaho

Allen RG, Tasumi M, Morse A, Trezza R, Wright JL, Bastiaanssen W, Kramber W, Lorite I, Robison CW (2007a) Satellite-based energy balance for mapping evapotranspiration with internalized calibration (METRIC)—applications. J Irrig Drain Eng 133:395–406

Allen RG, Tasumi M, Trezza R (2007b) Satellite-based energy balance for mapping evapotranspiration with internalized calibration (METRIC)–model. J Irrig and Drain Eng 133:380–394

Anderson T, Leal D (1998) The rise of the Enviro-Capitalists. Hover Digest, reprinted from The Wall Street Journal. http://www.hoover.org/publications/digest/3522876.html#

Bastiaanssen WGM, Noordman EJM, Pelgrum H, Davids G, Thoreson BP, Allen RG (2005) SEBAL model with remotely sensed data to improve water-resources management under actual field conditions. J Irrig Drain Eng 131:85–93

BOR (2009) Lower Colorado river accounting system evapotranspiration and evaporation calculations calendar year 2008. Report, Bureau of Reclamation Lower Colorado Region, September

Bowles J (2008) Colorado river farmers skip crops, send water to Southern California Instead. Riverside Enterprise Press, March 30

BPA (2001) www.bpa.gov/Corporate/KC/Mediacenter/pressconf/062901/customers.pdf

BPA (2002) Draft guide to tools and principles for a dry year strategy (Guide). http://www.bpa.gov/power/pgp/dryyear/08-2002_Draft_Guide.pdf

BPA (2006) Final dry year guide. http://www.bpa.gov/power/pgp/dryyear/11-2006_Final_Dry_Year_Tools_and_Principles.pdf

BPA (2008) Bonneville power administration Website. http://www.bpa.gov/corporate/

BPA/KC (2001) Keeping current. Taking on the energy crisis: providing tools for conservation and load reduction. May 2001. BPA

CDWR (2000) Preparing for California's next drought-changes since 1987–1992. http://watersupplyconditions.water.ca.gov/next_drought.cfm
CDWR (2002) DWR announces 2003 dry year water purchase program. Press release November 15, 2002. http://wwwowe.water.ca.gov/newsreleases/2002/11-15-02dryyear.doc
CDWR (2009). Issue no. 2 ETAW values. http://www.water.ca.gov/drought/transfers/issues.cfm
Clyde E (1965) Operations of the Utah Rental Market, delta area, U.S. Department of Agriculture, Economic Research Service
Colby B (1990) Transactions costs and efficiency in western water allocation. Am J Agric Econ 72:5
Colby B (2003) Water transactions as an urban water supply strategy. Managing urban water supply. Kluwer Academic Publishers, Dordrecht
Colby B, Basta E, Pittenger K (2011a) Temporary water transactions and climate change adaptation, Chap. 2. In: Bonnie C, George F (eds) Risk and resilience: the economics of climate-water-energy challenges in the arid southwest. Resource for the Future Press, Washington, DC
Colby B, Jones L, Pittenger K (2011b) Chapter 4, Economic tools for climate adaptation: water transaction price negotiations. In: Bonnie C, George F (eds) Risk and resilience: the economics of climate-water-energy challenges in the arid southwest. Resource for the Future Press, Washington, DC
DOI (2007) Secretary of the Interior, record of decision Colorado river interim guidelines for lower basin shortages and the coordinated operations for Lake Powell and Lake Mead, Washington DC, December 2007
DOI (2009) Department of the Interior land imaging report. Summary, Department of the interior remote sensing working group, Washington DC. http://remotesensing.usgs.gov/FY09DOILandImagingReport.pdf
Elhaddad A, Garcia LA, Chávez JL (2011) Using a surface energy balance model to calculate spatially distributed actual evapotranspiration. J Irrig Drain Eng 137:17–26
Garfin G (2012) Southwest climate assessment. Climate Assessment for the Southwest, Tucson
Garrick D, Jacobs K (2006) Water management on the Colorado River: from surplus to shortage in five years. Southwest Hydrol 5(3):8–9
Geerts S, Raes D (2009) Deficit irrigation as on on-farm strategy to maximize crop water productivity in dry areas. Agric Water Manage 96:1275–1284
Glenn EP, Neale CMU, Hunsaker DJ, Nagler PL (2011) Vegetation index-based crop coefficients to estimate evapotranspiration by remote sensing in agricultural and natural ecosystems. Hydrol Process 25:4050–4062
Grafton RQ, Landry C, Libecap GD, McGlennon S, O'Brien R (2010) An integrated assessment of water markets: Australia, Chile, China, South Africa and the USA. NBER working paper series working paper 16203
Hansen K, Howitt R, Williams J (2008) Valuing risk: options in California water markets. Am J Agric Econ 90(5):1336–1342
Hansen N, Holtzer T, Pritchett J, Lytle B (2010) Water-conserving cropping systems: lower south Platte irrigation research and demonstration project. Colo Water 27:7–10
Hansen K, Kaplan J, Kroll S (2013) Valuing options in water markets: a laboratory investigation. Environ Resour Econ 57:59–80
Hartmann HC (2005) Chapter 202, Use of climate information in water resources management. In: Anderson MG (ed) Encyclopedia of hydrological sciences. Wiley, West Sussex
Hass N (2006) Drought and the price of water. CLIMAS Update. 9(1). http://www.climas.arizona.edu/pubs/update/may2006.pdf
Hirsch RM, Helsel DR, Cohm TA, Gilroy EJ (1993) Chapter 17, Statistical treatment of hydrologic data. In: Maidment DR (ed) Handbook of hydrology. McGraw-Hill Inc, New York
Howe C (1996) Increasing efficiency in water markets: examples from the Western United States. In: Anderson TL, Hill PJ (eds) Water marketing – the next generation
Howitt RE (1998) Spot prices, option prices, and water markets: an analysis of emerging markets in California. In: Markets for water. Natural Resource Management and Policy, vol 15, Kluwer Academic Publishers, Boston, pp 119–140

Hunsaker DJ, Fitzgerald GJ, French AN, Clarke TR, Ottman MJ, Pinter PJ Jr (2007) Wheat irrigation management using multispectral crop coefficients: I. Crop evapotranspiration prediction. Trans ASABE 50:2017–2033

IDWR (2006) Attachment B IDWR water transactions program monitoring and evaluation protocol. Idaho Water Resource Board. http://www.idwr.idaho.gov/waterboard/WaterPlanning/Water%20Transaction%20Program/PDFs/Monprotocol.pdf

Jaeger WK, Mikesell R (2002) Increasing streamflow to sustain salmon and other native fish in the pacific northwest. Contemp Econ Policy 20(4):366–380

Jenkins M (2007) LA bets on the farm. High Country News 12 Nov 2007: n.p. Web. 27 May 2014

Jercich S (1997) California's 1995 water bank program: purchasing water supply options. J Water Resour Plann Manage 123(1):59–65

Jones L, Colby B (2010) Farmer participation in temporary irrigation forbearance: portfolio risk management, Western Regional Development Centre Rural Connections, May, 2010

Keplinger KO, McCarl B (1998) The 1997 irrigation suspension program for the Edwards Aquifer: evaluation and alternatives. Texas water resources institute, Technical report no. 178

KWAPA (2010) Klamath Water and Power Agency 2010 land idling program contract. Klamath Water and Power Agency, Klamath Falls, OR. PDF File

Lindenmayer RB, Hansen NC, Brummer J, Pritchett JG (2011) Deficit irrigation of alfalfa for water-savings in the great plains and intermountain west: a review and analysis of the literature. Agron J 103:45–50

Lyon SW, Troch PA, Broxton PD, Molotch NP, Brooks PD (2008) Monitoring the timing of snow melt and the initiation of streamflow using a distributed network of temperature/light sensors, ecohydrology. Ecohydrol 1:215–224. doi:10.1002/eco.18

Mayes L et al (2002) Urban water supply handbook. McGraw-Hill Inc, New York

Merchant J (2005) Making water available for Indian water rights settlements. In: Colby B, Thorson J, Britton S (eds) Negotiating tribal water rights: fulfilling promises in the arid west. The University of Arizona Press, Tucson, pp 95–103. Print

Michelsen A, Young R (1993) Optioning agricultural water rights for urban water supplies during drought. Am J Agric Econ 75:1010–1020

Morris M (2006) Soil moisture monitoring: low-cost tools and methods. Working paper no. IP277, ATTRA National Sustainable Agriculture Information Service. www.attra.ncat.org/attra-pub/PDF/soilmoisture.pdf

Mueller R (2011) Cropland mapping with satellite data. In: Remote sensing applications for U.S. – Mexico border water management. Water Education Foundation, San Diego

MWD (2003a) Metropolitan exercises second round of one-year water transfer options from Sacramento Valley. MWD Press Release, March 20, 2003. http://www.mwdh2o.com/mwdh2o/pages/news/press%5Freleases/2003%2D03/watertransfer.hym

MWD (2003b) Improved 2003 water picture allows Metropolitan to decline final Sacramento Valley transfer option. MWD Press Release, May 1, 2003. http://www.mwdh2o.com/mwdh2o/pages/news/press%5Freleases/2003%2D05/sacramento%5Fvalley.htm

MWD (2004) Metropolitan board takes steps to secure quality, reliability of Southern California water for 2005: Board agrees to store Colorado River water for Las Vegas area; authorizes pursuit of Central Valley water transfer options for 2005. MWD Press Release, Oct. 12, 2004. http://www.mwdh2o.com/mwdh2o/pages/news/press%5Freleases/2004%2D10/reliability.htm

Nagler PL, Morino K, Murray RS, Osterberg J, Glenn EP (2009) An empirical algorithm for estimating agricultural and riparian evapotranspiration using MODIS enhanced vegetation index and ground measurements of ET. I. Description of method. Remote Sens 1:1273–1297

NASS (2011) NASS launches new CropScape geospatial data service. News release. http://www.nass.usda.gov/Newsroom/printable/01_10_11.pdf, January

Piñón-Villarreal AR, Samani ZA, Bawazir AS, Bleiweiss MP, Skaggs R, Tran VT (2010) Estimating water use through satellite remote sensing. Technical completion report no. 357, WRRI, December

PVID (2004a) PVID-MWD Forbearance and Fallowing Program. Term sheet for Forbearance and Fallowing Program Agreement. August 16, 2004

PVID (2004b) Landowner agreement for fallowing in the PVID ("Landowner Agreement") between the MWD of Southern California and PVID
Rafn EB, Contor B, Ames DP (2008) Evaluation of a method for estimating irrigated crop-evapotranspiration coefficients from remotely sensed data in Idaho. J Irrigation Drain Eng 134:722–729
Salas JD (1993) Chapter 19, Analysis and modeling of hydrologic time series. In: Maidment DR (ed) Handbook of hydrology. McGraw-Hill Inc, New York
Saliba B, Bush D (1986) Water markets in theory and practice. Westview Press, Boulder
Samani Z, Bawazir AS, Skaggs RK, Bleiweiss MP, Piñon A, Tran V (2007) Water use by agricultural crops and riparian vegetation: an application of remote sensing technology. J Contemp Water Res Educ 137:8–13
Senay GB, Leake S, Nagler PL, Artan G, Dickinson J, Cordova JT, Glenn EP (2011) Estimating basin scale evapotranspiration (ET) by water balance and remote sensing methods. Hydrol Process 25:4037–4049
Stedinger JR, Vogel RM, Foufoula-Georgiou F (1993) Chapter 18, The frequency analysis of extreme events. In: Maidment DR (ed) Handbook of hydrology. McGraw-Hill Inc, New York
Taghvaeian S (2011) Water and energy balance of a riparian and agricultural ecosystem along the lower Colorado River
Teuling AJ, Hupet F, Uijlenhoet R, Troch PA (2007) Climate variability effects on spatial soil moisture dynamics. Geophys Res Lett 34 (6):Art. No. L06406
TNC (2010) Our woods, our waters, our way of life georgia update. www.nature.org/media/georgia/georgia_spring_2010_update.pdf, Spring
Trends (2004) 2004 Trends in agriculture land & lease values, region seven, American Society of Farm Managers and Rural Appraisers, California Chapter. http://www.aglandtrends.com/2004/pdfs/2004-reg07a.pdf
Troch PA et al (2008) Dealing with landscape heterogeneity in watershed hydrology: a review of recent progress toward new hydrological theory. Geogr Comp 2:10.1111/j.1749-8198.2008.00186.x
USDA (2006) Title 440 Programs Wetland Reserve Program Subpart G – Quality Control. Part no. 514, Natural Resource Conservation Service, http://directives.sc.egov.usda.gov/, September
USGS (2009) Irrigation water use. http://ga.water.usgs.gov/edu/wuir.html
Villinski M (2003a) A methodology for valuing multiple-exercise option contracts for water, Working paper WP03-4. University of Minnesota Center for International Food and Agricultural Policy. http://purl.umn.edu/14379
Villinski M (2003b) A framework for pricing multiple-exercise option contracts for water. PhD dissertation, Department of Applied Economics, University of Minnesota
Water Strategist (1997) "ESP" valuation of water contracts. Winter 10:4. Smith R, Vaughan R (eds)
Water Transfer Workgroup (2002) Water transfer issues in California. Report, Final Report to the California State Water Resources Control Board, June
Watson R, Scarborough B (2010) Flint river basin irrigation: wireless water for biodiversity. PERC case study, Property and Environment Research Center, September
Williams J (2007) Option markets for water in California: effective management of water supply risk
Willis D, Caldas J, Frasier M, Whittlesey N, Hamilton J (1998) The effects of water rights and irrigation technology on streamflow augmentation cost in the Snake River Basin. J Agric Resour Econ 23(1):225–243
Woo C, Horowitz I, Hoang K (2001) Cross hedging and value at risk: wholesale electricity forward contracts. In: Lee C (ed) Advances in investment and portfolio management, vol 8. Elsevier Science Ltd., Kidlington/Oxford, pp 283–302
Young M (2010) Environmental effectiveness and economic efficiency of water use in agriculture: the experience of and lessons from the Australian water reform programme. Report, OECD. www.oecd.org/water

Chapter 5
Are Lease Water Markets Still Emerging in California?

Richard E. Howitt

Abstract Over the past 15 years water markets in California have evolved, but not as fast as expected, and not between the agents who were initially expected to be active in the market. The chapter reviews the disappointing performance of the state-sponsored groundwater bank in the 2009 drought and advances some hypotheses as to why the trades were not larger. The growth in bilateral trades between urban and agricultural regions and the role of environmental constraints on restricting water trades is summarized and discussed. One source of problems for the emerging water market in California is the multiplicity of ways in which opponents can use valid environmental regulations to delay or block water trades until the window of opportunity for spot trades is no longer open. Two recent examples are analyzed.

The chapter concludes that water markets are still emerging California, but they are not yet fully emerged or formed. A policy conclusion that results from this review is that excessive environmental caution can provide a mechanism to increase transaction costs of short-term spot trades needed for drought management, to a point at which they are no longer viable. Some suggestions for simpler and more robust institutions that would enable short-term spot trades are suggested, and the recent recognition of the role of markets by water industry stakeholders gives rise to cautious optimism for future water markets in California.

Keywords Water markets • California water • Transaction costs • Market institutions • Drought management

R.E. Howitt (✉)
Department of Agricultural and Resource Economics, University of California at Davis, One Shields Avenue, 2116 Social Science & Humanities, Davis, CA 95616, USA
e-mail: howitt@primal.ucdavis.edu

K.W. Easter and Q. Huang (eds.), *Water Markets for the 21st Century: What Have We Learned?*, Global Issues in Water Policy 11, DOI 10.1007/978-94-017-9081-9_5,

5.1 Introduction

There are several reasons why California should have more active water markets than it currently does. First, there is a wide difference in the value marginal product of water both within the agricultural sector and certainly between agriculture and the urban sector. One glaring case is in the central San Joaquin Valley where, in water scarce years such as 2009, the highest going price in Westlands water district in the latter half of the summer was $500 per acre foot, while 50 miles away the effective cost of water was less than $40 per acre foot. The price and value variation within urban areas is less pronounced, but still enough to overcome reasonable transaction costs.

A second reason why one would anticipate an active spot market in California is the degree of interconnection between alternative locations of water demand. The state and federal water systems namely the California Water Project (CWP) and the Central Valley Project (CVP) have a well-developed canal structure that runs from the north to the south of the state connecting urban and agricultural regions over a 700 mile long linked river and canal system. In addition, many other irrigation districts are connected indirectly by river systems and local control structures. One exception to this interconnectedness is the difficulty in making water transfers from the East side of the San Joaquin valley to the West side. While canal capacity in the north-south conveyance system is often in surplus supply, due to the original planning anticipating more water development in the north than actually occurred, institutional rigidity and obstructionism make the practical implementation of water trades difficult.

There are at least six different types of agency who are responsible for conveying water in California. In Fig. 5.1, federally funded conveyance systems are marked in black, state funded conveyance is orange, joint federal and state are marked as pink, and conveyance that is funded by the many types of local agencies is marked in green. The blue colored lines represent the natural river flows in the Colorado, Sacramento, American, San Joaquin and many other smaller rivers that are an essential part of the supply system, particularly those located on the east side of the San Joaquin and Sacramento valleys.

Since California has Mediterranean climate, water years are predominantly distributed as bimodal wet and dry years. This means that there is a predictable frequency of dry years in which the demand for spot market water transfers is high. In addition, the California economy is perpetually developing and changing. The first change was from an agrarian and extractive economy that was dominant from 1850 to 1910, with a gradual movement of the locally developed water out of gold mining and into irrigated agriculture. The second stage was an agrarian and industrial economy from 1911 to 1980 when large state and federal inter-basin projects were developed for both irrigated agriculture and municipal/industrial water supplies.

The current phase of California's economic development can be characterized as an irrigated agriculture and post-industrial service economy. These past substantial

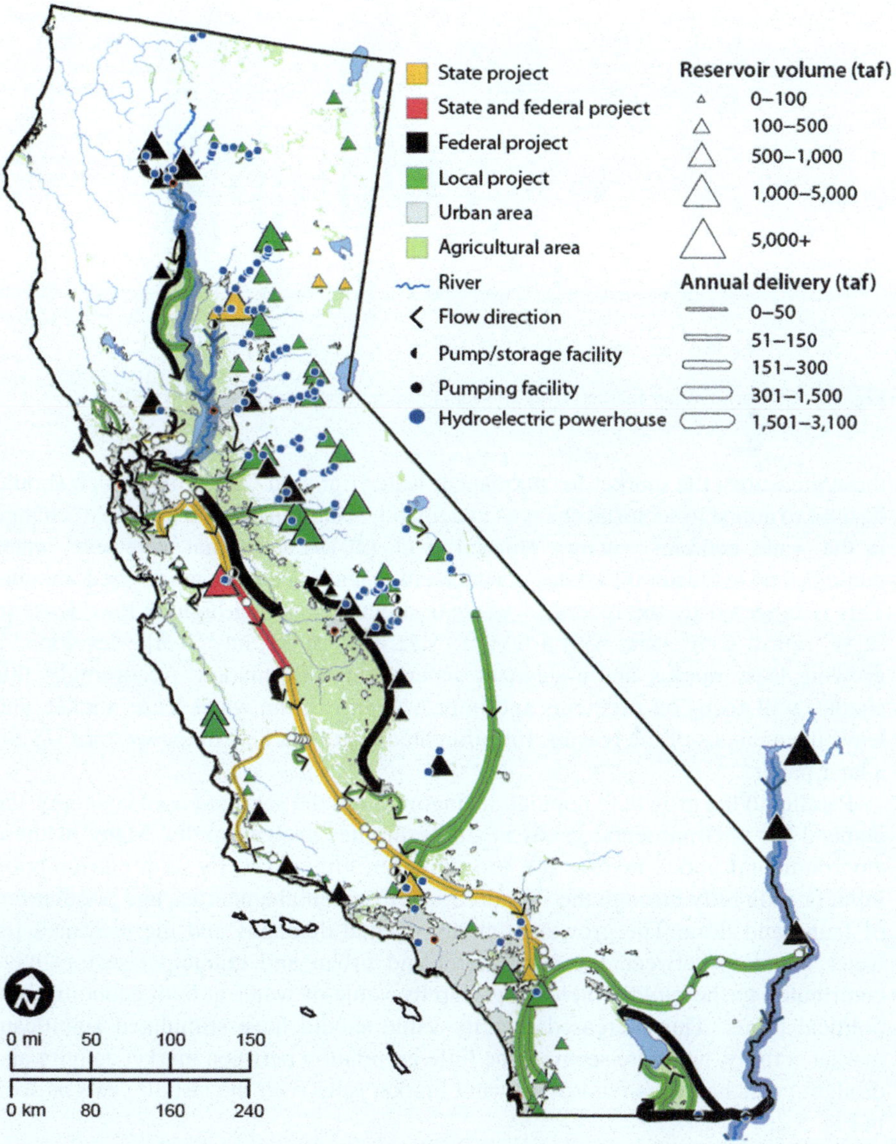

Fig. 5.1 California's interconnected water system (Source: Hanak et al. 2011)

shifts in the economic sectors have required parallel shifts in the development and allocation of natural resources, principally water. With the rapidly increasing economic and environmental cost of developing additional water supplies, the incentive to use water markets to adjust the allocation of the currently developed water supplies to changes in water using sectors of the economy is strong. Despite

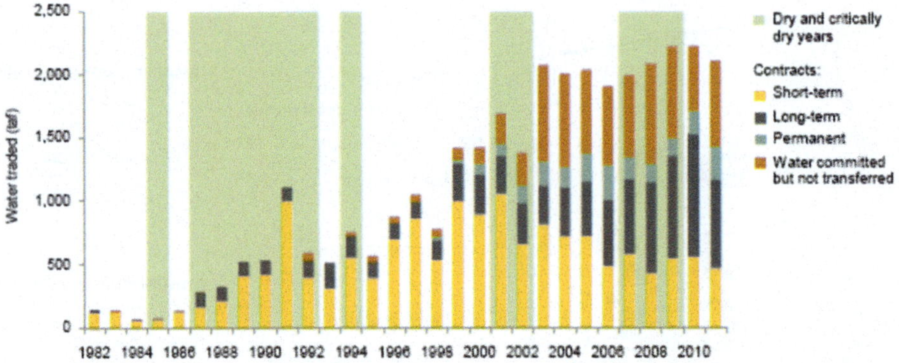

Fig. 5.2 California water markets over 30 years (Source: Hanak and Stryjewski 2012)

these pressures, the market for permanent water rights in California, which should be used to adjust to sectorial changes in demand, is lagging behind the rate of change in the water economy sectors. Hansen et al. (2013) show that state level water markets tend to cluster into states in which either lease or sales markets predominate. This is reflected by the record of water transactions having high or low “lease to sales” ratios. California, with a lease to sale ratio of 21, has clearly developed a growing lease market and a stagnant water rights sale market. Accordingly, this chapter will focus on problems and improvements to the water lease market, and leave the analysis of the reasons for bifurcated water markets in the western US for a later paper.

Parallel to the growth in population, income, and the service-based economy, the demand for environmental goods in California has grown rapidly. Many of these environmental goods involve the use of water, either directly as a consumption input or indirectly through the support of environmental amenities and populations of fauna and flora. The growth in environmental demands and the pressures for water transfers between the agricultural and urban and municipal sectors have contributed to the rapidly increasing scarcity value of water in both economic and political terms. This increased scarcity value should have stimulated significant market activity, but there seems to be little correlation between market activity and drought years based on records of water market sales over the last 30 years as seen in Fig. 5.2.

5.2 Water Market Response to Drought

The water market in California was inconsequential until the severe drought year in 1991. Faced with an extremely dry year in 1991 after a series of dry years, the state and governor were faced with the necessity of allocating the existing water supplies by directives or markets. Governor Deukmejian used emergency powers to suspend some water ordinances and create a state run emergency water bank. The water bank was not a market in the sense that purchase and sale prices were set at fixed rates that escalated slightly during the irrigation season. In addition, sales of water were initially allocated to users who could demonstrate that they had also taken some conservation actions. However, after late rains slightly improved the supplies, the water bank purchases exceeded the demand and 100,000 acre feet of water was carried over from the 1991 water year to 1992 when it was sold at a discount.

In both economic and political terms the 1991 Drought Water Bank was a success, with an estimated net return to the state of $105 million (1991$), and a total purchase quantity of 821,000 acre feet (Howitt et al. 1992). One of the reasons that this bank was so successful was that the governor gave it the highest priority, and a deputy director of the California Department of water resources was charged with the administration of the water bank. In no small measure, the success of the bank was due to the stellar reputation of the deputy director Bob Potter who had worked in the California Department of Water Resources for many years, and had the trust and respect of both the farmer sellers and urban buyer communities. This trust was an essential component of the 1991 water bank, and in real terms lowered the bank transaction costs which were estimated at 2.5 % of the value of the transactions.

Figure 5.2 shows the trends in the total volume of water sales, how the composition of the sales has changed, and also how dry years did or did not change the water market. Dry years are shown in Fig. 5.2 by tan highlighting. The market in 1991 jump-started with a large volume of spot trades which rapidly declined when the drought broke with a wet year in 1993. Spot trading volume increased slightly with below average precipitation years in 2001–2002, but trended downward from 2003 to 2010. This downward trend in spot trades was not broken by the 2007–2009 droughts. The state sponsored water bank that was reestablished in 2009 was ineffective for several reasons discussed below.

The 2009 Drought Water Bank was established in February 2009 to meet projected demands. Over the 2009 water season, the Drought Bank purchased 82,000 acre feet, falling far short of its initial target, which was, at one point, set at 600,000 acre feet. One of the main reasons for the lack of participation in the bank was the price that was offered to potential farmer sellers. At $275 per acre foot, a water purchase price that was just competitive with growing rice, but offered no real incentive to sell water to the bank, given the inevitable local skepticism about such sales. Additional problems with the 2009 water bank were restrictions on moving water across the Sacramento Delta, and the leadership of the bank which was given to a competent, but unknown, state agency manager rather than a known leader in the water industry as in previous years.

In short, despite the pressing need for an effective drought spot water market in California, far from growing from a promising start, the spot water market has dwindled downward, due the combination of layers of environmental legislation, and successful blocking action by local water trade opponents. More details of these actions are given later.

A small volume of long term trades emerged in 1992, but only started growing in 2000. Since then, the volume of long term and permanent water trades had grown steadily, and has offset the decline in spot market trading activity. In terms of actual water traded, the California water market volume has been stationary since 2003. Given the predictable periodic need for short term dry spot water markets in California, a detailed examination of the forces restricting spot water markets is used to suggest two potential solutions to the current torpor in water marketing.

5.3 Market Constraints

What went wrong in the development of the California water market? There are several reasons. Effective opposition in potential water exporting regions, increasing environmental restrictions in the Sacramento-San Joaquin Delta restricted conveyance, and a plethora of multiple agency regulations on conveyance capacity and the change of point of diversion, all combined to delay implementation of trades and increase their transaction costs. Timing is a critical factor for California drought spot market water trades. Despite good information on current dam capacity, the capricious nature of California precipitation means that potential market participants do not know if they are facing a drought until February. Since spot market water is mainly supplied from changes in agricultural operations, there is only a 2 month window in which spot market trades can be implemented before farmer crop timing prevents further trades. In the discussion that follows I will focus on the impediments to spot markets rather than long term or permanent transfers because the volume of spot transfers is in decline rather than growing as would be expected from the increasing and cyclical nature of California water scarcity.

Hanak (2003) has thoroughly analyzed the extent and growth of county ordinances on water exports that are designed to prevent or severely limit the export of water from a given county. Figure 5.3 shows the extent and type of ordinance. We conjecture that there are three dominant reasons for enacting these ordinances. The first concern is over the effect on groundwater depletion from water sales, directly by the sale of groundwater, or more commonly, indirectly by the sale of surface water and the substitution of increased groundwater pumping. The second concern is for the reduction in revenues received by local farm related businesses from increased crop fallowing due to water sales. Third, the prevention of environmental externalities from field fallowing, or from changes in river flows, temperature or return flows from crop irrigation.

Groundwater depletion is a valid concern for most potential water export regions, as they completely lack any form of quantified groundwater rights. With few exceptions, California's groundwater is governed by correlative water rights that only restrict overlying groundwater users to pump for beneficial use. The concept of beneficial use in agriculture is a very broad definition. One of the few recorded exceptions to beneficial use of groundwater was its use in gopher control by flooding the entire area that had holes. Groundwater is covered by an extensive "no injury" rule. While this is an equitable concept, there is no statewide criteria of "no injury" to groundwater, thus counties have defined fragmented and varying measures to

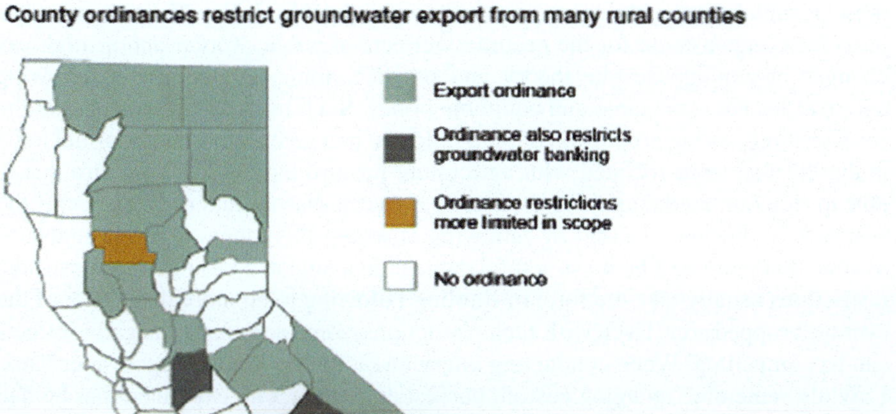

Fig. 5.3 County water ordinances (Note: Figure shows ordinance status as of 2002. To our knowledge, no additional county groundwater ordinances have been adopted since then. Kem County's ordinance is limited to the southeast portion of the county within the South Lahontan hydrologic region. Glenn County's ordinance was updated in 2000 and now relies on basin management objectives that do not automatically restrict groundwater exports. Source: Hanak 2003)

prevent injury. In some counties, any depletion of groundwater is assumed to be injurious to county residents, and water transfers are banned rather than having to compensate those county residents who are directly injured.

The effect of increased crop fallowing on local businesses is often cited as damage attributed to water exports. Few empirical studies have measured this impact. Howitt (1994) showed that the effect of fallowing and water exports on local businesses in 1991 ranged from a decrease of 4 % of agricultural business sales to 6.5 % of agricultural business profits. While no changes in water use can be free of these pecuniary externalities, economic theory and public policy does not provide a basis for compensation for the pecuniary effects of public works such as highway changes or zoning. Despite theory and practice, some restrictions on fallowing externalities are reasonable and equitable policy. Such restrictions usually take the form of limits on the proportion of land fallowed in a given area. The informal logic is that businesses associated with agriculture have to have a cost structure that is able to ride out the fluctuations in revenue between seasons and price cycles. If the reduction in business caused by fallowing is within this range, it is reasonable to assume that there will be no structural damage to associated businesses. These area restrictions usually take the form of limiting fallowing to no more than 20 % of the average cropped area. Even with such restrictions, some accounts of negative effects can be compelling. When conducting interviews after the 1991 drought water bank, I vividly remember an aging custom harvest contractor who explained that he had traded in all his combines for new larger models, only to have half of his custom contracts canceled due to water bank fallowing of local corn crops. Situations like this have no answers in economic theory, but do influence local politics.

A complete ban on exports to avoid environmental externalities is often short sighted, since water transfers can generate environmental benefits as well as costs. The "no injury" rule that is the keystone concept in regulating California surface water transfers is open to interpretation but, if taken literally, it sets a bar which would ban almost all economic activity. Clearly, the rule should be no significant injury in which the externalities from a water trade are balanced against the potential for compensation, and the cost of externalities versus the social benefits of the water trade. Given the lack of a statewide definition of a significant injury, the "no injury" rule is widely used as a delaying mechanism by those opposing water trades, and for spot markets the ability to delay a transfer is often the ability to prevent it.

5.4 Blocking Trades

An example of effective blocking of short run water trades is the attempt to trade water during the 2009 drought, between the Glenn Colusa Irrigation District (GCID) in the northern Sacramento Valley, and the Metropolitan Water District (MWD) in the Los Angeles region. Since 2003, GCID and MWD had agreed on a contract for water sales conditional on dry years as measured by the Sacramento River flow index. This type of conditional option contract should be an example of how spot market contracts can be negotiated to respond to drought year demands, and when such year occurs, they should be able to be implemented smoothly. In 2003 the option price was $13 dollars per acre foot, with the option condition being based on the Sacramento River index and a strike price of $102 per acre foot was agreed on in 2003. In 2009 the option condition of the Sacramento River index was clearly met in this drought year, and Glenn Colusa prepared for bids to fallow land and release surface water to fulfill the contract. The fallowing of land in the irrigation district was challenged by local groups on the basis that the fallowed land would make it difficult for an endangered species of snake, namely the Giant Garter snake, to make the dens in the dry soil. Given the restrictions of the Endangered Species Act, the water district proposed to fulfill the contract without fallowing land, but substituting groundwater irrigation for surface water that was normally used to irrigate the area. The landowners had correlative rights to groundwater which was from a comparatively shallow aquifer and normally recharged by natural recharge and deep percolation from surrounding irrigated fields. This substitution of groundwater was challenged by local organizations that opposed water trades, but the farmer's rights to pump groundwater were upheld. The opposition group then challenged the groundwater pumping on the basis of increased air pollution. They argued that pumping this additional groundwater would involve the use of temporary diesel powered pumps, and that such pumps would contribute toward the regional air pollution and should be prevented. When the district pointed out that many farmers already use diesel pumps to pump groundwater in this region, the ruling from the local air quality control board was that diesel pumps used pumped groundwater that was used for local irrigated agriculture were permissible under existing air pollution regulations, but the diesel pumps used to pump groundwater for irrigated crops when surface water was transferred could contribute to regional air pollution and thus had to be banned. At this stage, the water season was advanced to a time that alternative supply measures had been taken by the MWD, and this conditional, previously contracted; short-term water trade was no longer viable.

The experience of the GCID and MWD in 2009 shows that the ability to raise a sequence of challenges to short-term spot water trades is a very effective method of preventing the implementation of trades in a water spot market. It's not clear from an economic perspective why people would oppose such markets if the effect of environmental and local externality were adequately taken into account, which they were in the case of the conditional option agreement between GCID and MWD. However, it is reasonable to assume that any reallocation of the local natural resource will generate some degree of opposition. Certainly, the evidence is that the economic well-being of the County would show a net benefit. The problem seems to be, not in the existence of environmental constraints which are necessary, but in the presumption that trades have to be prevented until sequential objections are satisfied.

Another example of spot market trades creating problems between exporting growers in another region of the Sacramento Valley called Butte County started in the 1994 drought, and continued in the form of a water export ordinance. The crux of this water transfer problem in Butte County lies in the differences in farm size and profitability. The districts which wanted to transfer their surface water rights and use their legitimate groundwater rights to irrigate crops in 1994, are composed of a group of farmers who grow rice in the valley floor. The litigants were located in an area called the Cherokee Strip composed of small farms that could not grow rice and were dependent on groundwater. The Cherokee Strip is higher than the valley floor, and thus there could be a groundwater gradient and lateral flows between the exporters and the farmers in the strip.

As a drought year of 1994 progressed, wells in the Cherokee Strip started to go dry and run short of water. Landowners in this area claimed that this was due to additional groundwater pumping by water sellers lower in the valley. An investigation by the California Department of Water Resources found little hydrologic linkage between the aquifers, but landowners in the Cherokee Strip proceeded to court. The Cherokee Strip landowners failed to get an injunction and in 1995 claimed substantial impacts. The outcome of this controversy was that passage of the Butte County Groundwater Protection ordinance of 1996 which requires that permits were needed to export groundwater or to substitute groundwater for surface water exported out of the county. Permission would be refused if the exporters were deemed to cause any of following five types of injury. (1) Increased overdraft. (2) Saltwater intrusion. (3) Exceeding the safe yield of a basin or sub-basin. (4) Uncompensated injury. (5) Subsidence.

5.5 Removing Constraints

The five conditions above are very reasonable and in theory allow transfers to occur where they do not cause long-term damage to the aquifer or users that were not compensated. In practice, the political interpretation at the county level has been such that since 1996, no permits have been approved for the export of groundwater despite several severe drought years. In this case, my sympathies lie with the small farmers in the Cherokee Strip who should have been given the benefit of the doubt, even if for political reasons. At a minimum, they should have had subsidized well improvement to restore their groundwater service. The response in this case contrasts with the Glenn Colusa Irrigation District that has well improvement policy in exporting areas that can be described as "Fix the well first, for free, and ask questions afterwards". This approach by GCID reverses the normal response to complainers of "prove your damages" before talks of compensation can occur. It takes a politically more sophisticated view that recognizes that in dry years, the timeliness of response is critical. Just as transfers should be given the option under certain conditions to transfer without being blocked, compensation for damage to wells from transfers should be dealt with by repairing the well first, and then discussing the degree to which it was damaged by water transfer. It seems to me on a subjective basis, that the political goodwill generated by rapid response to these problems outweighs the cost of the inevitable free riders on the system.

The cost that agencies charge for using their facilities to convey, or wheel, transfers may indeed prove an obstruction to the transfer process. A case that has been litigated and argued over the past 13 years concerns the flat or "postage stamp" rate that the Metropolitan Water District (MWD) charges transferors using their facilities. The California water code in Section 1810 outlines the framework for transferring, or wheeling, water and states that ".. neither the state nor any regional or local public agency may deny a bona fide transferor of water the use of a water conveyance facility which has unused capacity of the period of time for which that capacity is available, if fair compensation is paid from a use". Of course, the whole controversy is over what is fair compensation for that use. The code defines fair compensation as "reasonable charges incurred by the conveyance system including capital, operation, maintenance and replacement costs, increased costs from any necessitated purchase of supplemental power, and including reasonable credit for any offsetting benefits from the use of the conveyance system". Discussions at the time of wording the new water code centered about the problems of pricing decreasing marginal cost systems, and how to approximate the marginal cost of a transfer. MWD interpreted this cost as requiring a full average cost charge for the whole system under a system called "postage stamp" pricing. As the name implies, the charge to wheel water through the system is exactly the same for 3 miles or 100 miles. In addition, the MWD wheeling charged is extremely high at $670 an acre foot just to move water any distance through the system. A recent breakdown of this charge shows costs of $195 per acre foot for facilities and $195 per acre foot in support costs and an additional administrative overhead cost of $279 per acre foot based on the entire cost of billing and meter reading averaged over the whole system. These latter costs seem to have nothing to do with the marginal cost of moving water through the system. This extraordinary high fixed cost, does however, have an effective dampening effect on any outside agencies wanting to make transfers independently of the MWD. The best example of this is the charge to the San Diego County Water Authority for moving 200,000 acre-feet of purchased water from Imperial County through the Metropolitan system. This charge is despite the fact that the San Diego County Water Authority is the largest of the member water agencies in the Metropolitan system.

The ability to make short-term dry year spot market transfers in California is significantly handicapped under current regulatory interpretations and environmental regulations. The cases discussed above all can be categorized under three groups. First, the use of environmental objections to transfers based on a number of area origin regulations on species preservation, air pollution, and land use. Second, restrictions on water exports are often successfully based on the impact on groundwater resources as reflected in the many local county level ordinances which flourish in the absence of a statewide ordinance. Third, the inability and cost of transferring water from areas of origin to areas of demand through the existing conveyance system may be prohibitive, due to excessive transfer charges, or a thicket of time-consuming regulatory inter-agency requirements.

5.6 Policy Options

The following section will outline two policy changes that could lower the transaction costs of dry year water spot markets. First, a generic definition of environmental impacts of water transfers would establish a statewide framework for the assessment of regional and environmental impacts of water transfers. In addition, a uniform Environmental Impact Report (EIR) could be the basis of a change in the legal presumption, to one of being able to make such transfers if the environmental standards are satisfied. I term this approach a generic transfer environmental impact criterion.

A second policy change would optimize the use of California's water conveyance system both for the current contractors and additional water buyers and sellers. The institutional structure of the new management system would mimic a system successfully used in electric power industry, and often termed an Independent System Operator or ISO for water distribution.

The connection between the doctrine of reasonable use and water markets is still under discussion, and has been so since Gray's seminal work in 1994. Despite the opinions that water markets, in terms of the price based incentives for conservation are entirely consistent with the reasonable use doctrine, many holders of water rights are yet to be convinced that selling some of their water does not put their reasonable use water rights at risk. It is now generally accepted that price signals induce more agricultural conservation actions than threats, or command and control policies, generally termed "best management practices". The conservation of urban uses of water shows opposite tendencies with the price effect on conservation being relatively slight given the inelastic demand for urban water, while the effect of public outreach campaigns designed to raise the consciousness of water conservation have been shown to be statistically significant methods to reduce urban water use by Renwick and Green (2000). Further reassurance on the security of reasonable use property rights in the face of transfers and market allocation can probably only be achieved through a series of test court cases, a long and arduous process.

The costs and complications of the environmental review of water transfers were characterized in the three cases discussed above. While long-term transfers are subject to review by the State Water Resources Control Board (SWRCB), short-term transfers of 1 year or less are exempt from review by the SWRCB. However, many transfers are not subject to review by the board, and thus not exempt from the EIR process which is often used to delay or disqualify short-term dry year transfers.

The key legislation that governs conditions under which water transfers can occur is the "no injury" rule that protects against damage from changes in place or purpose of use of surface water that is subject to regulation by the State Water Resources Control Board. The extent of injury covers all other water users and fish and wildlife associated with the water resource. One glaring omission is that since there is no state law governing groundwater use, the "no injury" rule cannot apply to groundwater, even when it is substituted for surface water to implement a trade. Another omission in the "no injury" rule is that it does not take account of the effects on both local economies and/or environmental amenities from fallowing land to generate tradable water, since this is not technically a change in place or use of water but merely an absence of using water. Even where there is a change in place or use of surface water injuries categorized under the "no injury" rule does not include third-party economic impacts on the local economy, unemployment or loss of tax revenues. This is not surprising as no public compensation system exists for this type of damage. All of these shortfalls mean that on a statewide basis the regulatory basis for mitigating the effects of water transfers is largely absent. Given this gap in the legislation it's not surprising that local regions concerned about the impact of water transfers have enacted local ordinances which are often used as methods to block transfers rather than to mitigate any injuries. In Section 1745 the California water code calls for public review of transfers involving more than 20 % of local water supplies. While this is a reasonable rule of thumb to limit pecuniary third-party impacts, it does not guarantee that any mitigation payments will be made. Despite this absence of legislation, just about every long-term transfer and many short-term transfers are accompanied by payments to local economies designed to offset any deleterious impacts. In the initial 1991 Emergency Drought Water Bank, payments to Yolo County to compensate for increased unemployment and public support were negotiated between the County and the contractors purchasing Water bank water. The amount that was agreed upon was $65,000. However, due to technical problems of one state agency paying another, the actual payment was never implemented. Later conditional transfers from Imperial County and the Palo Verde Irrigation district have regional economic mitigation payments incorporated as part of the agreement.

There is no question that water transfers should be subject to environmental constraints, but the problem is that many of them are based on the California Environmental Quality Act (CEQA) which is subject to local regional interpretation. Again, social values should be taken into account when assessing the environmental impact of water transfers, however the analysis should be done using a consistent statewide set of principles, and in a way that allows preparation for a potential case for transfers before the dry year in question occurs. There are several state programs where statewide impact preparation manuals exist which local agencies are required to use to prepare the EIR for a subsidy or program. For example, applicants for public assistance for water quality control projects have to follow a uniform set of guidelines in preparation of the case for public funding. While the preparation of a transfer case should reflect local priorities, infrastructure, and water availability, the approach and criteria for granting water quality control financial assistance are uniform across different agencies and regions. This statewide template for a generic EIR for water transfers should be applicable to all types of water rights which are currently covered by local regulations. Prior analysis for the EIR should be done over a range of different hydrologic scenarios so that it is applicable to a wide range of dry year situations with different levels of dam capacity, river flows, previous droughts, and groundwater stocks.

5.7 Key EIR Topics

Essential topics to be covered by an EIR can be grouped under four headings. First the effect of transfers on surface water. The first effect would be that the transfers do not diminish the legitimate uses by other service water users. For example, the California rule that only consumptive use of water can be transferred prevents most third party effects from surface runoff or deep percolation. One exception to this is if the transfer is achieved by fallowing in an area where the runoff goes in a different direction than the area being supplied surface water, then parties relying on the surface runoff would be deprived of some of the surface water supply. The effect of transfers on surface water quality must also be addressed, in particular with contaminants, timing, or river temperature. One case of the transfer of groundwater out of Yolo County to Kern County was denied because the quality of the groundwater had a higher salinity level than the Sacramento River into which it was discharged, and thus the use of the river to convey the transfers would have degraded the quality by increasing the aggregate salt load. The volume of stream flows is usually not a problem for most transfers which are made from upstream sources to downstream sources, and thus increase the flows in the river. This assumes that the timing and the quality of the flows are not degraded.

Second the effects on groundwater supplies. The set of impacts that could occur due to transfers on groundwater supplies are fully listed in the Butte County groundwater ordinance discussed above, and cover both quantity and quality aspects. The mitigation of adverse effects to groundwater should be required to regulate groundwater supplies on an equal basis with surface water supplies. However this raises an interesting question over the use of groundwater to supplement the surface water that was sold. This is essentially the core of the Butte County problem, since the farmers down-slope of the Cherokee Strip were using their groundwater supplies in a perfectly legal manner, even though they would not normally use them in the absence of water sales. There is no question that this additional pumping would have an effect on surrounding groundwater users, the problem is whether this effect can be categorized as damage due to water transfers if the additional groundwater pumping is within the normal safe yield criteria for the area from which water is being sold. A strict property rights interpretation of the law would conclude that while there was an impact due to transfers, it was not an injury outside current property rights to groundwater held by the farmers who decided to sell their surface water. In this case I would have to conclude that while it was a detrimental impact on other groundwater pumpers, it should not be classified as an injurious impact since the exporting farmers were exercising their groundwater rights in a responsible and balanced manner.

Third the effects of crop fallowing on the environment. The simplest mechanism to release transferable water from agricultural production is to fallow crops and sell the released consumptive water. Even when deep percolation and surface return flows are unchanged, the process of fallowing can have detrimental environmental impacts on the local fauna and Flora. One well-documented case is that of the giant garter snake discussed above. Other environmental impacts can result from changes in the riparian vegetation growing along the distribution canals and laterals. The degree to which farmers are responsible for these externalities is uncertain, as it is clear that in the normal course of farm income optimization, farmers have the right to switch crops and fallow without restriction. Again, like the problems associated with increased legal groundwater pumping, it seems that the current interpretation of California law is that water transfers are not viewed as normal farm operations, but are in a special class in which any externalities resulting from have to be fully mitigated.

Fourth the effects of fallowing transfers on the local economy. The inclusion of mitigation payments to offset economic costs to local government and even associated local industries should be included as part of the generic EIR. While the theoretical economic efficiency argument for such transfer payments is very weak, political expediency and equity considerations push strongly for such payments as part of the process of implementing water transfers.

A statewide template for a generic EIR as outlined above will provide the basis for reassurance to exporters of water of economic and environmental controls on the extent of transfers and mitigation of any significant damage that occurs. Water importers will also benefit since the conditions and costs of making water transfers will be more transparent, and more importantly, predictable in advance. Once these criteria are established in a consistent statewide formula this generic EIR can be satisfied in advance, thus allowing the rapid response that is necessary in order to implement a dry year spot water market under California hydrologic conditions.

5.8 Managing Water Distribution

At the start of this chapter we noted the extensive water transfer grid that links many water supplies and users in California. What was also obvious from Fig. 5.1 was the wide range of federal, state, and local agencies that had developed different parts of this network system and control its access, pricing, and maintenance. The classic approach to traditional water development in California has been one of vertical integration in the full supply chain. Ownership or control is normally established for the storage source, the canal or river linkages, and the end use by local water agencies. The older water districts on the East side of the San Joaquin Valley and parts of the Sacramento Valley have supplies based on local river systems and dams or on old established pre-1914 riparian water rights. This model of vertical development usually has the cost of storage, transfer, and distribution in a single charge based on cost recovery for the entire system. The structure was very successful in developing the current network of individual systems. The normal way to refer to a particular system including the dam, supply canal and distribution is to categorize it as state, federal or local. The state and federal systems were developed later in the previous century and are based on inter-basin transfers through large long-distance canal systems and pumping plants. Despite the federal and state basis of financing the systems, access to the systems by non state or federal contractors is jealously guarded. In recent years there has been a coordinated operating agreement in effect between the State Water Project and the federal Central Valley project. This agreement has significantly improved the efficiency of the system, particularly with respect to the operation of the shared Sacramento San Joaquin Delta and the San Luis dam. It could be regarded as a preliminary test of the advantages that an Independent System Operator (ISO) like structure could bring to the entire water distribution system. A shift to water markets can be envisaged as move from vertically integrated supply and delivery systems to horizontally integrated networks that enable the efficient reallocation between sectors and locations. A necessary condition for market transactions is to lower the transaction costs of water movement. I propose that an ISO structure that manages water transfers can achieve the same efficiency savings in water that have been demonstrated in the energy sector. A Water-ISO would likely be opposed by some interests, but if implemented in politically acceptable stages, it would open potential markets for water that are currently hamstrung by the lack of predictable access to water transfers.

In proposing an alternative system in which the ability to implement water transfers and sales without the standard thicket of regulations or exorbitant charges designed to discourage trade, is based on an institution developed in the electric power industry called an Independent System Operator (ISO). With the partial deregulation of electrical energy it became clear that the distribution grid can be operated more effectively if there is equal access to all parts of the grid by those wishing to move power across it. The ISO structure is based on the principle of horizontal integration rather than vertical integration. It has several key and critical characteristics. First, it does not own any of the facilities but does have the control of flows and operations conveying them. Second, the ISO is a public agency, but run on a nonprofit basis by a staff that are not civil servants and thus subject to both the benefits and costs of private market employment. The governing board of the ISO is appointed on a rotating basis by the governors in states in which it has jurisdiction. While an ideal system would be designed to be fully integrated, the shifts in ownership and control needed to do this are unrealistic in the California water sector. The advantage of the ISO institution is that it is designed to be grafted on top of existing institutions without changing the fundamental ownership structure. This design characteristic is a significant advantage from both the political and operational perspective. In addition, an ISO is not beholden of any one water wholesaler or environmental interest, and this increases the likelihood of an impartial allocation of an increasingly scarce resource between competing interests. A clear motivation and operating method for the electric power ISO system is that it uses market prices to provide an incentive for effective supply management of this network commodity. CAISO (2013).

The ISO system was installed to operate in an electrical market in which generation, transmission, and retailing of power were separated into horizontal layers of function rather than the previously vertically integrated utility-based system. It is clear to us that the water sector in California has very similar efficiency gains to be obtained by a shift from vertically integrated utilities to horizontally integrated functions. In the water industry the horizontal integration should be by storage, transmission and wheeling, and wholesaling through irrigation and water districts. Integration of the transmission system would greatly facilitate the ability to move water between regions of different scarcity. Under the ISO implementing legislation individual retail water districts would lose their exclusive franchise on the operation, but not the ownership, of certain canals and sources of water. Districts and agencies would also be liable for transfer charges on their own system, and might feel that the charges in the ISO market exceeded the current value of the water to them. However, since the water districts retained ownership of these canals and dams, they would be compensated by a gain in revenue from ISO operations, and thus could lower their retail rates from the higher charges rebated by the ISO for their share of the distribution network.

A Water-ISO, like the current electricity ISO, would be a nonprofit public benefit corporation with an independent board appointed by the governor and similar mechanisms to ensure stakeholders have input into the operation. It is important to stress that the Water-ISO would not own any canals, conveyance, or dam facilities. It would be important however that the ISO control sufficient proportion of the water market to form stable prices. Fortunately, the two major arterial water conveyances from north to south of the state are owned by the State Water Project and the federal Central Valley Project. If these two systems were exclusively operated by the Water-ISO, the majority of north-south water movement and East-west water movement in the San Joaquin Valley will be facilitated by a Water-ISO. Independent water districts and systems on the East side of the San Joaquin Valley and in the Sacramento Valley would have to use the Water-ISO for any trades outside their immediate district, and hopefully would see the benefits of combining with the larger system. Since they are still the full owners of dams and canals, the state and federal contractors will be responsible for the maintenance development and investment in the facilities. If the same efficiency gains from combined operation that have been realized in the electricity sector emerge from a Water-ISO, the additional revenues would justify further investment and development of the existing agency systems.

Ideally, the Water-ISO would be established with sufficient scale to form an effective market by requiring that all water conveyed through the federal and state systems is subject to operational control by the ISO. If the political will to do this is lacking, a phased in system could also act as a test of the ISO concept. A phased-in system would have two components, one would be an extension of the existing coordinated operating agreement between the state and federal systems to ensure consistency and some efficiency gains in the operation of the combined system. Current state and federal contractors would have priority and compose the majority of the water moved in the combined system, but excess capacity would be quickly identified and made available for those trades which can be agreed on between private or agency parties. Rather than directly make a market with different types of water contract over different periods of time, the phased-in Water-ISO would act as an efficient conveyor, and a market facilitator between independent parties. Under this phased-in system efficiency gains will be more muted, but the market value of water would be more transparent, and the ability to move it would be faster and more efficient than the current system.

There have been increased restrictions on moving water through the Sacramento San Joaquin Delta which were further enforced in 2009. These restrictions are in response to several suits under the Endangered Species Act in which fish species such as the Delta Smelt and winter run Salmon, which are listed species, could be harmed by excessive exports of water from the Delta. These seasonal restrictions, which also depend on the severity of the water year, significantly reduce the normal contracted exports from the Delta under the federal and state water systems. For example, districts with the most junior rights in the federal system had their allocations for 2012 cut to 40 %, and 2013 which is another dry year, resulted in an additional cut to 20 % of the contracted quantity. There is no question that the endangered species fish populations are at an extremely low levels, and that under the Endangered Species Act actions have to be taken. One problem is that the hierarchal nature of water rights, and the unwillingness of agricultural contractors in the San Joaquin Valley to trade with each other, exacerbate the problem and concentrate cutbacks in areas with the most junior rights, which paradoxically, are those with high production and high water use efficiency. These restrictions on contracted exports from the Delta make the opportunity to use the Delta facilities for water trades increasingly difficult and reduce the ability to move water and have water trades from the lower water value part of the state in the Sacramento Valley to the high water value parts in the southern San Joaquin Valley or the Los Angeles basin.

One recent development that is spurred by this cutback is a reassessment of the value of water trades by agricultural and other contractors. Water trades are now being seen as a required part of the adjustment process. As of writing in 2013, some typical responses have been as follows. "With a long-term average water supply of about 45 % for agricultural service contractors, there will always be a need for supplemental water supplies to meet demands," said Frances Mizuno of the San Luis-Delta Mendota Water Authority. She continued, saying that "Groundwater pumping and water transfers are the primary sources of supplemental supplies. We need to have in place long-term, programmatic environmental documents that include a cumulative effects analysis for water transfers."

The 2013 water crisis has resulted in an encouraging response from the California Governor. In May 2013, Governor Brown issued an executive order to streamline approvals for voluntary water transfers. The order directs the state Water Resources Control Board and Department of Water Resources to expedite review and processing of voluntary water transfers and water rights, consistent with current law. The State Water Resources Control Board currently has water transfer petitions totaling about 260,000 acre-feet, with 194,000 acre-feet included in petitions to transfer water between state and federal water projects. George Kostyrko, a State Board spokesman said that in normal years, the water board expects to process three or four water transfer petitions. In drought years, or when there is an executive order, the number of petitions processed increases to 15 on average and can go as high as 30. The water board currently lists 11 pending petitions to transfer water to entities south of the delta. "In terms of time, it depends on the type of transfer," Kostyrko said, reaffirming that. "It depends on the circumstances of the transfer and whether or not there are protests or comments. The Division of Water Rights has made the processing of a transfer petition its highest priority over other water-right permitting activities."

California Farm Bureau Federation reported that their President Paul Wenger thanked Governor Brown for recognizing the need to streamline California water transfer rules. "In a year like this, voluntary transfers of water from areas that have a surplus give our system more flexibility so that farmers facing water supply cutbacks, especially those with permanent crops, may find alternative sources," Wenger said.

5.9 Conclusion

While these very recent statements are encouraging for the reemergence of California water market from its current position of stasis and decline in short-term spot water market trades, many problems still remain. The lack of criteria for groundwater and its connection to surface water transfer remains a major problem, as does the fragmentation of environmental regulation of groundwater use in areas with several types of surface water rights, which in turn, leads to local ordinances that are often used to prevent water trades. The organization of the water wheeling network between federal state and many local agencies is also a major impediment to actually moving traded water, once the terms have been agreed. Adding to this is the current impasse over environmental standards in the Sacramento San Joaquin Delta. This chapter has suggested two principle institutional corrections for these problems. First a statewide adoption of standards and environmental criteria to pre-approve trades that can be rapidly implemented in drought periods. It is encouraging to see that water contractor's organizations are calling for this innovation. The second suggestion is more radical and involves significant reorganization and shift in power over the ability to wheel transferred water between places of diversion and use. The second innovation, namely, the concept of establishing a Water-ISO will take longer than a generic environmental impact report, but should greatly facilitate trades if it occurs. There is no doubt that California water markets have not fully emerged from an initially promising start, however the pressures of water scarcity and the wide discrepancies in its value of use between different locations in uses will stimulate change. We envisage that some version of the market reforms suggested here will evolve in the future, and recent statements by water leaders are encouraging along these lines.

References

CAISO (2013) http://www.caiso.com/Documents/CompanyInformation_Facts.pdf

Gray BE (1994) The market and the community: lessons from California's drought water bank. Hastings West-Northwest J Environ Law Policy 1:17–45

Hanak E (2003) Who should be allowed to sell water? PPIC Press, Public Policy Institute of California, San Francisco

Hanak E, Stryjewski E (2012) California's water market, by the numbers: update 2012. PPIC Press, Public Policy Institute of California, San Francisco

Hanak E, Lund J, Dinar A, Gray B, Howitt RE, Mount J, Moyle P, Thompson B (2011) Managing California's water: from conflict to reconciliation. PPIC Press, Public Policy Institute of California, San Francisco

Hansen K, Howitt R, Williams J (2013) Water trades in the Western US: risk speculation, and property rights. Book chapter. In: Maestu J (ed) Water trading and global water scarcity: international perspectives. 2012. RFF Press/Routledge, Abingdon, UK, chapter 4, pp 55–67

Howitt RE (1994) Effects of water marketing on the farm economy, chapter 5. In: Carter HO, Vaux HJ Jr, Scheuring AF (eds) Sharing scarcity gainers and losers in water marketing. University of California, Agricultural Issues Center, University of California, Davis, California, pp 97–132, June 1994

Howitt R, Moore N, Smith RT (1992) A retrospective on California's 1991 emergency drought water bank. California Department of Water Resources, Sacramento, March

Renwick ME, Green RD (2000) Do residential water demand side management policies measure to up? An analysis of eight California water agencies. J Environ Econ Manage 40:37–55

Chapter 6
Water Markets in Chile: Are They Meeting Needs?

Robert Hearne and Guillermo Donoso

Abstract Water markets in Chile have been enabled since the 1981 National Water Law. This law was designed to provide market incentives with the capacity to reallocate water toward more valuable uses. Although there have been many controversies, especially concerning the initial allocation of water-use rights, and the coexistence of consumptive and non-consumptive use rights, water markets have gradually expanded. Large scale trading has not occurred in many basins. Likewise, large scale intersectoral trading has not occurred because the agricultural sector has remained prosperous, with growth outpacing the rest of the economy. Market transfers have occurred throughout Chile. The 2005 Reform of the Water Law addressed many concerns, especially speculation in unused non-consumptive water-use rights. And some concerns remain. Groundwater has been depleted and new efforts need to be made to improve groundwater data and groundwater users associations. Registering water-use rights and recording transactions are inconsistent. Large dispersions in prices reflect imperfect market conditions. Nonetheless the expansion, geographically, and volumetrically, of market trading suggests that these markets are meeting the needs of most water using sectors.

Keywords Water markets • Water management • Water law • Water-use rights

---

R. Hearne (✉)
Department of Agribusiness and Applied Economics, North Dakota State University, 811 2nd Ave N., Fargo, ND 58102, USA
e-mail: Robert.hearne@ndsu.edu

G. Donoso
Department of Agricultural Economics, Pontificia Universidad Católica de Chile, Casilla 306, Santiago 22, Chile
e-mail: gdonosoh@uc.cl

K.W. Easter and Q. Huang (eds.), *Water Markets for the 21st Century: What Have We Learned?*, Global Issues in Water Policy 11, DOI 10.1007/978-94-017-9081-9_6,

6.1 Introduction

Chile has maintained a system of transferable water-use rights since the 1981 National Water Law. This system has led to active water markets in a few key river basins including some intersectoral water transfers. The uneven geographic distribution of sectoral activity and water resources led to intersectoral water transfers between mining, agriculture and urban areas in the North and predominantly intrasectoral, agricultural water transfers in the Central and South of Chile (GWI 2010).

Because Chile has been unique, among developing countries, in its adoption of privatized water use rights its water management and water markets have received attention in the international academic literature (Rosegrant and Gazmuri 1994; Cristi and Trapp 2003; Quentin et al. 2012). Chilean water markets were seen as emblematic of a new paradigm of water management. Part of this attention is due to Chile's adoption and maintenance of neoliberal economic policies – which includes defense of individual property rights. Many economists have focused on the benefits of market transactions, but noted that markets were not widespread (Hearne and Easter 1997; Donoso et al. 2002; Hadjigeorgalis 2004; Zegarra 2002). Other researchers have focused on the limitations of the 1981 Water Law including poor conflict resolution and the absence of river basin management (Bauer 1998, 2004; Dourojeanni and Jouravlev 1999).

Recent water law reforms and the gradual expansion of local market activity have led to a maturation of water markets in Chile. The long sought reform to reduce speculative hoarding of unused water rights is diminishing this constraint to progress. This chapter reviews Chile's experience with water markets with an emphasis on recent developments. The second and third sections of this chapter present a geographical and institutional review of Chile's water resources and management. This will be followed by a section reviewing empirical results that support the conclusion that Chile's water markets are continuing to expand.

6.2 Geographic Background

Chile's geography contributes to the divergence of opinions over its water management. Chile's land mass stretches from the Atacama Desert in the north through the humid forests of southern Patagonia and Tierra del Fuego. Chile's numerous rivers flow from the Andes to the Pacific and provide irrigation and drinking water and valuable hydroelectricity. Most of the precipitation falls in the cooler winter months and is stored in the Andean snowpack. This allows for high river flows during the early growing season. Within the global context, Chile as a whole may be considered privileged in terms of water resources. The average available water from

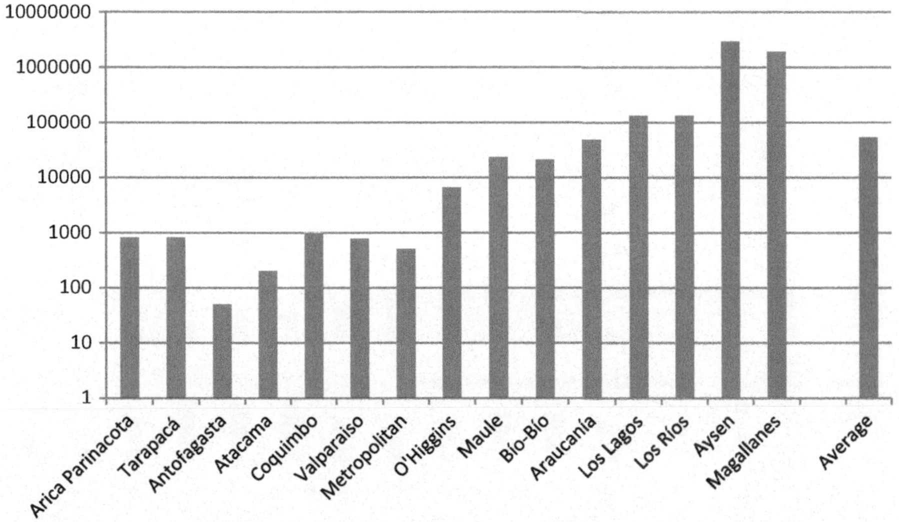

Fig. 6.1 Average water availability per person per year (m^3/person/year) (Source: World Bank 2011)

precipitation is equivalent to 53,000 m^3/person/year (World Bank 2011), a value considerably higher than the world average (6,600 m^3/person/year). However, there exist significant regional differences: from Santiago to the north, arid conditions prevail with average water availability below 800 m^3/person/year, while south of the capital Santiago the water availability is significantly higher reaching over 10,000 m^3/person/year (see Fig. 6.1).

Surface water flows in Chile are characterized by high annual variability due to influence of various phenomena, such as the presence of North Atlantic Oscillation and El Niño Southern Oscillation (ENSO). Southern rivers have greater stability of water flows, as shown in Fig. 6.2 (FCCyT 2012).

Chile's aquifers are generally unconfined or semi-confined, small, and shallow (less than 50 m). They are highly heterogeneous with respect to size (Instituto de Ingenieros de Chile 2011). It is estimated that Chile has a large volume of groundwater resources and the estimated average recharge is approximately 55 m^3/s in regions north of the Metropolitan Region (Ayala 2010; Salazar 2003). The Tarapaca, Antofagasta and Atacama regions have recharge rates of about 10 m^3/s, while the central Metropolitan and Valparaiso regions have recharge levels of 50–100 m^3/s (FCCyT 2012). South of the O'Higgins Region there is no detailed information on the recharge level; however, a first approximation of the General Directorate of Water (DGA), indicates that groundwater recharge is approximately 160 m^3/s between the regions of Maule and Los Lagos (Dirección General de Aguas 2010; World Bank 2011).

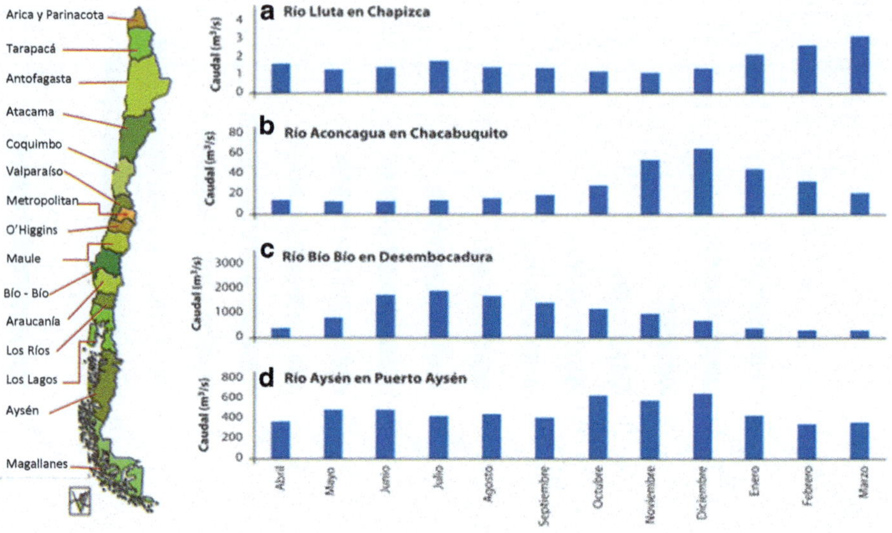

Fig. 6.2 Average monthly hydrographs of some Chilean rivers (Source: FCCyT 2012)

6.3 Water Use and Management

In the desert north, the sparse water resources are divided among Chile's principle mining operations, and a sparse population. In north central Chile irrigated agriculture is well suited for fruit crops grown for international markets. Central Chile is the nation's population and economic center and includes the Santiago metropolitan area with nearly one third of the nation's population. This area is semi-arid and irrigated agriculture is important. South central Chile is humid with large rivers and a substantial generation of hydroelectricity. Southern Chile is sparsely populated but has important forests, fisheries, and aquaculture resources.

Irrigation accounts for 73 % of consumptive water use and 40 % of cultivated land. Most of the higher value crops, including the exported fruit, are irrigated. Principle irrigated crops include fruits, vineyards, pasture, grains, and vegetables. Agriculture accounts for 7 % of Chile's exports and employs 9 % of its labor force. The remaining share of extracted water is used in industry (12 %), mining (9 %), and household use (6 %). Much of the mining sector is located in the arid northern areas of the country and is reliant on groundwater extraction (World Bank 2011).

Chile has prided itself with effective regional water supply and sanitation (WSS) companies that have provided extensive coverage with quality service. These regional WSS companies function as private enterprises although the state, through its investment corporation CORFO, still owns a large portion of these companies. These companies provide service under national regulation. Coverage is nearly universal in urban areas. Surface water accounts for 54 % of the WSS water supply and groundwater provides the remaining 46 %. From 1998 to 2011 the percentage of sewage water that is treated increased from 17 to 94.2 % (World Bank 2011; Superintendencia de Servicios Sanitarios 2012).

Chile has continued to develop its southern rivers for hydroelectricity. In 2012 hydroelectricity surpassed thermal sources as the primary source of electricity generation in Chile, although this is highly dependent on precipitation (Daugherty 2012). The Pangue Dam, completed in 1996 with a 4,687 M-watts (MW) capacity and the Ralco Dam, completed in 2004 with 690 MW capacity were both highly controversial and opposed by environmentalists and indigenous populations (Bauer 2004; Hearne and Donoso 2005). A current proposal to add an additional 2,750 MW of capacity by constructing five dams on the Baker and Pascua Rivers in Chile's remote and humid Patagonia has been favored by the Piñera government but opposed by environmentalists who have inspired nationwide protests (Nelson 2011).

The increased wastewater treatment has improved surface water quality, reduced biochemical oxygen demand, and reduced incidence of waterborne diseases. However, despite the increased wastewater treatment and the short length to Chile's rivers water quality has remained a problem. CADE-IDEPE (2005) concludes that water pollution in the northern area is mainly produced by mineral-rich soils due to the erosion of rock formations that deposit contaminants such as copper, chromium, molybdenum, boron and aluminum in the water (FCCyT 2012). Indeed in northern Chile the natural levels of arsenic and boron surpass standards for drinking water and irrigation respectively. In central Chile water pollution load originates from urban settlements and agriculture. There is evidence that in areas of intensive agriculture, non-point pollution has remained problematic. However, there is little information regarding pollution discharges and of the quality of the receiving water bodies (FCCyT 2012). The World Bank has identified the need to improve Chile's system of water quality monitoring (World Bank 2011). Finally, in the south the situation is different, due to the higher dilution capacity of water bodies the level of pollution is much lower (FCCyT 2012).

Until the 1990s, environmental and water management policies did not pay much attention to meeting water requirements for environmental purposes. Evidence shows that the river flows in the central and northern area of Chile were fully allocated, or over allocated, to consumptive uses and thus potentially unavailable for instream uses. The high water use has led to the deterioration of aquatic ecosystems in semiarid and arid regions of Chile. This has gradually changed with the introduction and continuous improvement of the System of Environmental Impact Assessment (SEIA) in 1994 along with the 2005 reform of the Water Law which imposed the obligation to establish minimum ecological flows. Before the 2005 reform, most river basins located north of Santiago, were over-allocated and, thus, the DGA has not been able to implement minimum ecological flows due to the lack of water. River basins that have protected minimum ecological flows are mainly located in the south of the country where water is more abundant and has lower use values (World Bank 2011).

Chile is increasingly exploiting its groundwater resources. In 2003 groundwater withdrawals were estimated at 88 m^3/s (Ayala, Cabrera y Asociados Ltda 2007).

Close to 50 % of groundwater use is for agriculture, with 35 % used for residential purposes and the rest in industry. Groundwater is especially important for the mining sector and potable water supply, providing 63 % and 46 % of the water used for these purposes, respectively. Groundwater accounted for nearly 11 % of total consumptive water use in 2003. Since 1990, requests for groundwater-use rights have increased substantially in areas where surface water resources have become scarce. Aquifers in the north and central regions of the country have been overexploited resulting in conflicts. Consequently, the DGA has concentrated its efforts at understanding groundwater quantities and recharge in the dryer regions north of the capital, Santiago where 112 separate restriction zones have been declared. Recently it has started to identify aquifers of concern in the nation's southern regions (World Bank 2011).

6.4 Water Rights, Water Law and Water Management

The clear enabling factor that allowed for the implementation of the water use rights market in Chile was Chile's tradition and culture, dating back to colonial times, of managing water resources with water use rights. The 1981 Water Law establishes that water right owners are responsible for water management. Chile has a tradition of private sector development of water infrastructure, user management of canals, and allocation of irrigation water based upon individual user rights that dates back to colonial times. The key concept that surface waters are national goods, but that individuals can obtain water-use rights was established in the 1855 Civil Code and reinforced with the 1930 and 1951 National Water Laws. The 1951 National Water Law strengthened the concept of water-use rights as property that could be used under the rules established by the state. This tradition provided incentives for private financing and development of irrigation canals (Hearne and Donoso 2005).

The private possession of water use rights was restricted by the 1967 National Water Law which stressed the state's role in managing water as a public good. This law was implemented during the governments of Frei Montalva (1964–1970) and Allende (1970–1973) and allowed the state to expropriate water and reallocate it toward beneficial uses as determined by regional plans. This law complemented the Agrarian Reform which reallocated large estates without compensation. The 1981 National Water Law reflected the shift in the political paradigm toward the defense of private property rights that accompanied the 1973 military government.

The 1981 National Water law maintained water as a national good with individual user rights. Water use rights were defined for consumptive and non-consumptive uses. Non-consumptive rights were designed to facilitate hydroelectric generation and required users to return flows to the river without interfering with consumptive rights. Water-use rights (WR) are, by law specified as a volume per unit of time. However, given that river flows are highly variable in most basins, these WR are recognized in times of scarcity as shares of water flows. This characteristic of WR, which combines volumetric maximum amounts per unit time in times of plenty, with shares in times of scarcity has proven to be appropriate, since the use of a system of use rights defined as pure shares precludes any excess water use for other purposes such as environmental objectives since it would lead to full use of water by the current holders of WR (World Bank 2011). Consumptive rights entitled users to completely use all water without an obligation to return any water to the source.

Water use rights were also specified for groundwater. Water use rights could be obtained through a petition to the Directorate General of Waters, or could be purchased. Article 79 of D.L. 2603 recognizes usufructuary WR that were conceded previous to the Water Law of 1981. A water user shall be the owner of a usufructuary water use right once their use over a certain amount of time is proven and ensuring that no third party effects or conflicts exist associated with this use. The specific details of this recognition are specified in the transitory second article of the Water Law. Usufructuary rights were respected but all users were encouraged to register their traditional rights.

Much of the direct management of rivers and canals has been the responsibility of individual water user associations. User management has existed in Chile since the colonial era, and currently there are more than 4,000 Water User Associations (WUAs) (Dourojeanni and Jouravlev 1999). There are three recognized types of WUAs. Small-scale water communities (*comunidades de aguas*) consist of users of the same water source. These water communities can be chartered, although many are not chartered. Canal associations (*associaciones de canalistas*) consist of the users of a common canal and can receive a charter which gives the association legal status. River management committees (*juntas de vigilancía*) are comprised of all users and canal associations along a portion of a river and are responsible for canal intakes along that portion. These WUAs maintain their own lists of water users and often these lists are more important than the official registries. The fact that much of the nation's canal infrastructure was operated by private sector WUAs complemented the system of privatized user rights. Many of these WUAs have professional management (Hearne and Donoso 2005).

The effectiveness of some of these institutions in managing irrigation systems and reducing transactions costs for water market transactions has been noted (Hearne and Easter 1995, 1997). WUAs have facilitated trade in the Limarí Valley, which is noted for active trading between irrigators. Yet WUAs mostly only provide information, and their role in initiating trade has been small. Market intermediaries, including real estate agents and attorneys, introduce stakeholders to trading, by for example, bringing them together and providing insurance for the transaction. Once markets are established, WUA can play an important role in increasing and facilitating trade (Macaulay 2009).

According to the DGA and the Directorate of Water Works (DOH), a large percentage of WUAs have not updated their capacity to meet new challenges. Many managers of these user organizations do not have technical capacity and do not effectively communicate with their members. Additionally, Bauer (1998) points out that vigilance committees have not been effective in resolving intersectoral conflicts. To address some of these concerns, the DOH and DGA have implemented programs to train WUA managers and directors since the mid-1990s (Peña 2000; Puig 1998).

The different user organizations have some common competences. First, their primary responsibility is the distribution of water resources between water users. Second, the management decisions are voted on in general meetings by shareholders in proportion to their WR shares. Third, under drought conditions, water is distributed proportional to the amount of WR each water user holds.

In 2010 the number of water communities was more than 10 times the number of canal associations; this is due to the fact that it is easier to form a water community than a canal association. Water communities and canal associations are responsible for both the management, maintenance and renovation of more than 40,000 km of primary and secondary canals, as well as dams built by the private sector or transferred to the user associations by the State (World Bank 2011). At present there is only one groundwater community in the country, in the region of Atacama. The Water Law of 1981 establishes that any aquifer that has been declared under restriction or protection must have a groundwater community. Compliance with this regulation is very low since several aquifers have been declared under restriction and protection and have not formed groundwater communities.

River management committees are different from the other two types of WUAs, since all their competences and legal powers are over surface water before it's withdrawn. Since the Water Law reform in 2005, river management committees also must integrate groundwater users into its jurisdiction, in an attempt to move toward integrated surface and groundwater management. The main responsibilities of these WUAs are:

(a) Generate hydrological information in order to improve user's understanding of the water system;
(b) Manage surface and groundwater water withdrawals;
(c) Surveillance and monitoring of surface and groundwater water withdrawals;
(d) Water extraction enforcement; and
(e) Application of sanctions to non-compliers.

However, Chile's water institutions present important limitations to effectively address integrated water resource management. First, Chile has sought to create institutional arrangements in which each economic sector has a defined regulatory framework, with appropriate incentives for the efficient management of resources in their particular area. This approach has not allowed for an effective management of the multiple interactions that arise between the public and private sectors present at a watershed level. Second, the fragmented approach lead to the lack of a strong agency that identifies, formulates and implements national water policy as well as gives coherence to the actions of the various other agencies. Third, OECD (2011) concludes that Chile's water institutions and organizations present obstacles to achieve effective horizontal co-ordination between public agencies at the central level as well as a vertical coordination. The most important of these obstacles is the excessive fragmentation of Chile's water institutions and organizations, the existence of overlapping and unclear allocation of responsibilities, a competition for powers between ministries, lack of an adequate budget for public agencies, and the lack of citizens' concern for water policy.

Table 6.1 Water flow (l/s) of regularized and un-regularized WR

Water basin	Region	Regularized WR flows (l/s)	Un-regularized WR flows (l/s)
Río Salado	Antofagasta	397	1
Río Copiapó	Atacama	2,740	8
Estero Pupío	Coquimbo	437	128
Río Quilimarí	Coquimbo	346	65
Río Petorca	Valparaíso	2,355	1,622
Río La Ligua	Valparaíso	3,531	2,738
Río Maipo	Metropolitan	82,473	34,247
Río Bio-Bio	Bío-Bío	62,236	38,852

Source: Rhodos (2010) and World Bank (2011)

Furthermore, the Instituto de Ingenieros de Chile (2011) points out that the current practice of independently managing water resources at the level of a river or aquifer section prevents the implementation of an integrated approach to water management. The Water Law of 1981 considers river sections and aquifer sections as independent bodies of water. Thus, each independent section has a WUA that optimizes water resources for its water users without considering downstream effects or impacts on groundwater users. For example, in the past 3 years that have been characterized by drought, several canal associations have lined their channels so as to reduce water percolation and deliver more water to their surface water users. This is an optimal decision for surface water users; however, it significantly reduces groundwater recharge. What is more worrying is the fact that most of these investments have been subsidized by the state. Thus, government funded investments have generated externalities on groundwater users.

Although the 1981 National Water Law was successful in promoting investments related to water and improving water use efficiency in many economic sectors, it also led to new difficulties which were partially addressed in the reforms of 2005. Key reforms initiated with the 2005 National Water Law were intended to address some concerns regarding speculative hoarding of unused, non-consumptive water-use rights by implementing a tax on all WR that were owned but remained unused (Yanez 2008).

Both the WUAs and the real estate registries (*Conservadores de Bienes Raices*) maintain cadasters of water use rights. Water-use rights need to be registered before they are sold. Since the promulgation of the Water Law of 1981, efforts have been made to regularize (ensure that WR have proper titles), grant title, and register WR in order to resolve overlapping claims to water. This is especially important for WR that were redistributed under the Agrarian Reform and might be contested by previous owners. Estimates of WR that are not registered range from 60 to 90 % (Dourojeanni and Jouravlev 1999). This can be in part explained by the fact that courts have protected unregistered rights, which has undermined the registration requirement (Bauer 1998).

Rhodos (2010) documents legitimate uses in different water basins that have been recognized for decades by WUAs which have not been regularized and registered; his findings are presented in Table 6.1. In the northern basins, water flows associated with recognized but undocumented (or in process of regularization) WR are small. However, the contrary is true for the central and southern water basins where there are many legitimately recognized WR which have not been regularized.

Thus the regularization procedure has not been very effective. This lack of regularization and registration can be explained by the following reasons (World Bank 2011):

(a) The lack of incentives and penalties for holders of WR to regularize and register customary WR. In particular, the second transitory article does oblige users to register their WR but there are no impediments to exercising their rights even though the WR are unregistered;
(b) Regularization procedures are complex and lengthy, due to the complexity and rigor of the verification process. However, it is also due to an excessive judicialization of proceedings. Of the customary WR that have been certified by the DGA since 1981, between 40 and 65 % are still awaiting a court ruling (World Bank 2011).

Only registered rights can be bought, sold, and mortgaged, and thus, the fact that most rights remain unregistered impedes the transfer of water. However, most WUAs maintain their own registries in order to effectively distribute water to rights owners. These do not imply legal title. The DGA is also responsible for maintaining the Public Water Registry (PWR) which contains information on all water-use rights that are granted by the DGA. This PWR also contains hydrological and water-quality data, information on WUAs and water withdrawals, and all transactions. However, this registry does not imply legal title, and is often incomplete.

The regularization procedures have generated an important proportion of the current water use conflicts that must be settled by the DGA and the judicial courts. Given that the regularization procedures were established 20 years ago, the difficulty of verifying the validity of the customary water use has significantly increased (World Bank 2011). According to Rhodos (2010) the regularization procedure has lent itself to many abuses. For example, several WR that were being regularized were not recorded in the cadaster of water users that was conducted by the DGA between 1981 and 1987 for surface water and in 1976 for groundwater.

A centerpiece of the information system on WR is the PWR which provides the DGA with the necessary information on WR to enable it to effectively fulfill its functions. Furthermore, the PWR should allow water users to obtain the data required for efficient water management and planning, as well as for WR market transactions. However, as discussed previously, the PWR is incomplete; only 20 % of all WR and 50 % of market transaction cases are registered (World Bank 2011). The main reason why the PWR is incomplete is that only regularized and formally inscribed WR can be registered. Moreover, the record is not completely updated because the real estate registries often do not forward to the DGA market transaction data, even though the 2005 reform of the Water Law stipulates that they should.

The 1981 law was designed to facilitate market transactions of water-use rights, and allow for market reallocation of water to higher valued uses. Consistent with the prevailing neo-liberal economic paradigm, the law's backers sought the efficiencies of market allocation. Thus the 1981 Water Law distinguished itself from the 1951 Water Law in upholding the private property characteristics of a water-use right but also fostering market reallocation (Buchi 1993; Hearne and Donoso 2005).

As Vergara (2010) asserts, there is a distinction between centralized, decentralized and autonomous water institutions. Centralized organizations comprise the administrative bodies of the State. These centralized institutions include water quantity and quality management bodies. Decentralized bodies are represented by user organizations. Finally, there are several autonomous institutions such as the judicial system that resolve most water use conflicts. Table 6.2 presents the institutional mapping of water policy roles and responsibilities. As can be appreciated in Table 6.2, multiple central authorities (ministries, departments, public agencies) are involved in water policy making and regulation at the central government level. In Chile the number of actors involved in water policy making is 15; one of the highest of OECD countries that were surveyed in an OECD (2011) study on water governance.

Under the Water Law of 1981, the State reduced its intervention in water resources management to a minimum and increased the management powers of water use rights holders that are organized in water user associations (WUAs). The water resource management roles assigned to centralized institutions are the following:

(a) To measure and determine the availability of water resources and to generate the necessary data and information that allow for a well-informed management of water resources on the part of WUAs;
(b) To regulate the use of water resources avoiding third party effects and their overexploitation. For that purpose the State must analyze water resource availability and potential water use conflicts before granting new water use rights, authorizing water use right transfers and other authorizations such as changes in water distribution infrastructure; and
(c) To conserve and protect water resources, by means of an environmental impact assessment of investment projects, establishment of minimum ecological flows and environmental policies.

The 1981 Water Law provided the DGA with broad powers and responsibilities in managing waters, especially in times of a drought. The DGA does have certain discretionary authority, especially in periods of drought. But it has maintained a limited role in direct management of national waters, and limited its interference with private water-use rights during periods of drought (Peña 2002).

The Water Law of 1981 paid little attention to the sustainable management of groundwater because at that time, groundwater extraction was marginal. Recognizing the need to improve groundwater management regulation due to increased groundwater pumping, the 2005 amendment of the Water Law of 1981 introduced procedures to reach a sustainable management of underground water resources. The main provisions are: (a) extraction restrictions when third parties are affected, (b) the authorization for the DGA to require the installation of extraction measurement equipment in order to effectively monitor extractions, (c) the establishment of areas

Table 6.2 Institutional mapping of water policy: centralized state agencies and autonomous institutions

Ministries		**Sub-Ministerial agencies**	
Acronym	Portfolio	Acronym	English name
MOP	Public works	DGA	General Directorate of Water
		DOH	Directorate of Public Works
		SISS	Superintendence of Water and Sanitation
		INH	National Hydrology Institute
MNE	Energy	CNE	National Energy Commission
MINAGRI	Agriculture	CNR	National Irrigation Commission
		SAG	Agricultural Service
		INDAP	National Development Institute
		CONAF	National Forestry Service
MINDEF	Defense	DMC	Defense Media Center
MINVU	Urbanism and housing	DDU	Urban Development Division
		SERVIU	Urban and housing service
MININT	Interior	ONEMI	National Emergency Office
		GORE	Regional Governments
MINSAL	Health	ISP	Institute of Public Health
MINECON	Economy	SUBPESCA	Subsecretary of Fisheries
		SERNAPESCA	National Fisheries Service
MMA	Environment	SEA	Environmental Impact Assessment Service
		SMA	Subsecretary of Environment
MM	Mining	SERNAGEOMIN	National Geological and Mining Service
Autonomous institutions			
		PJ	Judiciary
		TDLC	Defense Tribunal for Competition
		TA	Environmental Tribunals
		CONADI	National Indigenous Commission
		CMS	Sustainability Ministers Council
		CGR	General Comptroller
		CBR	Real Estate Registries
			Municipalities
Decentralized institutions/water user associations			
		JdV	River Management Committees
		COD	Water Communities
		ASCAN	ASCAN: Canal Associations
		COMAG	COMAG: Groundwater Communities

subject to extraction prohibitions and restrictions, and (d) the need to consider the interaction between surface water and groundwater when analyzing a petition for new surface or groundwater WR.

World Bank (2011) concludes that these groundwater regulations have not been fully implemented over time and thus, there exist various problems associated with groundwater management. A major concern is the general lack of information about groundwater and insufficient knowledge about its dynamics, in particular its interaction with surface waters. There are significant gaps in the registry of wells, extraction and quality measurements, recharge balances, and identification of pollution sources. In general, information systems are not linked to the measurement and monitoring of aquifers to estimate groundwater withdrawals. An effective information system is a prerequisite to be able to control and sustainably manage an aquifer.

An additional challenge for a sustainable groundwater management is the fact that at present ground and surface waters are managed independently despite their recognized interrelations. This implies that there is no conjunctive management of surface and groundwater, which has proven to be an effective adaptation mechanism for climate change.

There are, in general, no WUAs that manage groundwater user rights; the only exception is in some sections of the over-exploited Copiapo aquifer in the Atacama Region. There should exist a groundwater WUA at least for all aquifers that have a restriction or prohibition declaration by the DGA. The fact that users have not yet organized themselves in groundwater WUAs to take over the management of groundwater may reflect the lack of understanding of a large proportion of users of the long term effects that uncontrolled exploitation of aquifers may cause. In the absence of groundwater WUAs, the Water Law of 1981 establishes that the DGA is responsible for controlling and monitoring groundwater withdrawals. Evidence has shown that the DGA does not have the necessary resources (human, technical, and financial) to monitor all groundwater extractions.

A number of additional practical issues were of concern to the DGA and Chilean water managers. The 1981 Water Law specified that the DGA would grant water use rights upon petition. But did not specify the circumstances under which these rights were to be denied to petitioners. Because of this many non-consumptive use rights were granted to speculators who maintained ownership of these rights without using them. In 1997, the DGA estimated that 87 % of the granted non-consumptive use rights, which are mostly in the humid southern half of the country, were unused. There was also concern that these unused rights were concentrated in the possession of a few hoarders who maintained market power over future development of non-consumptive use rights. Additional concerns included the inconsistent registration of rights. Although local DGA offices and real estate cadasters maintain records of registered water-use rights, rights do not need to be registered in order to be respected by the law. Although market transactions for water-use rights are supposed to be maintained at the local real estate registries, these records have inconsistently reported the quantities transacted and exchange prices (Yanez 2008; Bauer 2004; Cristi 2011).

These concerns were partially addressed in the 2005 reform of the 1981 Water Law which: (i) mandated a fee to be paid for unused water-use rights; (ii) directed the DGA to protect minimum ecological streamflows; (iii) established a national registry for all water-use rights; (iv) established open auctions for the granting of water-use rights that have multiple petitioners; (v) required water-use right petitioners to state the use of the water; and (vi) provided the DGA increased powers to restrict granting water-use rights. The 2005 reform did not reverse the market orientation of the 1981 Water Law. Indeed this reform diversified the use of economic incentives with auctions for new contested rights and fees to maintain unused rights. It did adopt the principle that water should be used (Yanez 2008; Donoso et al. 2010).

The most noteworthy provision of the 2005 reform of the Water Law was the establishment of a fee for maintaining water–use rights without using the water. The non-use tax rate for consumptive and non-consumptive WR is calculated $\tau = \gamma Qf$ and $\tau = \gamma QHf$, respectively, where γ is a constant that equals 0.1 for all regions between Magallanes and Los Lagos, 0.2 for Regions between O'Higgins and Araucania, and 1.6 for all regions north of the Metropolitan Region, Q represents the average flow of water that is not used, measured in m^3/s, f is a temporal factor that increases over time if the water right remains unused ($f = 1$ for years 1–5, $f = 2$ for the years 6–10, $f = 4$ for over 11 years without an effective use of the waters), and H, which applies only to the non-consumptive WR is the difference between the water extraction level and the level of restitution of water, with a minimum of 10 m ($H \geq 10$).

According to Peña (2010), as a result of this reform, combined with the performance of the Antitrust Commission, the monopolistic distortion due to speculation and non-consumptive WR hoarding has been reduced. In turn, Jouravlev (2010) notes that as a result of the reform of 2005 (together with other measures), WR that still are not used are generally no longer a major obstacle to the development of river basins, and it is likely that non-use of WR will continue to decline in the future due to the projected increase in the non-use tax. Similarly, Valenzuela (2009) finds that the non-use tax has operated as an incentive for the increased offering of WR in the WR markets; between 2007 and 2008, an equivalent of 65 m^3/s additional rights have been offered, which represents 1 % of the total WR affected by the non-use tax. Cristi and Poblete (2010), finds evidence that during the year 2009 that 2.08 % of the WR subject to the non-use tax were offered in the market or began to be used. Thus, the effectiveness of the non-use tariff has been increasing over time, as would be expected.

It is important to point out the major flaws with respect to the design of the non-use tax:

(a) The non-use tax is only applied to WR for which the water intake infrastructure has not been constructed. However, the mere existence of water intake infrastructure does not necessarily ensure that water is used in practice;

(b) It can be applied only to the registered WR and regularized customary WR that are contained in the records of the PWR. As was previously mentioned, the majority of WR are not registered;
(c) The calculation formulas are defined in the 2005 Reform of the Water Law, which makes it difficult to change, in particular to reflect the increased economic value of water over time. It is foreseeable that the economic value of water will increase significantly in the future;
(d) The non-use tax for consumptive WR is associated with the relative scarcity of water as indicated by the regional multipliers in the tax rate determination. This does not represent the real opportunity cost of water for all economic sectors and thus will not act as an incentive when the sector's opportunity cost is greater than that of agriculture.

The protection of minimum ecological flows has also been a long-time concern of the DGA. In 2004 Humberto Peña, then Director of the DGA suggested that the current law did not leave sufficient protection for water in certain dry areas such as the northern desert oases (Peña 2002, p 20). However, there has been very little conflict between environmental and consumptive uses of water in most of Chile.

A number of initiatives have recently been implemented to facilitate market trading through reduced transactions costs. In 2006 the use of specialized consulting services to facilitate trading was advanced with the formation of the first enterprise to specialize in these services (Global Water Intelligence 2010). In 2012, a new internet platform for spot market transaction in the Limarí basin canal system was launched. This electronic market for water is an initiative to develop and test a mechanism to reduce transactions costs and improve market transparency. This electronic market is sponsored by the National Irrigation Commission, with technical assistance from two universities and the full cooperation of the major WUAs. Both volumes of water and options are transacted. During the markets initial 3 months over 450,000 m^3 were bought and sold at an average price of US \$0.24 per m^3.

6.5 Analyses of Water Markets

With the exception of few localized markets, trading has been traditionally limited in Chile (Peña 2002). Despite a legal separation between land and WR, many Chilean farmers maintain that water and land should not be separated. This traditional integration of land and water has kept many farmers from offering water for sale without selling the corresponding land. Also, the agricultural sector in Chile has continued to grow, often at a rate greater than the rest of the Chilean economy (World Bank 2003; ODEPA 2004). Because of this growth, the value of water in irrigation has remained high and farmers have little incentive to sell water. And, many farmers maintain surplus WR in order to mitigate the risk of drought.

In addition, backed by Chile's free market policy, the government has no particular objective to improve water markets, other than providing a framework in which the market can develop. The promotion of information mechanisms is not considered by the government. Some real estate agencies have expanded their services to facilitate WR exchanges. Most buyers and sellers still rely on newspaper posts or word of mouth to find each other. The lack of quantitative data on Chile's water markets, limits the analysis of its evolution.

Since 1990 research has documented functional water markets in a few important river basins. In the Limarí Basin, for example, a semiarid water basin located in the northern region of the country, water is a scarce good with a high economic value (especially for the export oriented agricultural sector). This scarcity generates strong competition for water between users, which in turn causes the temporary and permanent water market to be very active. In this basin, an integrated system of dams, canals and WUAs have reduced transactions costs, such that during the 1993–1999 period, 6,000 WR were traded (Hearne and Easter 1997; Hadjigeorgalis 2004; Zegarra 2002; Donoso 2006). Between the years 2000 and 1981, 27 % of the total WR in the Limarí Basin were reallocated via market transfers (Cristi et al. 2001).

Studies have also shown intersectoral trading in the Elquí and Maipo-Mapocho basins where urban growth has increased the demand for potable water supplies (Hearne and Easter 1997; Donoso et al. 2002). In the Maipo system, in the central region of the country, water supply is greater and demands from the agricultural sector lower. In the first section of this river basin 793 WR were traded in the period 1993–1999 (Donoso 2006). Also residential developers will purchase land and WR together, and thus alleviate the need to purchase WR independently as urban areas expand into previously irrigated acreage.

However, these studies concluded that most of Chile's river and canal systems did not have active trading and water markets have not been institutionalized throughout much of the country. A general conclusion of these studies is that trading is relatively active in basins where water has high values and where transactions costs are low. Even in the Limarí basin, where trading is common, this trading is more frequent in dry years when water has increased scarcity and value (Peña 2002).

Yet Chile's system of privatized water-use rights was a key element in the expansion of high valued export crops. With secure property rights farmers had incentives to develop improved irrigation systems and invest in perennial fruit crops. Even in the absence of active trading, the ownership of WR provided farmers with an appreciation that they were using an asset that had market value in alternative uses. In addition the presence of market transactions could reduce conflicts that would otherwise occur in the reallocation of water to growing municipal and industrial demands (Hearne and Easter 1995; Donoso 2006).

Chile's use of privatized WR was clearly designed to facilitate market reallocation of water (Buchi 1993). Because of this confidence in markets, Chile's 1981 Water Law has always had its skeptics. Dourojeanni and Jouravlev (1999) have been critical of the water law's neglect of river basin management. Bauer (1998, 2004) identified key concerns and especially involving non-consumptive use rights and inadequate conflict resolution institutions. Romano and Leporati (2002) assessed the distributional aspects of the water markets in the Limarí Basin and concluded that asymmetric information and barriers to entry have led to a concentration of WR toward the most wealthy and powerful segments of society. Contrary results were found by Hadjigeorgalis (2008) who shows that water markets in the basin have been successful in moving water and WR from low- to high-valued uses and that resource-constrained farmers of the Limarí Basin use temporary WR markets as a safety net. Additionally, the results indicate that WR markets have not been inequitable with respect to prices offered; resource-constrained farmers receive the same offer prices for their water and WR as wealthier farmers.

Recent research has further documented the expansion of water markets throughout Chile. Most of this analysis has focused on price variability in different localized markets. This research has often upheld the conclusions of previous studies. Hadjigeorgalis and Riquelme (2002) assessed transaction in the Cachapoal River basin in Chile central valley. Using data from real estate registries, a hedonic price analysis of 126 transactions during the 1990s was conducted. WR from zones with less variability in water delivery commanded a premium price. Water in canals with better infrastructure and more organized WUAs also had higher sales prices. Buyers who purchased large quantities of water were shown to pay a price premium. Water-use rights prices were shown not to have increased over time during this period, despite the increased relative scarcity of water. The authors suggest that improved irrigation technology might be one reason for constant prices.

Cristi et al. (2001) and Cristi (2007) used transactions data from 1981 to 2000 and survey data for 1998–2000 to analyze both the spot and permanent WR markets in the Limarí Basin. The results demonstrated that both markets had active trading. The spot market was active in both wet and dry years. The interconnected canals and the well managed WUAs contributed to the low transactions costs that fostered market trading. Cristi found that differences in attitudes toward risk acceptance led to heterogeneous values for water across farmers and to water market transactions. Unequal access to credit led certain farmers to sell water to others. Cristi observed that there were many transactions for small quantities of water-use rights. This implies that transactions costs are relatively low and that irrigators can use market transactions to achieve marginal changes to the quantities of water they receive. Despite the good market information and low transactions costs Cristi found heterogeneous market prices across and within canal systems.

Jordán (2007) studied the determinants of water-use rights prices in the first section of the Maipo River, a key water source for the Santiago Metropolitan area. Agriculture and real estate are the most important participating sectors in this market. Results of an econometric analysis show that there is relatively little intersectoral transactions. Key price determinants include the experience of the buyer or seller, the location of the point of withdrawal from the river, and the profitability of the sectors involved. The author concludes that prices are highly variable because the markets have few buyers and sellers which result in prices based on one-on-one bargaining.

Donoso and coauthors (2010) analyzed water markets in the Aconcagua basin. This is a large important basin north of the Santiago Metropolitan area. Focus groups with farmers and interviews with key water managers complemented an econometric assessment of price variability. Based upon a dataset of 1,675 water-use rights market transactions, for the period of 2000–2008 gathered from real estate registries the authors concluded that the markets in the separate segments and aquifers of the Aconcagua basin, remained thin. Prices varied substantially across and within these river segments, and mean prices were highly skewed due to high-end outliers. Farmers paid less for surface water-use rights than real estate and WSS companies. Despite perceived increased water scarcity, especially among industrial, mining, and residential sectors water-use rights prices have not increased. The authors attribute this to the reluctance of farmers to participate in the market. Farmers groups were concerned about the infrastructure costs necessary to move water across canals. They feel the need to keep water with land and prefer informal exchanges of volumes of water among themselves. In contrast industrial users and WSS companies favored water markets as a way to meet their future needs. These stakeholders expected markets to continue to develop and identified the need to improve cadasters of water users and complementary water market information.

In a review of national water market trends Cristi (2011) reviewed the database of transactions in the registry of water-use rights that is maintained by the DGA (see Table 6.3). This is the registry that was established by the 2005 reform of the Water Law and intended to provide transparent market information. Information for this registry is forwarded by local real estate registries throughout the nation. However, Cristi determined that only 60 % of the real estate registries reported transactions. Furthermore, many of the recorded transactions have incomplete information as to the price and the quantity of water transacted. Cristi concluded that despite efforts to identify and remove clear outliers in terms of price per volume, water market prices remained highly variable.

Data collected by Cristi are for water market transactions independent of land transactions, for the years 2005–2008. These transactions are spread throughout the country and not limited to the most frequently studied Limarí Basin (IV Region) and the Maipo and Mapocho Basins (in the Metropolitan Region). Approximately 88 % of these transactions were for consumptive use rights (Cristi 2011). Although water markets are most active in central Chile from the Coquimbo to the Maule regions, transactions occur throughout the nation. The active trading in the Limarí basin has not made the Coquimbo region an outlier, as the Metropolitan, Maule, and Valparaiso regions all witnessed greater value of water-use rights traded. The Santiago metropolitan region, with 40 % of national population and 44 % of GDP accounted for 48 % of the value of water market transactions. This data also supports the conclusion that water markets are more active in dry years, since 2007 was a relatively dry year and accounts for 50 % of the value of water trading in this 4 year period. Moreover, the later years witnessed more water market trading than the first 2 years.

The total number of granted consumptive surface water and groundwater use rights is 81,818 (DGA 2012a, b). Of the total consumptive WR, 54.2 % are for surface water and 45.8 % groundwater use rights. The majority of the granted consumptive WR are concentrated in the Bío-Bío, Araucanía and Los Lagos

Table 6.3 Water market transactions as reported to the national water-use rights registry

Region[a]	2005	2006	2007	2008	Total
Water market transactions					
Arica y Parinacota and Tarapacá	92	179	197	96	564
Antofagasta	13	7	63	48	131
Atacama	4	10	1		15
Coquimbo	775	1,231	1,155	287	3,448
Valparaiso	513	732	926	668	2,839
Metropolitan	585	1261	1210	1170	4,226
O'Higgins	465	568	513	464	2,010
Maule	968	1,471	1,678	2,042	6,159
Bío-Bío	300	643	934	285	2,162
Araucanía	145	200	29	113	487
Los Ríos and Los Lagos	28	131	39	25	223
Aysén		11	47	10	68
Magallanes		4	2		6
Total	**3.886**	**6,448**	**6,794**	**5.208**	**22,338**
Value of water market transactions (1,000 constant 2005 US $)					
Arica y Parinacota and Tarapacá	523	1,264	7,118	9,598	18,503
Antofagasta	43	121	3,327	205,906	209,397
Atacama	5,629	41	6	0	5,676
Coquimbo	25,675	194,483	234,095	3,757	458,010
Valparaiso	6,149	11,621	81,584	387,771	487,125
Metropolitan	6,392	52,439	1,451,153	555,291	2,065,275
O'Higgins	2,469	11,826	331,431	107,443	453,169
Maule	23,571	148,686	32,679	355,634	560,569
Bío-Bío	1,653	8,560	10,781	3,577	24,570
Araucanía	1,830	1,589	432	3,035	6,886
Los Ríos and Los Lagos	165	10,889	807	8,082	19,943
Aysén	0	95	27	7	128
Magallanes	0	385	0	0	385
Total	**74,098**	**441,999**	**2,153,440**	**1,640,100**	**4,309,638**

Source: Cristi (2011), Banco Central de Chile
[a]This data precedes the 2006 changes to Chile's regions. Therefor Arica and Parinacota remains combined with Tarapacá, and Los Ríos and Los Lagos remain combined

regions, with 11.8 %, 16.9 % and 16.7 % respectively (see Fig. 6.3). Total traded consumptive WR during 2011 corresponds to 5.6 % of total granted WR; 4.4 % of granted consumptive surface water and 7 % of total consumptive groundwater use rights. Figure 6.4 shows consumptive WR transaction data based on data of the PWR of the DGA, for the period 2005–2008.[1] The results for this 4-year period show that there were 24,177 WR transactions of which 92.3 % were independent

[1]The PWR of the DGA has data only for the period 2005–2009. The data for the year 2009 is incomplete and thus not included in this analysis.

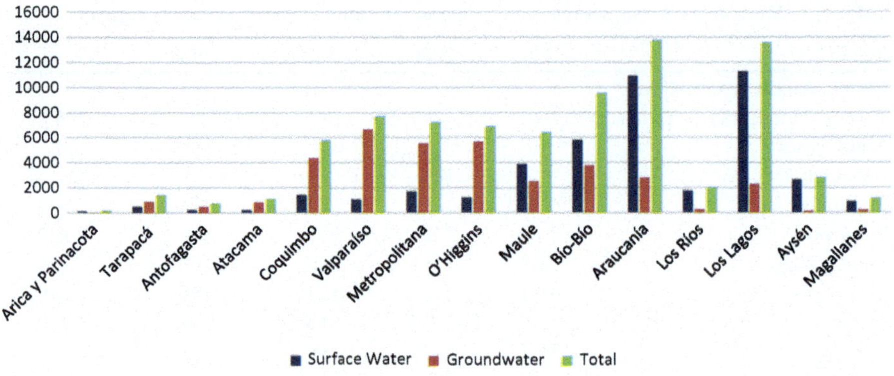

Fig. 6.3 Granted water rights (Source: Dirección General de Aguas 2012a)

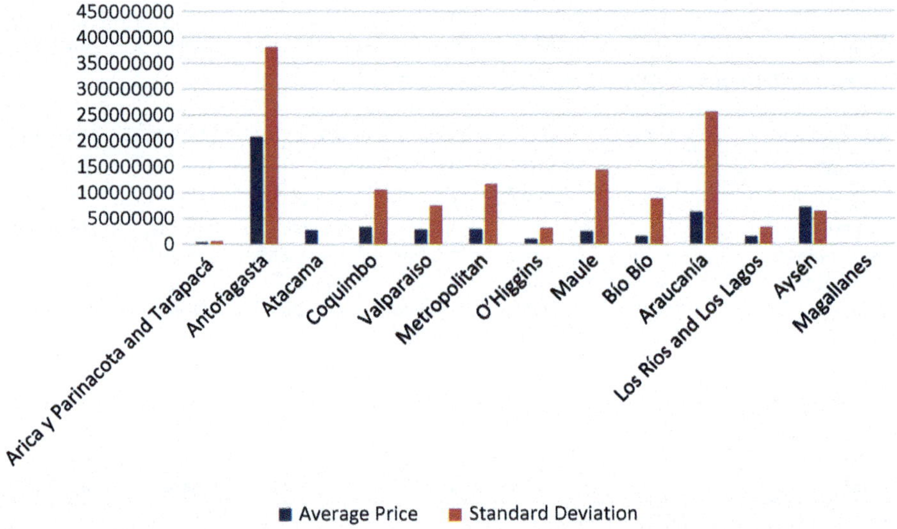

Fig. 6.4 Average regional surface WR price (US\$/m^3/s) (Source: Dirección General de Aguas 2012b)

of other property transactions, such as land. The total value of WR transactions independent of other property transactions is U.S. \$ 4.8 billion, representing an annual transaction value of U.S. \$ 1.2 billion.

The average WR price is US \$ 41,572,531/m^3/s. WR prices in the north of the country are, in general, greater than in the South, which indicates that the market, at least in part, reflects the relative scarcity of water. WR prices have a coefficient of variation of 245 %, and price dispersion is lower in the more active WR markets. Thus, Chilean WR markets are characterized by large price dispersion for homogeneous WR (Cristi and Poblete 2010).

This large price dispersion is due, in great part, to the lack of reliable public information on WR prices and transactions. Given the lack of reliable information, each WR transaction is the result of a bilateral negotiation between an interested buyer and seller of WR where each agent's information, market experience and negotiating capacity is important in determining the final result (Jordán 2007). Bjornlund (2002) in a study of WR markets in the Goulburn-Murray Irrigation District in South Australia, found similar results, where factors that influence the negotiating process and the agent's negotiating power significantly influenced WR prices. First, Bjornlund (2002) found that the agent's awareness of prevailing market prices is a significant factor explaining the WR price; hence, an important variable that influences the negotiating process is each agent's expected price, which is based on previously traded WR prices. Second, Bjornlund (2002) concludes that a major determinant of WR prices is the bargaining strength of the buyers and sellers.

6.6 Conclusions

Water markets in Chile have been enabled since the 1981 National Water Law. This law was designed to provide market incentives with the capacity to reallocate water toward more valuable uses. Although there have been many controversies, especially concerning the initial allocation of WR and the coexistence of consumptive and non-consumptive use rights, water markets have gradually expanded. Large scale trading has not occurred in many basins. Likewise, large scale intersectoral trading has not occurred because the agricultural sector has remained prosperous, with growth outpacing the rest of the economy.

The 2005 Reform of the Water Law have addressed the issues of unused water-use rights held by speculators. This has reduced what many critics of the 1981 Water Law have considered to be a significant constraint to water management and development. It has also provided for auctions of forfeited water-use rights which may provide additional stimulus and transparency to local markets.

Many water-use rights have remained unregistered. Given that unregistered water-use rights cannot be sold in markets this remains an impediment to market trading. Additional efforts are needed to ensure consistent titling and registration of all water rights. The 2005 Reform of the Water Law requires that records of all WR transactions be forwarded to the National Water Registry. Efforts to ensure that complete transactions records, including data on price and volume, may provide improved market information to buyers and sellers, and reduce price dispersion.

Also with the 2005 Reform, the DGA has been allowed to protect minimal environmental flows in rivers. This protection is expected to be minimally effective in the arid north where all water has been allocated to water-use rights. Additional efforts are needed to protect marine and riverine ecosystems. Additional regulation may be required in order to maintain water quality.

Groundwater resources have become increasingly exploited. And many aquifers have become over-allocated. Improved hydrological data is required.

And, responsible user management of aquifers would be fostered with efforts to facilitate the development of WUAs for aquifers.

The 1981 Water Law was designed to bring market incentives into water allocation decisions. It was also designed to reduce the role of government bureaucracies in water management. In general this system has been effective. Recent data reveals water market transactions throughout Chile. Markets are more active, and prices are higher in basins with relative water scarcity. Many traditionally recognized WR remain untitled, and outside of trading. But they remain protected by courts. Market prices are highly dispersed, due to the influence of bilateral negotiations between parties with different experience and information levels. New information systems, such as electronic water trading and the PWR will likely help to reduce transactions costs and price variability. Nonetheless, Chile's system of water markets has expanded geographically and quantitatively in the last decades. Agriculture, mining, hydropower, and urban development have all been able to expand without serious constraints imposed by water management.

References

Ayala L (2010) Aspectos Técnicos de la Gestión Integrada de kas Aguas (GIRH). Primera etapa: Diagnóstico. Santiago: Informe preparado para el diagnóstico de la gestión de los recursos hídricos

Ayala, Cabrera Y Asociados Ltda (2007) Estimaciones de Demanda de Agua y Proyecciones Futuras. Zona I Norte. Regiones I a IV. Informe Final. Report To the Direccion General De Aguas. S.I.T. N° 122. Available at http://documentos.dga.cl/USO4854v1.pdf. Accessed June 2013

Bauer C (1998) Against the current: privatization, water markets and the state in Chile. Kluwer, Boston

Bauer C (2004) Siren song: Chilean water law as a model for international reform. Resources for the Future Press, Washington, DC

Bjornlund H (2002) Signs of maturity in Australian water markets. N Z Prop J 2002(July):31–46

Buchi H (1993) La transformación económica de Chile. Del Estatismo a la libertad económica. Norma, Bogota

CADE-IDEPE (2005) Diagnóstico y Clasificación de los Cursos y Cuerpos de Agua según Objetivos de Calidad. DGA

Cristi O (2007) The influence of heterogeneous risk preferences on water market activity: an application to the Paloma System of the Limarí Water Basin, Chile. Ph.D. Thesis, University of Maryland, College Park. Available at http://drum.lib.umd.edu/handle/1903/6695. Accessed June 2012

Cristi O (2011) Análysis del Mercado De Agua. Paper presented at the Conference Proyecto:Mercado Electronico Del Agua (MEDA) Santiago. April 2011

Cristi O, Poblete C (2010) Que nos dicen los conservadores de bienes raices del mercado de aguas en Chile. Por publicarse en Serie Working Paper, Escuela de Gobierno, Universidad del Desarrollo. Santiago, Chile

Cristi O, Trapp A (2003) "Mercado de Agua para Irrigación y Uso Urbano: una Aplicación a la Cuenca del Río Elqui, Chile". World Bank-Netherlands Water Partnership Program (BNWPP), Water Rights System Window, July 2003

Cristi O, Vicuña S, Azevedo L, Baltar A (2001) Mercado de agua para irrigación: una aplicación al sistema paloma de la cuenca del Limarí, Chile. Presented at the III Encuentro de las Aguas, Santiago de Chile, 24–26 Oct 2001

Daugherty A (2012) Hydroelectricity surpasses thermoelectricity in Chile. The Santiago Times. 20 June 2012. Available at http://www.santiagotimes.cl/business/energy/23953-hydroelectricity-surpasses-thermoelectricity-in-chile. Consulted February 2013

Dirección General de Aguas (2010) Estimación de Recargas de Aguas Subterráneas en Cuencas de las Regiones VII. VIII, XIV y X. Gobierno de Chile, Santiago

Dirección General de Aguas (2012a) Catrasto Público de Aguas. Available at http://www.dga.cl/administracionrecursoshidricos/Paginas/default.aspx. Consulted August 2013

Dirección General de Aguas (2012b) Información histórica recibida de Conservadores de Bienes Raices en relación a derechos de aprovechamiento de tipo Consuntivo. Available at http://www.dga.cl/FNE/CBR/Paginas/default.aspx. Consulted August 2013

Donoso G (2006) Water markets: case study of Chile's 1981 water code. Cien Inv Agr 33(2):157–171

Donoso G, Montero JP, Vicuña S (2002) Análisis de los Mercados de Derechos de Aprovechamiento de Agua en las Cuencas del Maipú y el Sistema Paloma en Chile: Efectos de la Variabilidad de la Oferta Hídrica y de los Costos de Transacción. Páginas 109–151 In: Embid A (ed) El Derecho de Aguas en Ibero América y España: Cambio y Modernización en el Inicio del Tercer Milenio, Madrid, Civitas

Donoso G, Cancino J, Melo O, Rodriguez C, Contreras H (2010) Informe Final "Análisis Del Mercado Del Agua De Riego En Chile: Una Revisión Crítica A Través Del Caso De La Región De Valparaíso." Oficina de Estudios y Políticas Agrarias (Odepa) del Ministerio de Agricultura. Available at http://www.odepa.gob.cl/odepaweb/servicios-informacion/publica/Analisis_agua_riego_Valparaiso.pdf. Consulted June 2013

Dourojeanni A, Jouravlev A (1999) El Código De Aguas De Chile: Entre La Ideología Y La Realidad. S E R I E, Recursos Naturales e Infraestructura No. 3. Santiago, Chile. CEPAL

FCCyT (2012) Diagnóstico del Agua en las Américas. Red Interamericana de Academias de Ciencias. . México, Distrito Federal: Foro Consultivo Científico y Tecnológico, AC

Global Water Intelligence (2010) A step in the right direction. 11(10). October 2010. http://www.globalwaterintel.com/archive/11/10/market-insight/step-right-direction.html

Hadjigeorgalis E (2004) Comerciando con Incertidumbre: Los Mercados de Agua en la Agricultura Chilena. Cuadernos de Economía 40(122):3–34

Hadjigeorgalis E (2008) Distributional impacts of water markets on small farmers: is there a safety net? Water Resour Res 44(10). doi:10.1029/2007WR006527

Hadjigeorgalis E, Riquelme C (2002) Análisis de los Precios de los Derechos de Aprovechamiento de Aguas en el Río Cachapoal. Ciencia e Investigación Agraria 29(2):91–99

Hearne R, Donoso G (2005) Water institutional reforms in Chile. Water Policy 7:53–69

Hearne R, Easter KW (1995) Water allocation and water markets: an analysis of gains-from-trade in Chile. Technical paper #315. World Bank, Washington, DC

Hearne R, Easter KW (1997) The economic and financial gains from water markets in Chile. Agric Econ 15:187–199

Instituto de Ingenieros de Chile (2011) Temas Prioritarios para una Política Nacional de Recursos Hídricos. Available at: http://www.iing.cl/index.php?option=com_content&view=article&id=95&Itemid=234. Accessed May 2014

Jordán C (2007) Analisis del mercado de derechos de aprovechamiento de aguas de la primera seccion del Rio Maipo, region metropolitana: comportamiento de los diferentes agentes economicos y variabilidad en precios. MS. Thesis, Universidad Catolica de Chile. Fac. de Agronomia e Ingenieria Forestal

Jouravlev A (2010) Código de Aguas de Chile: Desafíos pendientes Paper presented at the Reunión de trabajo con el Comité Técnico del Banco Mundial (DGA, Santiago de Chile, 15 de julio de 2010). Available at http://www.eclac.cl/drni/noticias/noticias/5/40355/DGA_Chile_15_07_2010.pdf. Accessed June 2012

Macaulay S (2009) Advancement of progressive management strategies to promote regional and statewide water supply reliability in California, United States of America. Consultancy report for Rijkswaterstaat Centre for Water Management, Dutch Ministry of Transport, Public Works and Water Management. West Yost Associates, Davis

Nelson A (2011) Chile dams its rivers to unleash its economy. Time. May 11, 2011. Available at http://www.time.com/time/world/article/0,8599,2070816,00.html#ixzz2LNoyJsuO. Consulted February 2013

ODEPA (Oficina de Estudios y Políticas Agrarias) (2004) Agenda Agroalimentaria Y Forestal 2004–2006. http://www.odepa.gob.cl/

OECD (2011) Water governance in OECD countries. doi: 10.1787/9789264119284-en. OECD, Paris. Available at http://www.oecd-ilibrary.org/environment/water-governance-in-oecd-countries/chile_9789264119284-12-en. Consulted July 2013

Peña H (2000) Desafıós a las Organizaciones de Usuarios en el Siglo XXI. III Jornadas de Derechos de Agua. Universidad Católica de Chile, Santiago

Peña H (2002) Los Mercados Del Agua: La Experiencia Chilena. Presented at the international conference of basin organizations. Madrid, Spain. Nov. 2002

Peña H (2010) —Copiapó: realidades, desafíos y lecciones. Presentación para "¿Existe sobre-explotación del agua en Chile?, Seminario Nacional AHLSUD Capitulo Chileno A.G

Puig A (1998) El fortalecimiento de las organizaciones de usuarios para una gestión integrada de los recursos hídricos. Paper presented at the international conference water and sustainable development, Paris, 19–21 March 1998. http://www.oieau.fr/ciedd/contributions/at2/contribution/aurora.htm

Quentin Grafton R, Libecap G, McGlennon S, Landryj C, O'Brien B (2012) An integrated assessment of water markets: a cross-country comparison. Rev Environ Econ Policy 5(2):219–239

Rhodos Asesproas y Proyectos Ltda (2010) Resultados de estudios de levantamiento de información sobre derechos no inscritos susceptibles de regularizar. Santiago

Romano D, Leporati M (2002) The distributive impact of the water market in Chile: a case study of the Limarí. Q J Int Agric 41:41–58

Rosegrant WM, Gazmuri R (1994) Reforming water allocation policy through markets in tradable water rights: lessons from Chile, Mexico and California. EPTD discussion paper N° 6, IFPRI

Salazar C (2003) Situación de los Recursos. Hídricos en Chile, Reporte de Investigación, SISS, Santiago

Superintendencia de Servicios Sanitarios (2012) Informe anual de coberturas urbanas de servicios sanitarios. Available at: www.siss.gob.cl/577/articles-9886_recurso_1.pdf. Accessed May 2014

Valenzuela, C (2009) La patente por la no-utilización de las aguas en Chile: origen, diseño y primeras experiencias en su implementación, Santiago de Chile, Universidad de Chile/Facultad de Ciencias Agronómicas. Available at: http://www.eclac.cl/cgi-bin/getProd.asp?xml=/drni/noticias/documentosdetrabajo/0/39270/P39270.xml&xsl=/drni/tpl/p38f.xsl&base=/drni/tpl/top-bottom.xsl. Accessed May 2014

Vergara A (2010) Diagnóstico de Problemas en la Gestión de Recursos Hídricos: Aspectos institucionales para una propuesta de modificaciones legales, reglamentarias y/o de prácticas administrativas. Informe preparado para el diagnóstco de la gestión de los recursos hídricos, Santiago

World Bank (2003) Chile at a glance. Available at http://www.worldbank.org/data/

World Bank (2011) Chile: Diagnóstico de la gestión de los recursos hídricos. No. 63392. Available at http://water.worldbank.org/node/83999. Accessed June 2013

Yanez R (2008) El mercado de las aguas en Chile. Informe Especial. Ideas para Chile. Vol XIX N° 200. ISSN 0717–7933

Zegarra E (2002) Water markets and coordination failures: the case of the Limarí valley in Chile. PhD Thesis, University of Wisconsin, Madison

Chapter 7
Water Markets in Spain: Meeting Twenty-First Century Challenges with Twentieth Century Regulations

Dolores Rey, Alberto Garrido, and Javier Calatrava

Abstract Water scarcity is a growing reality in many Spanish basins which creates the need for more flexible and efficient market-based allocation instruments. This chapter critically analyzes water markets' strengths and weaknesses, evaluates some recent trading experiences, and assesses some recent reforms in the Spanish water legislation. Formal and informal trading, and variants in between, have facilitated temporary and permanent water exchanges, with and without explicit support of public agencies. Based on our analyses and other literature findings, we propose a number of reforms that we consider necessary to upgrade water markets in Spain, including some innovations such as optioning rights, and quality-graded water exchanges.

Keywords Water markets • Spain • Market reform • Informal trading • Trade barriers

D. Rey
CEIGRAM, Universidad Politécnica de Madrid, Senda del Rey 13, 28040 Madrid, Spain
e-mail: dolores.rey@upm.es

A. Garrido
Department of Agricultural Economics and Social Sciences, Research Centre for the Management of Agricultural and Environmental Risks, Centro de Estudios e Investigación para la Gestión de Riesgos Agrarios y Medioambientales, CEIGRAM, Universidad Politécnica de Madrid, E.T.S. Ingenieros Agrónomos, Avda. Complutense s/n, 28040 Madrid, Spain
e-mail: alberto.garrido@upm.es

J. Calatrava (✉)
Escuela Técnica Superior de Ingeniería Agronómica, Universidad Politécnica de Cartagena, Paseo Alfonso XIII, 48, 30203 Cartagena, Spain
e-mail: j.calatrava@upct.es

K.W. Easter and Q. Huang (eds.), *Water Markets for the 21st Century: What Have We Learned?*, Global Issues in Water Policy 11, DOI 10.1007/978-94-017-9081-9_7,

7.1 Introduction

Water scarcity is a reality in many Spanish basins, and it will be exacerbated in the foreseen future by climate change and increasing water demands. This creates the need for more flexible and efficient market-based allocation instruments (Stefano and Llamas 2012). Markets facilitate the reallocation of water resources among users, improving water use efficiency and allocating water to high-value uses.

The 1999 Reform of the Water Act of 1985 introduced the legal possibility of voluntary exchanges of public water rights (water concessions, as they are called in the Act). Initially, the formal trading activity was limited to a few isolated cases across the country (Garrido et al. 2012a). The 2005–2008 drought gave rise to an increase in water exchanges that significantly improved the conditions in those areas where water scarcity was most severe. Since 2005 the water trading activity has been more frequent in Spain, although traded volumes in dry years represent less than 1 % of all annual consumptive uses. Various water trading mechanisms were defined in the 1999 Reform, to which one more was added in 2012 to address problems of groundwater overexploitation. A specific market regulation in the Water Law of the Andalusian region (see Fig. 7.1) enabled differentiated options to be used to exchange water in internal basins of the Andalusian region.

In parallel with formal trading operations, and going back at least three decades, informal water markets of a very different nature have evolved and developed extremely diverse and innovative mechanisms (Hernández-Mora and De Stefano 2013), mainly in the Southeast of Spain and in the Canary Islands (whose water law is different from that in Iberian Spain). Some of the exchanges within this informal category eventually gave rise to formal agreements or adjudications. Still many others are in a legal limbo, but provide a wealth of services and water supply to otherwise thirsty users, showing that the regulatory framework in force is not sufficiently rich or encompassing to include the many market variants and approaches. This chapter also reviews the informal or quasi-informal water trading in Spain.

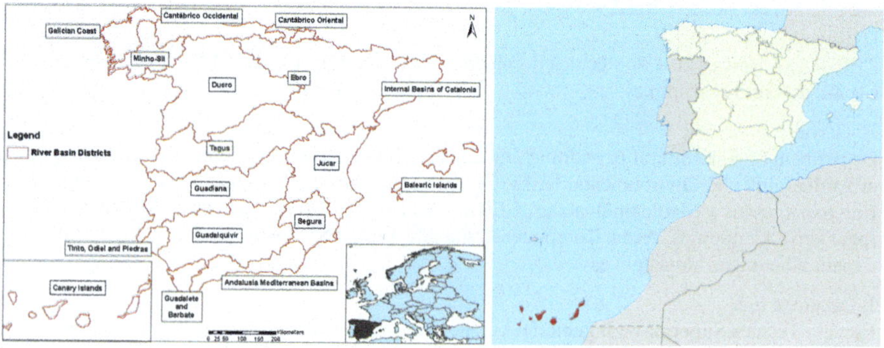

Fig. 7.1 Spanish river basins (*left*), including the canaries (*right*)

Under this diverse institutional landscape, the threat of climate change gives support to the development of water supply policies and institutions that are sufficiently flexible, adaptive, and robust to deal with an uncertain and changing water future (Adler 2009). According to the CEDEX (2011), precipitation will decrease in Spain by 7–14 % between 2010 and 2040 depending on the GHG emission scenario considered. In semiarid areas, decreases in available water resources may be equivalent to 50 % of the potential resource in Iberia (Iglesias et al. 2005; Moreno 2005; Garrido et al. 2012b). Water reallocation is seen as one pillar of water demand management, making a better use of existing resources as opposed to supply augmentation options (Molle and Berkoff 2006).

This chapter claims that the institutional design of water markets should be improved and new types of trading mechanisms should be developed in order to overcome the drawbacks of the current water market regulation. We proceed with a short review of the Spanish water market regulations and their most recent reforms. Then, the chapter summarizes all of the different trading formats and approaches, which have been recently documented in the literature (Ariño and Sastre 2009; Garrido et al. 2012a, 2013), including "informal" exchanges (Hernández-Mora and De Stefano 2013); and analyzes the causes of low participation in the market. At the end, some recommendations and conclusions are provided, that could be applied in Spain and other countries facing similar water scarcity problems.

7.2 Water Market Legislation in Spain

7.2.1 General Approach[1]

In Spain, there are public and private water rights. Public water rights are concessions granted by the Water Authorities for 30–75 years. According to the 1985 Water Act, rights can be granted to pump groundwater or divert water resources directly from surface water bodies. Water use rights are defined by the point of withdrawal, type of use, date of withdrawal (calendar), plots to be irrigated and irrigation technologies, usable volume or flow and return flows. The type of use, location, withdrawal prerequisites or return flow points cannot be changed without an explicit approval by the River Basin Agency (RBA). Rights differ in the priority of their access to water depending on the type of use (domestic, environmental, agricultural, hydropower or industrial). Holders of private groundwater rights, before the 1985 Law came into force, were given the choice of keeping their rights as a private right or else converting them into temporal water concessions. A vast majority (more than 80 % of right holders according to Llamas et al. 2001) opted for the first option.

[1]This section borrows heavily from other authors' work (Garrido et al. 2012a).

The differences between water rights and public rights are the following: public rights are use permits granted by the State for a duration of 30 years; they can be revoked, transformed, amended or interrupted by the Basin Agencies if conditions advise such decisions; their legal foundation stems from the 1985 Water Act, which declared all water resources to be part of the public domain; they are registered in a separate section of the section on private rights. Private rights, in contrast to public rights, have a longer maturity, existed before the 1985 Water Act came into force and are considered private property that can be sold, leased and form part of a company or cooperative assets. Maintaining the status of water rights requires that the technical conditions of use (depth and location of wells, power of pumps, pumped volume) not be altered.

Swapping private rights with a concession was in principle stimulated by the rigidity with which the former were defined. Since the legislators preferred to have most users under the public regimes, the Act preserved the private rights under the exact conditions established in the registry, forcing anyone wishing to change them to request a change to the public section and have it transformed into a concession. Unexpectedly to the legislators, most preferred to keep rights private.

The 1999 Reform of the Water Act introduced the legal possibility of voluntary exchanges of public water rights (concessions), but with many restrictions. Before this reform only private rights could be formally traded; water flows pumped from private wells could be leased, auctioned or sold.

There are various types of barriers to exchanging water rights: market regulation barriers, barriers related to water rights' definition, institutional barriers, and environmental barriers. All of them make trading activity quite difficult by raising transaction costs as well as preventing certain types of trades. In Sect. 7.4, this issue is analyzed in depth.

The 1999 Reform identified only two ways to exchange public water use rights: i) right-holders that voluntarily agree on specific terms of trade and jointly file a request to the Agency, or lease-out for a number of years the water to which right-holders are entitled; ii) water bank operations (or *water exchange centers*, as they are called in the 1999 Reform of the Water Law). Users of private groundwater rights, individually or as firms or cooperatives, can sell, lease or rent pumped water, although such trading is subject to specific restrictions.

Initiated by the RBAs, water banks are set up as public tenders for interested right-holders who are willing to relinquish their water rights temporally or for the remaining maturity period. The bank's water supply operation involves procuring volumes from voluntary sellers, and making them available for other users and uses, including environmental restoration purposes. They may also acquire permanent water rights. Water Banks are supposed to be administered by the RBAs and operate in exceptional situations of drought or overexploitation of aquifers (WWF 2005). In practice, these *water exchange centers* have only functioned as buyers of water concessions or leased water use rights just for 1 year. Water has not been sold to other users. Instead, purchased water has been made available to other users in the form of new water concessions or devoted to maintaining environmental river flows and/or raising water tables in overdrafted aquifers.

There is a great diversity in the ways exchanging systems have evolved since the 1970s, primarily in the most water stressed areas (Segura and Jucar basins and the eastern part of the Andalusian Mediterranean Basins[2]). In Sect. 7.3, we provide a realistic overview of water markets and trading in Spain, including formal and informal trading, and the middle ground between the two.

7.2.2 Subsequent Reforms

At the national level, the last reform of the Spanish Water Law of May 2012 highlights the need to simplify and accelerate the administrative procedures, and to add more flexibility and efficiency to the water management system. The reform focuses mainly on groundwater resources. It proposed several measures to deal with water availability problems, including the encouragement of transformation of private water rights into public water concessions. Although this reform is meant to improve water management, there are also some details that could threaten groundwater resources sustainability, and be in breach of the mandates of the European Union Water Framework Directive (WFD), one of which is to avoid any further deterioration of a water body already heavily damaged. The new regulation establishes the possibility of recharging aquifers with external water resources in order to avoid the risk of not achieving a good quantitative status for these aquifers. This could potentially persuade water users that the best solution for declining groundwater tables is always to provide external resources, and thus it is not necessary to change the exploitation rate of aquifers. Also, the Reform grants new water concessions under certain circumstances in groundwater reservoirs at risk, which presumably will cause a higher overexploitation of groundwater resources.

The regional government of Andalusia passed more advanced legislation in 2010. This new Andalusian Water Law includes some differences from the National Law that result in more flexible trading mechanisms. However, the water market regulation in Andalusia is only applicable in the Andalusian Mediterranean Basins

[2]The Andalusian Mediterranean Basins are a series of basins on the southern Mediterranean coast of Spain that are completely within the boundaries of Andalusia and thus water management is the responsibility of the Regional Government of Andalusia. The Spanish Mediterranean basins include the Analusian Mediterranean Basins, the Segura basin, the Júcar basin, the Ebro basin and the Catalonian Internal basins (basins on the Mediterranean coast that are completely within the boundaries of Catalonia). The Andalusia region has other basins that are on the Atlantic coast including the Guadalquivir and Guadiana basins that empty into the Atlantic. The Guadalquivir basin includes territories in three regions different from Andalusia. More than 80 % of the Guadalquivir basin is in Andalusia and its climate is markedly Mediterranean but it is not included in the Mediterranean basin.

Table 7.1 Main differences between the National Law and the Andalusian Law related to water markets

Andalusian Law[a]	National Law
Agriculture, industry and tourism are considered at the same level in the water uses priority range	Agriculture is in a higher level, so farmers cannot sell their water rights for industrial or touristic activities
Water Banks are conceived as a mechanism to trade water under every circumstance	Water Banks are conceived as a mechanism to trade water only during drought periods
For acquiring water through a water bank, there is no need to be a water user with formal rights	Only users with formal water rights have access to the Water Bank or to purchase from other user

Source: Authors' elaboration

[a]The Andalusian Water Law take precedent over the National Law only in the basins that are contained entirely within Andalusia's borders as its regional government has jurisdiction over all water management in these basins

(see Fig. 7.1). The main innovations introduced by this reform are summarized in Table 7.1.[3] This approach could hopefully serve as a precedent for future amendments to the market regulation in the rest of Spain.

The differences in the Andalusian Law from the National Law provide flexibility for the water market system, allowing farmers (the main water rights holders) to sell water to industries, renewable energy plants (thermo-solar installations) or to the tourist sector. The most relevant criterion to determine the priority among these uses are: the impact on sustainability of the resource, maintenance of territorial cohesion and the higher added value in terms of job and wealth creation for the region. As in the National legislation, the Andalusian Law always guarantees the primary water requirements for the urban sector, and also for environmental purposes in order to achieve a good ecological status for all water bodies.

Water banks are considered an important tool not only for solving drought or environmental problems in Andalusia, but also to create a water stock for future purposes, to sell water use rights to users for a given price, and to avoid imbalances in the distribution of water resources. Through water banks the regional government can make offers for public purchase of rights, and expropriate or revise water concessions. The possibility of purchasing water through the water bank without previously being a right holder allows users facing new emerging water demands to obtain water. Currently there is an initiative to establish three water banks in three different basins in Andalusia.

[3]BOJA num. 155. Law 9/2010, July 30th. Andalusian Water Law.

7.3 Overview and Evaluation of Past Experiences

Two canonical water trading formats exist in Spain: one involves right holders exchanging registered water rights, using formal procedures and in full compliance with water law. The other canonical extreme involves two agents (persons or firms) agreeing verbally on purchasing a given volume pumped from an unregistered and unapproved borehole, leaving no written document or contract. This is the typical illegal type of exchange (no permit to pump, no water right, no records). This other extreme could also require, in some cases, the use of a pipe that connects points several kilometers away from each other. There are all kinds of middle ground in between these two extremes (formal and informal). Figure 7.2 attempts to sketch them out.

While formal trading has been thoroughly documented in the literature, very little has been published about the different types of informal trading. We will review some of the formats that have been documented in the gray literature or found in the authors' own field work.

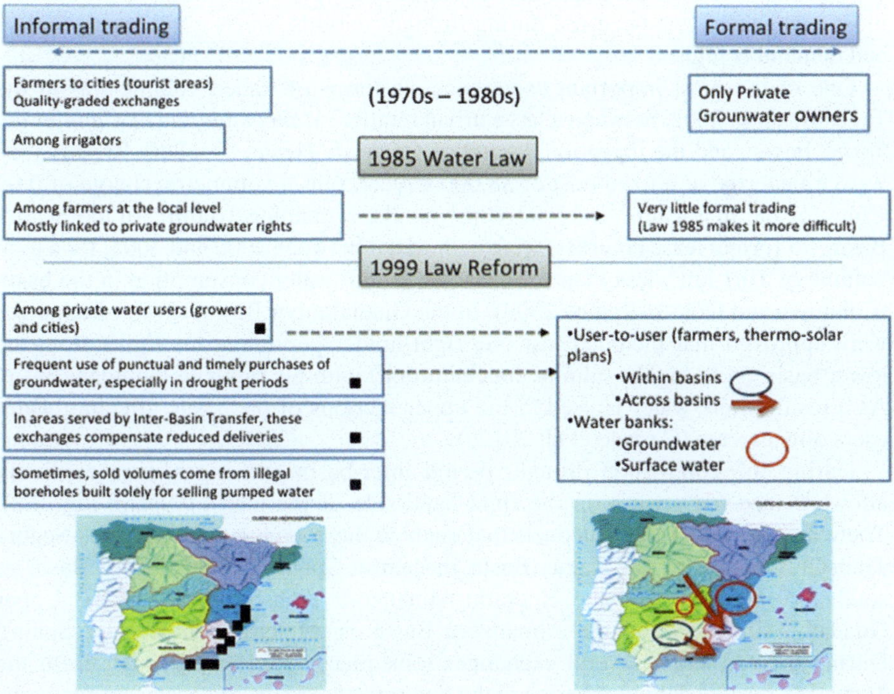

Fig. 7.2 Formal and informal trading in Spain (Source: authors' elaboration)

7.3.1 *Formal Trading Mechanisms*

Under this heading we review trading mechanisms that are situated on the extreme right, or close to it, in Fig. 7.2. Note, however, that two arrows connect trading schemes that begin on the left, (informal qualification) and end up being on the right side. These involve exchanges that are initiated and made effective without any legal support, but eventually are filed with the water authority and adjudicated. We are not aware of any reverse changes in trading format from right to the left.

7.3.1.1 Bilateral Agreements

The number of formal lease contracts were expected to increase significantly upon the approval of the 1999 Reform, especially between different areas of the same basin, but in practice they declined significantly. Temporary leases are predominant, whereas permanent exchanges of water rights are less common. Irrigation districts have been the main water sellers, with other districts, urban suppliers and thermo solar plants being the main purchasers. In general, prices have been high because most exchanges have occurred during drought periods, when water supply is low and demand is high.

One of the most important experiences in terms of traded volume was in the Tagus Basin in 2002, between a large urban retailer (*Mancomunidad de Canales del Sorbe,* buyer) and the irrigation district of *Canal de Henares* (seller). Twenty hm^3 were transferred at a fixed cost of 38,000 €/year, plus a volumetric charge of 0.04 €/m^3 for the first 4 hm^3 and 0.02 €/m^3 for the remaining 16 hm^3. In the Segura Basin, 35 formal lease contracts were authorized between 2000 and 2005, for a total volume of 10.1 hm^3, less than 1 % of total annual water consumption in the basin (Calatrava and Gómez-Ramos 2009). In the Guadalquivir Basin, several exchanges were approved that included just one right-holder permuting his rights from the lower basin (with higher salinity concentration) with his rights in the upper basin. As a result, more water is used in the upper sections of the basin, affecting water users downstream (Garrido et al. 2012a).

During the 2005–2008 drought period inter-basin exchanges were explicitly allowed[4] (see red arrows on the right-hand-side of the map of Spain, Fig. 7.2). There are two important aqueducts that enabled these exchanges: the Tagus-Segura Transfer (connecting the Tagus Basin in central Spain and the Segura Basin in southeastern Spain) and the Negratín-Almanzora Transfer (between the Upper Guadalquivir Basin and the Almanzora Basin in Almería, southeastern Spain). During these 3 years several exchanges took place, transferring water from the Tagus and the Guadalquivir basins to the Segura and Almanzora basins respectively.

[4]This required four annual Royal Decrees that permitted inter-basin exchanges, using pre-existing infrastructures, on the basis of drought situations in the recipient basins (Segura, Júcar and Andalusian Mediterranean Basins).

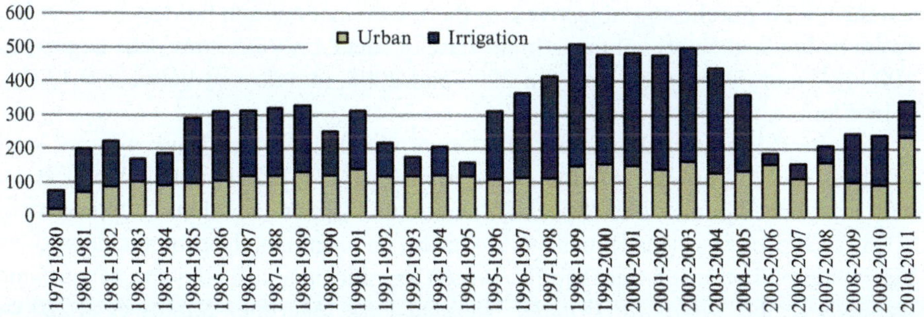

Fig. 7.3 Transferred water volume (hm^3) for irrigators and urban suppliers through the Tagus-Segura transfer, 1979–2011 (Source: San Martín 2011)

These were annual agreements for specified volumes, at prices net of transportation costs that ranged from 0.15 to 0.28 €/m^3. The severe drought situation that the country was suffering at that time led the Government to allow water users to use the aqueduct for these exchanges without paying any transportation fee (Garrido et al. 2012a). In the case of the Tagus-Segura Transfer, sellers were farmers from the Tagus Basin, and buyers were the major urban water supplier in the Segura Basin (Taibilla's Canals Commonwealth, *Mancomunidad de Canales del Taibilla)*, and the Central Association of the Irrigators of the Tagus-Segura Aqueduct (*Sindicato Central de Regantes del Acueducto Tajo-Segura*) (Garrido et al. 2012a).

In the Segura Basin, several large irrigation districts and a majority of municipalities depend on the water resources from the Tagus Basin, delivered through the Inter-basin Tagus-Segura Transfer,[5] which it is not a water market, but rather an institutional arrangement between both basins. The transferred volumes through this Aqueduct vary considerably from year to year, as they depend on the stock level in *Entrepeñas-Buendía* reservoir in the Upper Tagus Basin. One market driver with profound effects is the instability of these transferred volumes (see Fig. 7.3). In fact, short falls in deliveries were compensated for, in part, by water purchases referred to above in Sect. 7.3.1.1.

7.3.1.2 Water Banks

The 1999 Reform of the Water Law established the option of creating publicly run and administered water banks. Normally, water banks in Spain have been established to solve environmental problems. This was the case in Guadiana, Jucar and Segura basins, where water banks had different budgets, features, procedures and results (Garrido et al. 2012a). The Jucar and Segura water exchange centers did

[5]The Tagus-Segura Transfer also serves users in the Andalusian Mediterranean Basins and the Jucar Basin.

not meet their purchasing objectives, as there were not enough bidders to cover the entire budget and target volumes. In the case of Jucar, only 77.9 hm^3 (the target was 100 hm^3) were purchased between 2006 and 2008, at a cost of between 0.13 and 0.19 €/m^3.

Despite the large budget spent, the Upper Guadiana Water Bank was not well managed and did not provide the expected results (WWF 2012). This Water Bank was established under the "Special Plan of the Upper Guadiana", approved in 2008, as the primary instrument to solve the environmental problems caused by the overexploitation of one aquifer, which significantly affected the remarkable wetlands in the *Tablas de Daimiel* National Park (Martínez-Santos et al. 2008; Llamas et al. 2010). Its goal was to acquire water rights to reduce pumping rates by 250 million m^3 by 2027 and raise the aquifer's water table (Garrido et al. 2012a). The initial idea was to purchase water rights to be re-allocated to other farmers (30 %) and to the environment (70 %) (López-Gunn et al. 2012).

Although this Special Plan established several requirements and conditions for the performance of the Water Bank (defined a priority area near the aquifer, only allowed farmers that had been using water for the last 3 years to sell their rights, etc.), the truth is that these conditions were not always applied. This impeded the achievement of a better ecological status for the aquifer, and even increased water consumption in some cases. According to WWF (2012), groundwater extractions have only been reduced in 1.1 hm^3 at a cost of around 6 million euros in public funds.

7.3.2 *Informal Trading*

The combination of scarcity, intensive agricultural production and the urban expansion to accommodate newcomers and tourist capacity has provided the ideal conditions for "informal" water exchanges. Before and after 1999, informal water exchanges at the local level have taken place frequently in many Spanish basins, primarily in East and Southeast Mediterranean areas (Segura, Jucar and Mediterranean Andalusian Basins; Fig. 7.2).

Transactions normally occur when water scarcity problems arise and water users need a rapid solution in order to obtain enough water to irrigate tree crops or to supply other critical water uses. Water exchanged in these informal markets usually comes from groundwater sources and mostly from private groundwater rights. The price in this type of exchanges is quite high compared to formal lease contracts and public purchases, and is often of a speculative nature. The prices also vary by location, water quality, alternative sources of supply and, to a larger extent, the scarcity level. Prices have been documented to reach 0.7 €/m^3, although in general there will always be a ceiling marked by the charges for desalinized water, plus the transportation costs (0.33 €/m^3 in coastal areas, and 0.39 €/m^3 in inland areas, with a total of 0.45–0.47 €/m^3 at the point of use), in those coastal areas where the resources are available. Quality graded water fetches different market prices with growers combining different sources to raise water quality to levels crops can tolerate. In addition, in some areas farmers or water companies desalinize deep saline groundwater, which is sometimes traded.[6] In some cases, water sold comes from illegal pumping.

It has been documented that even municipalities have participated in informal exchanges with farmers, mainly to meet the water demands derived from the tourist activity. That was the case of Benidorm (Alicante), with a seasonal population of 400,000 inhabitants and a regular one of 70,000. The resulting agreement was to swap fresh sources originally owned by horticulturalists for treated urban waste water (Martí 2005). In some cases, informal exchanges eventually become legalized or exchanged rights adjudicated by the Water Agency.

7.3.3 The Case of the Canary Islands

A very emblematic case of Spanish water markets is the one in the Canary Islands. This market has been active for a very long time, mainly for groundwater resources, and it is seen as an example of efficiency. Despite this, Canaries' water trading system has some problems and abuses: water is concentrated in a few hands (which determine the price and the conditions of the exchanges); there is a lack of transparency and information; water quality is not guaranteed by pipe owners and the owners are not responsible for water losses (Aguilera-Klink and Sánchez García 2005; Custodio and Cabrera 2012).

Some buyers prefer to purchase public water rather than private water, even when the price is higher than the price of private water in the market, mainly because it is more reliable, water quality is higher, and there are no charges for water lost in conveyance (Custodio 2011). Prices paid for irrigation water during high-demand periods can reach or exceed the price of seawater desalination; so only very competitive water users with high valued uses can afford to purchase it (Custodio and Cabrera 2012). However, the water market plays an important role for some agricultural areas and cities when there is no other available water source and it encourages economic and social development in the islands (Custodio 2011).

---

[6]A distinction has to be made between desalination of brackish waters and desalination of sea water. In some coastal areas of Southeast Spain, individual farmers (commonly larger ones) desalinize and use deep brackish water, about which hardly any reliable documentation can be found. Eventually, in drought periods, some of these volumes are sold in informal markets, mostly to smaller farmers that have shallower wells and no desalinization facilities. There are also water companies that sell desalinized/brackish water. We only know of one irrigation district desalinizing brackish water, as districts in Southeast areas more commonly rely on desalinized sea water, when available, of which some information exists about cost, contracts, and volumes used.

7.4 Reasons Behind Limited Success of Water Markets in Spain

Several reasons can explain the limited development of water markets in Spain. First, there are a number of restrictions and pre-requisites before a water exchange is approved that certainly add transaction costs and red tape (Garrido and Calatrava 2009). These are meant to avoid speculation and water rights hoarding, and to protect third-parties from negative effects; but result in low market activity.

There are a number of regulatory elements, identified by Ariño and Sastre (2009: 100–101), that can restrict the functioning of water markets including: (i) rights to consumptive uses cannot be sold to holders of non-consumptive use rights (hydropower) and vice versa; (ii) there are restrictions on potential water buyers, such as that rights can only be leased out to other rights holders of an equivalent or higher category in the order of preference established by river basin planning or in accordance with the National Water Act; (iii) there are limits to the spatial extent of trading: licenses for the use of public infrastructure connecting different river basin areas may only be authorized if they come under the National Hydrological Plan or other specific laws; (iv) there are limits on prices; regulations may determine maximum prices for water licenses. Competitive pricing can be superseded by administrative intervention. Unlike the Australian differentiation of entitlements and use rights, in Spain only a formal right, in the sense of entitlement, is defined. It was decided that the market should only be available for pre-existing and fully legally supported users.

Second, environmental limits are those enforced by public agencies responsible for the stewardship of the ecological quality of rivers and water bodies. In general, these limits, such as minimum environmental river flows, are based on modeling evidence, and are seldom contested. Occasionally, an *environmental tax* is imposed as a proportion of the volume/flow to which the traded right is entitled and which should be left in the natural source.

Third, most water in Spain is currently allocated through public water concessions, rather than private water rights, which still exist because their owners had rights before 1985. Water markets do not always work efficiently because water concessions were not designed for market transactions. Consider the situation of a drought. One would expect that shortages would trigger more market activity, but in fact water authorities effectively reduce the volumes accessible to the right-holders in areas facing scarcity, thereby reducing any incentive to purchase a water right. In a sense, the Agency still has a major role in allocating water under scarcity conditions. But decisions are agreed upon by all represented stakeholders, in meetings of formal committees with executive power. So the market is not deeply ingrained in Spanish water culture, and more collective responses to drought are common and widely accepted.

Again, this is not the case in Australia or in Chile. Moreover, there is also a problem of poorly defined water rights in some areas. It is not a coincidence that most of water trading in Spain has been inter-basin trading because scarcity situations have been different across basins and buyers and sellers have been able to trade different percentages of the volume or flow established in their formal right.

Fourth, with some exceptions, the potential for water trading between users in the same basin is limited, as differences in willingness to pay/willingness to accept are usually not significant. In addition, inter-basin water trading has only been allowed in drought periods as an emergency relief tool. The largest exchanges of water in the 2005–2008 drought period took place among users in different basins (Sect. 7.3.1.1).

Fifth, a significant proportion of agricultural users are grouped in Water Users Associations (WUAs) that in Spain usually take the form of communal entities. If their users agree, the WUA can become the right-holder of all the resources assigned to their members individually, but this implies the termination of the individual water rights. Under this case, WUAs rather than individual farmers are the ones participating in water trading, and they are less likely to participate as sellers in a water market. Furthermore, decisions to buy or sell are taken in Assembly or Commissions, rather than individually.

Finally, in spite of the functioning of formal water markets for more than a decade in Spain, there are still uncertainties. Criteria for approving or denying applications for water exchanges are not clear. Consequently, market participants rely more on previous experience than on a clear public definition of the circumstances under which trading is allowed (origin of water, area of destination, tradable volumes, fees to be paid, environmental restrictions, etc.). Similarly, the potential for inter-basin markets is hampered by the uncertainty about whether or not the Spanish Government will allow exchanges.

These and other barriers to trade result in other markets taking the role of water markets. The market for agricultural land (lease or purchase) and informal water markets substitute, to some extent, for formal water trading with a significantly higher cost. Consider the real case of a thermo-solar power plant, which needs water for cooling and replenishing vapor losses. If its owners do not hold water rights, the only way they can obtain water is by purchasing irrigated land and its attached water rights, and then request a change of use from the water authority. Furthermore, technologies and management practices, both on-farm, on site and at the district levels, have had a significant impact on reducing water application rates in Spain and deterring consumption. The energy cost component in many areas with abundant surface water supplies, on top of the financial and operating costs of recently modernized districts, have increased irrigation cost by 400 % (Hardy et al. 2012).

7.5 Possible Reforms

There are a number of shortcomings in water markets found in the Spanish system as well as in other countries: high transaction costs, slow administrative procedures, difficulties in finding buyers/sellers, high prices, rigid legislation, etc.

(Garrido et al. 2012a). However, markets in Australia, US and Chile are much more liquid and agile. As mentioned before, traded volumes in the water market have never represented more than 1 % of all annual consumptive uses. Furthermore water markets are mainly used during drought periods, except for a few water bank initiatives launched by basin authorities to buy-out groundwater rights. In the following points we offer some insights and ideas that would improve the functioning of water markets in Spain.

7.5.1 *Option Contracts*

Some of these barriers that affect the water trading activity could be avoided with option contracts. A formalized option contract gives the holder the right to acquire a prearranged water volume if needed, paying to the seller a premium at the beginning of the year. There are a lot of benefits derived from the establishment of this type of contracts. Among them, the reduction of transaction costs (Garrido and Gómez-Ramos 2009); less regulatory requirements than permanent transfers (Hansen et al. 2008); more certainty about the amount of water available in each irrigation season (Garrido and Gómez-Ramos 2009); provides reliability independently from the water rights owned; gives farmers opportunity to budget their costs and plant crops early in the season knowing that water will be available later at a given or even cheaper price (Cui and Schreider 2009); and secures urban drought water supplies at a lower cost than water rights purchases while maintaining agricultural production (Michelsen and Young 1993). The gains from trade are on average higher when options can be traded, by 46 % in competitive markets and by 63 % in dominant buyer markets (Hansen et al. 2008). A group of stakeholders and experts were consulted about introducing option contracts in the Spanish water markets. All of them agreed on the idea that option contracts could solve some of the inefficiencies of the current system. Option contracts may allow basins and users to manage drought and shortage risks much more effectively than spot markets. (See Chap. 4 for discussion of U.S. option markets for water).

7.5.2 *Water Saving Certificates*

In order to increase water use efficiency, an interesting alternative is the creation of water savings certificates. The most efficient water users who do not have easy access to other water sources would pay the less efficient ones to reduce their water losses. For that, they would get extra water corresponding to the water volume saved. These arrangements could increase water use efficiency in a given river basin, have beneficial impacts in the long-term and could help the recovery of overexploited aquifers (provided some of the conserved water is allowed to recharge the aquifer). For instance, the new Water Law in Peru foresees that users which individually or collectively obtain certificates of 'efficient use' are granted fee rebates and given preferential access to water in times of drought.

There are numerous ways with which saving certificates can be defined and measured. Satellite images and proper field records, coupled with the inspection of infrastructure and metering, can provide accurate evaluations of water consumption. Creating certificates is one indirect way to provide market incentives, without necessarily having market transactions, and would put the focus on the technical measurement of consumption by independent auditors. Moreover, it would also help in dissociating the notion of water right as a rigid formal right from the actual consumption, which is an hydraulic and environmental relevant variable.

7.5.3 Improvement of the Water Market Legislation

Based on the above, the regulatory framework of water markets in Spain needs profound reforms to make them more effective, secure and sustainable. Pending a serious legal assessment, we believe that the Water Law itself must be reformed. The following elements could help overcome its major weaknesses:

- Introduce a formal and effective separation of water rights and allocations, the latter being made also tradable (following the Australian system). This will require a redefinition of water rights in Spain (see Chaps. 9, 10 and 11 on Australia's water markets). This could be fostered indirectly with water savings certificates issued by independent technical auditors.
- Remove the hierarchy of use priorities, except for minimum volumes or allotments for urban suppliers and ecosystems. Once basic human and environmental needs are secured, the rest of economic or productive uses should have the same status. This will also require redefining water rights, and make the market more efficient and less distortive.
- Allow water exchanges only of the volume irretrievably lost from a given use, not of the total volume diverted. Irretrievable losses amount to water lost due to evaporation, crops' evapotranspiration or direct incorporation in manufactured products. Develop certification and statutory rules to ensure that this can be made effective.
- Adopt regulations for inter-basin and inter-regional trading, with the objective of reducing the political interference and arbitrariness. The idea is to define most possible contingencies and clarify when and how much water can be traded on pre-specified rules that all parties – users and administrations – commit to go by.
- Allow non-right holders to purchase water resources, removing an artificial impediment that prevents more efficient users from accessing water rights, which often is avoided by loopholes and costly paper work.

7.5.4 Water Management Improvements to Promote Efficient and Sustainable Water Markets

Some other improvements would not require a change in water legislation, but would certainly improve the functioning of water markets:

- Define and approve the major allocations for all water basins, finalizing the planning mandates of the Water Framework Directive (WFD),[7] including the environmental flow regimes and other restrictions. This would clarify, a lot, what amounts are subject to trade by all water rights holders at any given moment and location.
- Implement cost-recovery levels that are considered to be in full compliance with article 9 of the WFD. This would remove historical distortions that are no longer appropriate under current legislation in force concerning water prices.
- Ban any type of market operation request for users whose status falls short of being in full compliance with the Law, reducing concerns for spurious use of exchanging options.
- Implement a pre-registration and screening procedure for users interested in becoming market participants, with the intention to monitor and review market operations much more quickly. The idea is to implement a system in which pre-registered users can exchange water, and make the market operations more robust, agile and environmentally safe.

7.6 Conclusion

Water trading is a tool to cope with water scarcity and to improve water use efficiency. As water availability in the Mediterranean region is expected to diminish because of climate change (among other reasons), markets will have greater importance in the coming years. Since the approval of the 1999 Reform, water markets have helped water users mainly during drought episodes. It is important to start thinking about water markets as a tool to be used in every circumstance and not only during drought periods.

As important as trying to improve and encourage water markets is, there is also a need to achieve a fuller knowledge and understanding of how water is actually used in each Spanish basin and to control the effective use of this water while reviewing water concessions and increasing control of illegal extractions. Better control of the existing water resources and their final destination will lead to a much more efficient use of water.

---

[7]Directive 2000/60/EC of the European Parliament and of the Council of 23 October 2000 establishing a framework for Community action in the field of water policy.

After reviewing the latest reforms in the water legislation in Spain, it is clear that water regulation should move toward a more flexible, agile and dynamic management system. But equally important is to think about the good ecological status of our water bodies and establish sustainable exploitation rates. Although it is difficult to try to serve all water demands and at the same time maintain a good ecological status for water resources, that is the path that should be followed.

The existence of informal water markets of a very different nature along the Mediterranean basins proves that there is a demand for the reallocation of water resources among users and for improving supply reliability. Not only that, but there is also a demand to manage differently quality graded waters and allow each user to meet their requirements at the least possible cost. This demand is not met within the current regulatory framework, which is too limited and lacks provisions to cope with extremely diverse, quality graded, poorly monitored groundwater users. There is clearly a need for a new improved regulatory framework that provides sufficient flexibility for users in the most water-stressed basins, while at the same time allowing for protection of the public interests. Our proposed reforms could help to make the market more flexible and to overcome most of the difficulties found in the current system.

References

Adler JH (2009) Warming up to water markets. Regulation, winter 2008–2009: 14–17, CATO Institute

Aguilera-Klink F, Sánchez-García J (2005) Water markets in Tenerife: the conflict between instrumental and ceremonial functions of the institutions. Int J Water 3(2):166–185

Albiac J, Hanemann M, Calatrava J, Uche J, Tapia J (2006) The rise and fall of the Ebro water transfer. Nat Resour J 46(3):727–757

Ariño Ortiz G, Sastre Beceiro M (2009) Water sector regulation and liberalization. In: Garrido A, Llamas MR (eds) Water policy in Spain. Taylor & Francis, Leiden, pp 95–106

Arriaza M, Gómez-Limón JA, Upton M (2002) Local water markets for irrigation in southern Spain: a multicriteria approach. Aust J Agric Resour Econ 46(1):21–43

Blanco M, Viladrich M (2013) Estimación de las ganancias mínimas que debe garantizar la creación de un centro de intercambio de derechos de agua en la cuenca del Segre. Presented at the IX Spanish National Congres of Agricultural Economics, Castelldefels

Blanco M, Navarro D, Viladrich M (2010) ¿Pueden los instrumentos económicos contribuir a la sostenibilidad del regadío en las regiones mediterráneas? Anales de Economía Aplicada, ISBN 978-84-92954-15-5

Calatrava J, Garrido A (2005) Spot water markets and risk in water supply. Agric Econ 33(2):131–143

Calatrava J, Gómez-Ramos A (2009) El papel de los mercados de agua como instrumento de asignación de recursos hídricos en el regadío español. In: Gómez-Limón JA, Calatrava J, Garrido A, Sáez FJ, Xabadia A (eds) La economía del agua de riego en España. Fundación Cajamar, Almería, pp 295–319

Calatrava J, Martínez-Granados D (2012) El Valor de Uso Del Agua En El Regadío de La Cuenca Del Segura y En Las Zonas Regables Del Trasvase Tajo-Segura. Economía Agraria y Recursos Naturales 12(1):5–32. doi:10.7201/earn.2012.01.01.1

CEDEX (2011) Evaluación del impacto del cambio climático en los recursos hídricos en régimen natural. CEDEX, Madrid

Corominas J (2011) Aplicación a los regadíos del Guadalquivir del intercambio dederechos del agua en situaciones de dotación restringida. Botin Foundation Seminar on Water Markets in Spain. Madrid, June 2011

Cui J, Schreider S (2009) Modelling of pricing and market impacts for water options. J Hydrol 371:31–41

Custodio E (2011) Comentarios sobre el comercio y mercados del agua subterránea en Canarias. In: Cabrera MC, Jiménez J, Custodio E (eds) El conocimiento de los recursos hídricos en Canarias cuatro décadas después del proyecto SPA-15. AIH-GE, 2011

Custodio E, Cabrera M del C (2012) The Canary Islands. In: Lucia de Stefano L, Llamas MR (eds) Water, agriculture and the environment in Spain: can we square the circle? CRC Press, Leiden, The Netherlands, pp 281–289

Elorrieta JI, Castellano E, Martinez de Anguita P, Pellitero M, Rey C (2003) Establecimiento de un precio óptimo ambiental y social para el agua de riego mediante la internalización de los costes ambientales y el uso de matrices de contabilidad. Aplicación a la Comunidad Foral de Navarra, Economía Agraria y de los Recursos Naturales 3:139–166

Garrido A, Calatrava J (2009) Trends in water pricing and markets. In: Garrido A, Llamas MR (eds) Water policy in Spain. CRC Press/Taylor & Francis, Leiden, pp 129–142

Garrido A, Gómez-Ramos A (2009) Propuesta para la implementación de un centro de intercambio basado en contratos de opción. In: Gómez-Limón JA, Calatrava J, Garrido A, Sáez FJ, Xabadia A (eds) La economía del agua de riego en España. Fundación Cajamar, Almería, pp 321–341

Garrido A, Rey D, Calatrava J (2012a) Water trading in Spain. In: Lucia de Stefano L, Llamas MR (eds) Water, agriculture and the environment in Spain: can we square the circle? CRC Press/Balkema/Taylor & Francis, Leiden, pp 205–216

Garrido A, Willaarts B, López-Gunn E, Rey D (2012b) Considerations on climate variability and change in Spain. In: Lucia de Stefano L, Llamas MR (eds) Water, agriculture and the environment in Spain: can we square the circle? CRC Press/Balkema/Taylor & Francis, Leiden, pp 191–204

Garrido A, Maestu J, Gómez-Ramos A, Estrela T, Yagüe J, Segura R, Calatrava J, Arrojo P, Cubillo F (2013) Voluntary water trading in Spain: a mixed approach of public and private initiatives. In: Maestu J (ed) Water trading and global water scarcity: international experiences. RFF Press/Routledge, Abingdon, pp 180–193

Gomez-Limon JA, Martinez Y (2006) Multi-criteria modelling of irrigation water market at basin level: a Spanish case study. Eur J Oper Res 173(1):313–336

Hansen K, Kaplan J, Kroll S (2008) Valuing options in water markets: a laboratory investigation. DARE CSU working paper 001–08

Hardy L, Garrido A (2010) Análisis y evaluación de las relaciones entre el agua y la energía en España. Papeles de Agua Virtual 6, Botin Foundation, Madrid 2010

Hardy L, Garrido A, Juana L (2012) Evaluation of Spain's water–energy nexus. Int J Water Res Dev 28(1):151–170

Hernández-Mora N, De Stefano L (2013) Los mercados informales de aguas en España: una primera aproximación. In: Embid Irujo A (ed) Usos del Agua Concesiones, Autorizaciones y Mercados del Agua. Thomson Reuters Aranzadi, Madrid, pp 375–401

Iglesias A, Estrela T, Gallart F (2005) Impactos sobre los recursos hídricos. Impactos del cambio climático en España. Ministerio de Medio Ambiente, Madrid

Llamas MR, Fornés JM, Hernández-Mora N, Martínez L (2001) Aguassubterráneas: retos y oportunidades. Fundación Marcelino Botín and EditorialMundi-Prensa, Madrid

Llamas MR, Varela-Ortega C, de la Hera A, Aldaya MM, Villarroya F, Martínez-Santos P, Blanco-Gutiérrez IMM, Carmona-García G, Esteve-Bengoechea P, De Stefano L, Hernández-Mora N, Zorrilla P (2010) The Guadiana basin. In: Mysiak J, Henrikson HJ, Sullivan C, Bromley J, Pahl-Wostl C (eds) The adaptive water resource management handbook. Earthscan, London

López-Gunn E, Dumont A, Villarroya F (2012) Tablas de Daimiel National Park and groundwater conflicts. In: Lucia de Stefano L, Llamas MR (eds) Water, agriculture and the environment in Spain: can we square the circle? CRC Press/Balkema/Taylor & Francis, Leiden, pp 45–66

Martí TT (2005) Use competition and water exchange in Marina Baja district, Alicante, Spain. Water Sci Technol: Water Supply 5:265–272

Martin-Ortega J, Berbel J, Brouwer R (2009) Valoración económica de los beneficiosambientales no de mercado derivados de la mejora de la calidad del agua: una estimación en aplicación de la Directiva Marco del Agua al Guadalquivir. Revista de Economía Agraria y de los Recursos Naturales 9(1):65–89

Martínez-Santos P, De Stefano L, Llamas MR, Martínez-Alfaro PE (2008) Wetland Restoration in the Mancha occidental aquifer, Spain: a critical perspective on water, agricultural and environmental policies. Restor Ecol 16(3):511–521

Mesa-Jurado MA, Martin-Ortega J, Ruto E, Berbel J (2012) The economic value of guaranteed water supply for irrigation under scarcity conditions. Agric Water Manag 113:10–18

Michelsen AM, Young RA (1993) Optioning agricultural water rights for urban water suppliers during drought. Am J Agric Econ 75:1010–1020

Molle F, Berkoff J (2006) Cities versus agriculture: revisiting intersectoral water transfers, potential gains and conflicts. Comprehensive assessment of water management in agriculture. Research report 10, IWMI

Moreno JM (2005) Evaluación preliminar de los impactos en España por efecto del cambio climático. Centro de Publicaciones. Ministerio de Medio Ambiente, Madrid, 822 pp

Pujol J, Raggi M, Viaggi D (2006) The potential impacts of markets for irrigation water in Italy and Spain: a comparison of two study areas. Aust J Agric Resour Econ 50:361–380

Ramajo-Hernandez J, del Saz-Salazar S (2012) Estimating the non-market benefits of water quality improvement for a case study in Spain: a contingent valuation approach. Environ Sci Policy 22:47–59

Rigby D, Alcon F, Burton M (2010) Supply uncertainty and the economic value of irrigation water. Eur Rev Agric Econ 37(1):97–117

San Martín E (2011) Un análisis económico de los trasvases de agua intercuencas: el trasvase Tajo-Segura. Thesis, UNED, Madrid

Stefano L, Llamas MR (2012) Water, agriculture and the environment in Spain: can we square the circle? Botin Foundation, CRC Press, Boca Raton

WWF (2005) Los mercados de aguas y la conservación del medio ambiente: Oportunidades y retos para su implantación en España. WWF

WWF (2012) El fiasco del agua en el Alto Guadiana. Report 2012, WWF Spain

Chapter 8
Century Old Water Markets in Oman

Slim Zekri, Dennis Powers, and Abdullah Al-Ghafri

Abstract Water rights in Oman were instituted centuries ago and are still active. These water rights are treated similarly to real property rights since they are sold, rented, and inherited independently of the land on which water is used. In most of the observed irrigation schemes, water auctions are present. The common water rights are mostly leased. The paper discusses the efficiency of these auction markets as well as the revenue generation and equity. The paper also provides an analysis of quantitative data on Falaj auction markets through a case study of Falaj Belfae. During periods of low supply water prices increase drastically. It has been observed that water prices might increase 200-folds during periods of scarcity. The benefits of water markets revert to the farmers' community through a well maintained irrigation system. However during periods of water abundance observed water prices are below their long run marginal value. This is explained mainly by the absence of engineering mechanisms to prevent the groundwater from flowing to the surface and prevents its conservation for drought periods. In this sense the water market has not been sufficient to trigger the adoption of water saving technology at the supply level.

Keywords Water auction • Water demand • Falaj • Community market • Irrigation maintenance

S. Zekri (✉)
Department of Natural Resource Economics, College of Agricultural & Marine Sciences, Sultan Qaboos University, PO Box 34, Postal Code 123 Muscat, Oman
e-mail: slim@squ.edu.om

D. Powers • A. Al-Ghafri
Aflaj Research Unit, The University of Nizwa, Birkat Al Mouz, PO Box 33, PC 616 Nizwa, Sultanate of Oman
e-mail: dennis@unizwa.edu.om; a.alghafri@unizwa.edu.om

K.W. Easter and Q. Huang (eds.), *Water Markets for the 21st Century: What Have We Learned?*, Global Issues in Water Policy 11, DOI 10.1007/978-94-017-9081-9_8,

8.1 Introduction

Water markets in Oman are quite singular when compared to water markets in the US and Australia. A Falaj is an irrigation system in which the water is conveyed to the agricultural area powered only by gravity. For this reason, the water cannot be stored and thus the supply is taken as given by nature. Inside the Falaj community, the water is divided into time shares called *athars*. Typically athars entitle the owner to 30 min of water. The majority of athars are privately owned by members of the Falaj community, however, approximately 10–15 % of the athars are commonly owned by the Falaj management. The purpose of these commonly owned athars is to raise revenue for the expenses incurred in the Falaj operation and maintenance. The Falaj has two methods to use the common athars to raise revenue. It may combine the water with land to produce a crop, and sell the agricultural production for revenue. Alternatively, it may rent the water to farmers in the Falaj community using auctions. When renting the water, annual auctions, semi-annual auctions, and weekly auctions have all been observed. However, the most frequently observed are the weekly auctions. It is the weekly auction of these commonly owned athars that is the subject of this paper. In particular, the efficiency, revenue properties, and equity of these auction markets will be discussed. The paper is organized as follow, after the introduction, a description of the falaj and Omani traditional water rights and water markets is presented borrowing from Zekri and Al-Marshudi (2008). We then discuss the efficiency and equity of the water markets. Finally, quantitative evidence is provided on the elasticity of water demand implied by the auction markets.

8.2 Physical Structure and Importance of Falaj Systems

A Falaj is a traditional water network comprising man-made underground water galleries transporting groundwater to a village and an irrigation area. In most cases a single falaj provides water for one village. In Oman the three main types of falaj are (1) Ghaili falaj– water is drawn from perennial flows in the surface gravels of the river; (2) Aini falaj – water is drawn from one or several natural springs; (3) Daudi falaj – water is extracted from an underground aquifer through a tunnel dug in the upper layers of the aquifer. In all cases the water is conveyed by gravity and thus the supply of water is determined by nature. The Daudi aflaj (plural of falaj) have the most stable water flow rate throughout the year compared to the other two types. The renewable groundwater volume is estimated at 900 MCM/year, 93 % of which is used in agriculture (Zabet 2005). Thirty-eight percent of groundwater is used via falaj systems, and the remaining 62 % is exploited through individual wells (MRMEW 2002).

Table 8.1 Water supply and irrigated areas per falaj type in Oman

	Daudi	Aini	Ghaili	Total
Number of falaj per type	967	1,152	1,993	4,112
Percent of total number	24	28	48	100
Irrigated area	13,946	6,973	5,579	26,498
% of irrigated area by falaj systems at national level	20	10	8	38
Annual water supply (mm^3)	243	119	97	459
% of annual water supply	53	26	21	100

After Zekri and Al-Marshudi (2008)

Table 8.1 summarizes annual water supply and cropped irrigated area for the three types of falaj at the national level. Traditional falaj irrigation accounts for 38 % of the total cropped area. Forty eight percent of the falaj systems are ghaili, 24 % are daudi and aini represent 28 %. Ghaili falaj systems irrigate only 8 % of the total cropped area due to their seasonality. Daudi and aini irrigate 20 % and 10 % of the area, respectively. Falaj systems play a major role in the irrigation sector, as well as for recreation and tourism (Zekri et al 2012a, b, 2011).

8.3 Water Rights in Falaj Systems

8.3.1 Types of Water Rights

Regardless of the type of right, the initial allocation of water rights is in proportion to the participation of each farmer in the construction of the Falaj, since land is not a scarce resource. Water rights are classified into three categories. (1) Private water rights which are explicit rights with formal licenses or implicit rights without written licenses. The implicit water rights can be inherited but in no case traded or rented. The explicit rights can be freely traded. (2) Common water rights are rights owned by the falaj's community. They generate income for falaj maintenance and operations expenditures. The common water rights are rented weekly, semi-annually, or on an annual basis through water lease auctions. (3) Quasi public water rights are owned by charity institutions. In the past returns from these rights were used to finance mosques, schools and to assist people in need. These water rights are managed in the same way as common water rights. The quasi-public water rights can only be leased. Table 8.2 shows the distribution of water rights ownership for a sample of 8 falaj systems. The most significant category of water rights is privately owned rights, ranging from 70 to 98.7 %. Common property rights represents approximately 4–11 % and finally, quasi-public rights ranges from 0 to 18 %. The daudi falaj system has the highest share of common property right due to the fact that daudi operations and management costs are much higher than they are for the ghaili or aini falaj systems.

Table 8.2 Water rights distribution for a sample of eight falaj systems

Falaj type	Falaj name	Private rights (%)	Common rights (%)	Quasi public rights (%)	Total number of owners
Daudi	Al Malki	78	10.5	11.5	116
	Al Hali	77.8	11.1	11.1	98
	Al Maiser	70	11.1	18.9	167
Mean daudi		75.2	11	13.8	
Ghaili	Samdi	98.7	1.3	0	1,500
	Farsaki	95.4	3.3	1.3	450
	Dykali	90	10	0	150
Mean ghaili		94.7	4.9	0.4	
Aini	Al Hajeer	95.5	0	4.5	336
	Al Kasfa	97.2	0	2.8	N.A
Mean aini		96.3	0	3.7	
Overall mean	87.7	6.0	6.3		

After Zekri and Al-Marshudi (2008)

8.3.2 Characteristics of Water Rights

8.3.2.1 Separate from Land, Long Term and Secure

Ownership of water is often independent of land ownership. However, to reallocate water from those with relatively greater water supplies to those with relatively little water, markets exist. Water rights are perpetual and are transferred from one generation to another through inheritance. Falaj water rights are recognized at the national level and have the same legal aspects as any other private asset.

8.3.2.2 Proportional

The water right is expressed as a timeshare of the resource since the falaj flow rate fluctuates among seasons and years. The most common time share is the athar, which corresponds to approximately half an hour share per water cycle.

8.3.2.3 Seniority or Priority

A Falaj's highest priority is given to domestic water uses. Free access to water for household uses is granted at the main channel. Common bathing facilities are often located after the main source. Diversion of falaj water for private purposes is not permitted in the residential area. However, diversions for public institutions such as for a fort, a mosque or a school are common. Irrigation water is distributed according to the water right duration. Since wells and falaj systems exploit the same groundwater, seniority has been established to protect falaj owners. This seniority is implemented through an exclusion zone around every falaj's mother well, main canal and secondary canals.

8.3.2.4 Transferability

The most frequent market for water rights is among farmers within one single falaj system. Water transfer to other irrigation communities downstream is practiced when existing infrastructure allows it. No transfers of water for domestic or industrial uses are reported or observed. The absence of water transfer to urban uses may be explained by the fact that cities are relatively new, dating back to the 1980s and were planned based on government intervention with heavily subsidized desalinated sea water provided.

8.3.2.5 Water Rights Management

The falaj is a non-government entity administered by a manager elected from the water rights owners. He manages the falaj with the support of a cashier, a water distribution agent and a technician for maintenance (COA 1995). The manager organizes the bids, makes decisions regarding operation and maintenance, reports to water rights' holders on expenses & returns and resolves water conflicts between farmers. The manager is paid for his service 5–7 % of water lease returns.

8.4 Water Market Characteristics in Falaj Systems

Markets are divided into three main categories: (1) markets for the sale of water rights that usually take place in cases of inheritance, and thus are not frequent, (2) lease markets for common water rights which are the most frequent and (3) lease markets for private water rights which occur infrequently. These later two markets are based on the auction of the Athar. The timeshare arrangement makes it difficult to compare prices among falaj and even within a single falaj, as the water flow varies from one season to another. A number of studies have mentioned large variations in both sale and lease prices. Wilkinson (1977) mentioned that the price of water depends on falaj flow, weekly versus yearly leases, seasonal conditions and whether it is a day or night timeshare.

8.4.1 Efficiency of Falaj Auction Markets

A typical auction will rent 24–48 athars. In all auctions observed, the Falaj management institution uses a multi-unit ascending bid auction. The auction of these units is done sequentially. That is, the first athar is auctioned, then the second, etc. until the last athar is auctioned. The auctioneer calls out a price for the athar being auctioned. Those willing to buy at the price indicate to the auctioneer their willingness to buy. If more than one is willing to buy the auctioneer raises the price.

This process continues until only one buyer is left. This is then repeated for the other athars until all athars have been auctioned. Powers et al (2013) showed that in such an auction that if there are m athars being auctioned, the equilibrium auction price is equal the marginal value of the (m + 1)th athar.[1] Hence, the equilibrium price is equivalent to the price that is attained under competitive conditions. That is, the farmers that value the m athars the most and are willing to pay the most for them will receive those athars. Thus, the Falaj community has constructed a means of raising revenue that appears to preserve the efficiency properties of markets.

All auctions are not necessarily efficient. If supply can be controlled then auction prices could be manipulated, and inefficiency could result. Michelsen et al. (2000) argued that speculated changes in the supply and demand, future expectations of population growth and economic growth can bring about change in the farmers' willingness to pay and can also cause shifts in supply and demand. This is similar to how speculation and future expectations work in the housing and asset market. Any anticipation regarding the prices in the future would actually end up pushing prices up in the present. Zetland (2013) stated that the participation effect is observed when sellers are not present, leading to poor results in the water auction. However, as explained above, nature determines supply and neither the seller nor a buyer can affect supply. Hence there is no monopoly power or speculation possible in the Falaj auction markets.

In addition, bidders have clear expectations of the volume to be traded weekly as they themselves own athars in the Falaj and have a very good idea of how much water is flowing per hour. This possession of information is crucial to lowering the cost to bidders. The lower the cost of bidding the higher is the incentive for the farmers to engage in water use efficiency (Heaney and Beare 2003). Furthermore, when a farmer buys water, he is actually buying a volume to be used exclusively during the same day as the auction, as water cannot be stored. Thus there is no possibility for strategic bidding through Falaj auctions.

Even though auction markets are efficient in allocating the commonly owned athars, there is little evidence to suggest there is any short-term trading of privately owned athars. Indeed, it may be the case that the Falaj institution has suppressed private trading so as to support the weekly auctions of commonly owned water. If there are differences in the marginal value of water for farmers in the community, such a suppression of private trading would clearly lead to inefficient outcomes. Thus, while the auction markets themselves are efficient, any suppression of private trading to support the auction markets would lead to inefficiency. Again, no evidence on private trading appears to exist, hopefully future research can fill this void.

---

[1]Any differences in athar prices result from the fact that since the falaj is always flowing, some athars implies water is received by the farmer at inconvenient times, while others are received at convenient times. As a result, those with convenient times tend to be rented at a premium above the referred to equilibrium price.

8.4.2 *Revenue of Falaj Auction Markets*

While the Falaj auction markets are efficient in allocating water, the purpose of these markets is to raise revenue for Falaj operation and maintenance. Hence it is important to evaluate these auction markets on these grounds. Powers et al. (2013) has argued that because the auctions are repeated weekly, the demands of participants become known, and thus the winners can be identified prior to the auction. The argument is that since auctions are repeated weekly, the bidding pattern of agents (including the quantity of athars rented and when they rent them) will become known to other agents. That is, each agent is following a pattern, and this pattern can easily be seen by other agents. Hence the winners are identified by the other agents in the community. As a result, if the winners are known before the auction, and one agent is not a winner, then there is no reason for him to participate. Participants are thus limited to winners not by a community restriction, but by the choice of the individuals. That is, if there is a small cost to attending an auction, then if one is not going to win, he will choose to not participate. For this reason, participation in these auctions might be restricted to the winners only. As a result, the prices would likely drop. Instead of the equilibrium price equaling the marginal value of the $(m + 1)$ athar to the *community*, the price equals the $(m + 1)$ athar to the *participants*. In other words, prices drop below what would be the equilibrium amount if there was full participation. Hence the prices collapse, but not to zero (unless the value of the $m + 1$ athar is zero, as it may be when the falaj flow rate is high). It should be noted that Powers et al. demonstrates that the auctions are still efficient conditional on current water allocations since those participating are those that value the water the most. Nevertheless, the drop in prices implies auction revenue is reduced due to the chosen method of raising revenue. Of course, this is not necessarily a problem. It may still be that even with reduced prices the auction revenue is sufficient to meet the expenses of the Falaj. Indeed, the fact that the Aflaj have operated for centuries is sufficient proof that Aflaj communities have been able to raise enough revenue from auctions to ensure operation and maintenance.

Revenue may also be affected by collusion among bidders. If participation is low, as argued above, it would be possible for collusion among participants to occur, in an attempt to drive down prices. However, the Falaj manager is paid as percentage of the revenue collected at the end of each year, which is another incentive to make sure that the bidding process is run efficiently.

8.4.3 *Equity of Falaj Auction Markets*

An equitable scheme of funding Falaj maintenance can be defined as one in which the contribution of an individual is proportional to the number of athars owned. However, an individual's contribution to Falaj maintenance is determined by their demand for rented water. That is, those with the highest demand for rented water will win the auction, and thus contribute to Falaj maintenance, while those with low demand will not win the auction, and will thus make no contribution to Falaj maintenance. However, those that own the most athars (relative to their land holdings) will have the lowest demand for rented water, whereas those who own the fewest athars (relative to their land holdings) will have the highest demand for rented water. Thus, if there is a significant difference in athar holdings among Falaj community residents, then it may be the case that those who own the most athars contribute the least toward Falaj maintenance, and those that own the fewest athars contribute the most toward Falaj maintenance. If true, this implies this method of raising revenue shifts the burden of Falaj maintenance from those who own the most athars to those who own the least, and thus is probably not equitable.

Al-Abri (1992) argues that the initial allocation of athars was in proportion to the participation of each farmer in the construction of the Falaj. Usually the number of athars owned depends on the farm size as a risk minimizing strategy. Any increase in the farm land should be accompanied by either buying athars or depending on the weekly lease market. Thus, those who invested more in the beginning are in some ways sheltered from paying the maintenance fees compared to those who invested less or those who expanded their farm land later. This per se does not constitute ground for unfairness despite the observation that currently those who own the fewest athars per hectare are contributing the most to the Falaj maintenance. On the other hand, those who own less athars are risk takers since the Falaj flow varies considerably. They will pay medium to low water prices during average and high rainfall seasons and very high prices during drought periods. But drought does affect equally those who own few and those who own a lot of athars/ha. This is due to the fact that during drought all farmers will compete for water to maintain their plantations and all athars will be lower flows. It is during the drought periods that the highest income is collected for the Falaj.

8.5 Quantitative Analysis of Falaj Auction Markets

8.5.1 Analysis of Volume Traded

The Falaj water leases are obtained from the Falaj book. The manager of Falaj keeps record of all transactions including the name of the lessee, the time share and price. The Falaj water flow is used to change the time share into a volume of water. Falaj flows are measured on a monthly basis for a sample of Aflaj by the Ministry of regional Municipalities and Water Resources. Figure 8.1 shows the quantity-weighted monthly average prices and volumes of water leased during the period 1997–2012 for Falaj Balfae located 40 km south Muscat. Since a time series data is used, the prices are adjusted for inflation using the Consumer Price Index and are reported in 2011 US$. The figure shows how prices go up when supply is

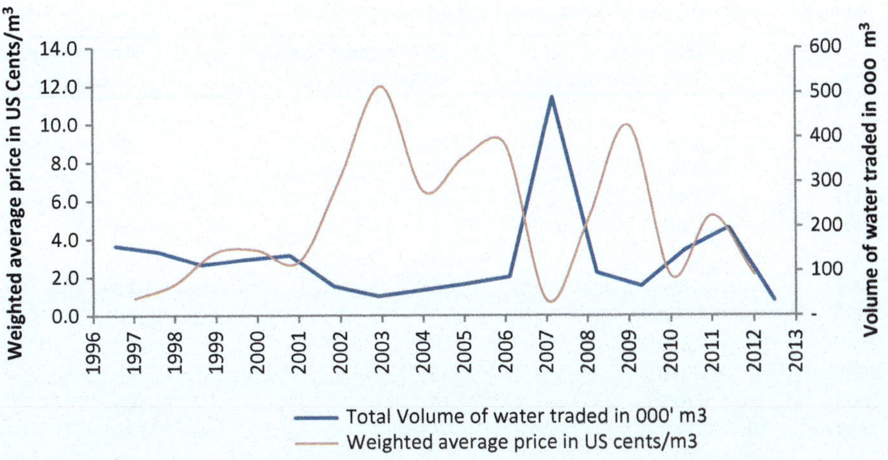

Fig. 8.1 Monthly weighted average price in US cents/m^3 and volume of water traded in 000' m^3

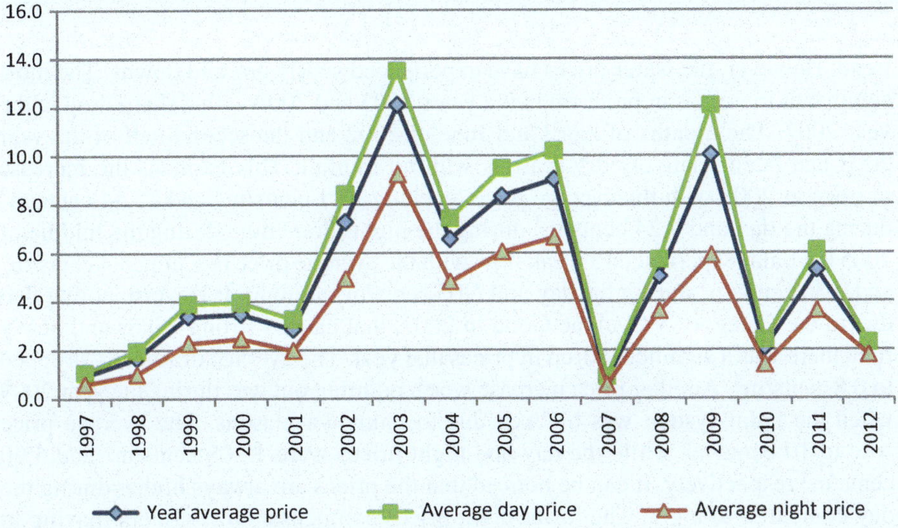

Fig. 8.2 Yearly weighted average price in US cents/m^3

low such in the period 2002–2006 and vice versa for year 2007. In most of the years the price and volume traded go in opposite direction due to the absence of storage and inability to control the flow. In other words, the flow of water is continuous year round, unless some natural conditions restrict the flow. There are no mechanisms to either stop or control the flow and this applies to all Aflaj in Oman.

Figure 8.2 shows the average price of traded water as well as the average day and night prices from 1997 to 2011. The reason both day and night prices exist is due to the natural Aflaj flows. The price trend over the years is a gradual increasing

Table 8.3 Yearly volume of water traded and revenue collected

Year	Total revenue from water sale	Total volume of water traded in 000' m^3	% of the average revenue (%)
1998	2,410	145	53
1999	3,863	115	85
2000	4,393	127	96
2001	3,867	136	85
2002	4,848	67	106
2003	5,308	44	116
2004	3,754	58	82
2005	5,906	71	129
2006	7,829	87	172
2007	3,910	491	86
2008	4,827	97	106
2009	6,663	67	146
2010	3,044	148	67
2011	3,225	198	71
Average	4,560	132	100

trend. However, the water prices have experienced yearly ups and downs. The most significant increases in price are in the years 2003 and 2009 with a sharp drop in the year 2007. The months of April and June of 2002 and the second half of the year 2003 had been relatively dry seasons with no rainfall. This explains the increase in price in 2003 with the average price reaching 12.1 cents/m^3 and 13.54 cents/m^3 during the day and 9.24 cents/m^3 during the night. Sometime during the middle of 2005 the Falaj experienced a weak flow with the average price reaching 8.4 cents/m^3 and 9.50 cents/m^3 during the day and 6.05 cents/m^3 at night. It is worth noting that due to the Category 5 Cyclone Gonu in 2007, that caused strong winds and heavy rain, there was a significant drop in prices that year. The average drastically dropped to 0.8 cents/m^3. Another price increase worth pointing out was during the year 2009 when the Falaj system was blocked due to unknown reasons. The average price rose to 10 cents/m^3 while the day and night prices were 12.08 cents/m^3 and 5.90 cents/m^3 respectively. It can be noticed that the prices are always higher during the day compared to night. The lower night prices compensate for farmers' having to work at night (Zekri et al. 2006a). Finally, during periods of abundance observed water prices were well below their long run marginal value, and even close to zero. This indicates that the Falaj water management is not appropriate during periods of abundance when water is inefficiently used, while it could be stored and supplied during droughts.

Table 8.3 shows that the revenue collected is negatively correlated to the volume traded. The highest revenue collected was during 2006, 72 % higher than the average. During low flow periods, most farmers, if not all, participate in the auction bringing up prices in order to protect their plantations from drying up. It is during such periods that most of the Falaj revenue is collected. Thus the concern about the burden of Falaj maintenance costs being mainly paid by those who own the least athars/ha may be somewhat reduced. However, further statistical analysis is required on who bought how much water, and at what price, to fully understand the equity issue.

8.5.2 *Falaj Water Demand*

In this section the demand function of Falaj water is estimated based on the lease of commonly owned water rights of Falaj Balfae. Each falaj has its own bundling of athars which reflects in number of packages being offered. No artificial recharge is undertaken on Aflaj. Most falaj have a bundling of the athars which might differ slightly. For Falaj Balfae the lease consists of 48 athars bundled in 7 packages, one package of 4 athars, 3 packages of 6 athars, 2 packages of 8 athars and 1 package of 10 athars. Thus, every week only seven bidders at maximum can lease water. However in practice sometimes more than one buyer cooperates and bid together for one package. Most water markets take place once a week, after Friday prayer, during week end to attract the maximum number of participants. Given the absence of channels to transport water from a village to another, only farmers from the same community participate in the bid and compete for the supplied water. The data of the weekly market allows one to conduct a detailed analysis of prices and quantities leased and the associated responsiveness of demand to price changes (Zekri et al. 2006b; Brooks and Harris 2008).

Several studies have been conducted since the 1970's to estimate the irrigation water demand. In the absence of water markets the demand for irrigation water is usually estimated using mathematical programming (Griffin 2006). Schoengold et al. (2006) used a panel estimation of an agricultural water demand function. Previous studies (Bontemps and Couture 2002; Hooker and Alexander 1998; Ogg and Gollehen 1989) showed that irrigation water demand is price inelastic with the value generally of less than −0.5.

Zekri et al (2006b) used observed water market data to estimate the water demand function. They found that the water price elasticity (−0.5) was in the high end of estimates from previous studies. Fruits and vegetables represent the major crops in the study area. Bjornlund et al. (2008) estimated both the supply and demand function of irrigation water based on Australian water market. The elasticity of demand was higher than that of supply, and varied between seasons. More recently Davidson and Hellegers (2011) used water market prices to derive the demand for irrigation water in India. On average the elasticity was found to be (−0.64). They also found that the demand was highly elastic (−2.1) for the high valued water used to produce vegetables and fruits.

The water demand for Balfae Falaj system was estimated using a Log-Log function of price and the supplemental volume traded. Actually most farmers own water shares and are buying extra volumes of water. Three correlations were estimated based on farmer level data weekly averages (see Table 8.4). The first one uses average bidding prices, the second uses average day bidding prices and the

Table 8.4 Irrigation water demand estimation for Balfae Falaj

	Average real price	Average day price	Average night price
α_0	5.519903248	6.263683624	6.255773601
α_1	−0.51496832	−0.520629327	−0.5059147
Significance F	1.33794E−50	8.36501E−51	3.86041E−47
t Stat	−21.5952229	−21.67653225	−20.23704155
Adjusted R Square	0.731281498	0.732759073	0.704937305

third one is based on the average night bidding prices. The econometric estimations were made using the following model

$$\text{Ln (V)} = \propto_{\circ} + \propto_1 Ln\ \text{(P)} + \varepsilon$$

where,

V = Volume of water
P = Average real price
ε = Error term
$\propto$ = Intercept ($\propto_{\circ}$) and coefficient ($\propto_1$)

The price elasticity of irrigation water is estimated at –0.51 using average prices. The average day and night prices show the same behavior with respective price elasticities of –0.52 and –0.50. This implies that water demand remains price inelastic throughout the day. The estimated elasticities are in line with the results of most of the studies that have been mentioned in this paper.

8.6 Conclusion

This paper set out to describe the auction markets in traditional water management systems in Oman known as a Falaj and characterize the performance of these auctions in terms of efficiency, revenue, and equity. With respect to efficiency, it was argued that while the auction markets are efficient in the allocation of the commonly owned athars, this may come at the expense of shutting down trading of private shares. Since no data on private trading exists, this is an area for future research. As a means to raise revenue for Falaj maintenance, it was shown that participation is likely to be low in these auctions resulting in lower revenue during years of more than average Falaj supply, while the participation is high during below average Falaj supply. Future research should compare auction revenue to other schemes for raising revenue. Concerning equity, it was argued that there is the potential for inequity in the funding of Falaj maintenance.

Finally, the paper analyzed the quantitative data on Falaj Belfae. The results showed that prices are correlated with supply. During high supply periods prices were very low and vice versa. During periods of abundance observed water prices were low implying an inefficient use of water. This is a possible indicator of the lack of innovation in the management of Falaj to store water during periods of high supply for use during droughts. The econometric estimation of demand showed that the price elasticity of irrigation water was −0.51, in line with previous studies that used water market prices to estimate the elasticity of irrigation water demand.

References

Al-Abri BBS (1992) Explanations of some Aflaj of Oman. Golden Press, Muscat/Ruwi (In Arabic)

Bjornlund H, Wheeler S, Shanahan M (2008) Supply and demand elasticity of irrigation water – implications for water college for regulation and market analysis. University of South Australia and University of Lethbridge, Alberta

Bontemps C, Couture S (2002) Irrigation water demand for the decision maker. Environ Dev Econ 7(4):643–657. doi:10.1017/S1355770X02000396

Brooks R, Harris E (2008) Efficiency gains from water markets: empirical analysis of watermove in Australia. Agric Water Manag 95:391–399

COA, College of Agriculture (1995) Traditional agriculture and fishing in the Sultanate of Oman. Sultan Qaboos University, Muscat

Davidson B, Hellegers P (2011) Estimating the own-price elasticity of demand for irrigation water in the Musi catchment of India. J Hydrol 408:226–234

Griffin R (2006) Water resource economics: the analysis of scarcity, policies, and projects. The MIT Press, Cambridge, MA

Heaney A, Beare S (2003) Improving water use efficiency. Aust Commod 10(2):226–277

Hooker MA, Alexander WE (1998) Estimating the demand for irrigation water in the central valley of California. J Am Water Resour Assoc 34(3):497–505

Michelsen AM, Booker JF, Person P (2000) Expectations in water-right prices. Int J Water Resour Dev 16(2):209–219

Ministry of Regional Municipalities, Environment and Water Resources (MRMEW) (2002) Aflaj in the Sultanate of Oman. International Printing Press, Muscat

Ogg CW, Gollehon NR (1989) Western irrigation response to pumping costs: a water demand analysis using climatic regions. Water Resour Res 25(5):763–773

Powers D, Ghafri AA, Schettino T (2013) An economic analysis of Falaj Auction markets, Aflaj Research Unit, University of Nizwa, working paper WP0113

Schoengold K, Sunding DL, Moreno G (2006) Price elasticity reconsidered: panel estimation of an agricultural water demand function. Water Resour Res. doi: 10.1029/2005WR004096

Wilkinson JC (1977) Water and tribal settlement in Southeast Arabia. Oxford research studies in geography. Clarendon Press, London

Zabet T (2005) Water demand management options in Oman. Paper presented during the world water day workshop. Ministry of housing electricity and water & ministry of regional municipalities, environment and water resources. 22–23 March, Muscat

Zekri S, Al-Marshudi A (2008) A millenarian water rights system and water markets in Oman. Water Int 33(3):350–360

Zekri S, Kotagama H, Boughanmi H (2006a) Traditional indigenous water market in Oman and the response of farmers to water prices. Int Symp Irrig Modern 28(31):74–84

Zekri S, Kotagama H, Boughanmi H (2006b) Temporary water markets in Oman. Agric Mar Sci 11:77–84

Zekri S, Mbaga M, Fouzai A, Al-Shaqsi S (2011) Recreational value of an Oasis in Oman. Environ Manage 48(1):81–88

Zekri S, Fouzai A, Helmi T (2012a) Damage cost in dry Aflaj in the Sultanate of Oman. Agric Mar Sci 17:9–19

Zekri S, Mbaga M, Fouzai A (2012b) Complementarity between agriculture and tourism towards sustainability. Int J Agric Res 7(10): 482. doi:10.3923/ijar.2012.482.493

Zetland D (2013) All-in-Auctions for water. J Environ Manage 115:78–86

Chapter 9
The Evolution of Water Legislation in Australia

John Tisdell

Abstract As water at its core still remains fleeting and its supply independent, water rights and the associated trade in such rights depend on a set of institutional and legislative frameworks. In this light, it is important to appreciate the history of water legislation. This chapter, in complementing Chap. 10, gives a short history of water legislation in Australia – from its place in the colonization of the country to modern multiagency management of the Murray Darling Basin. As the demand for water moves from an immature phase of abundance to a mature phase of scarcity and degradation, water law must also evolve. There has been a raft of progressive legislation in the development of water management in Australia. This chapter will touch on those key pieces which molded the formation of water law in the eastern states of Australia.

Keywords Water management • Australian water law • Environmental water • Deakin • Environment law

9.1 Introduction

Water entitlement regimes in Australia were developed in the late nineteenth and early twentieth centuries and replaced the English riparian doctrines initially established in the colonies. Those responsible for developing water entitlement regimes in Australia, such as Alfred Deakin, benefited from observing the prior appropriation and riparian doctrines in use in the U.S.A, and developed the non-priority permit

J. Tisdell (✉)
Tasmanian School of Business and Economics,
University of Tasmania, Hobart, Tasmania 7001, Australia
e-mail: john.tisdell@utas.edu.au

K.W. Easter and Q. Huang (eds.), *Water Markets for the 21st Century: What Have We Learned?*, Global Issues in Water Policy 11, DOI 10.1007/978-94-017-9081-9_9,

doctrine as the preferred institution for Australia.[1] The non-priority permit system in Australia operates on the allocation of entitlements granted at the discretion of the government, rather than an individual right of ownership to water. The resource itself is nationalized by vesting the control of it with state government water authorities. Entitlements are defined in standard units, usually megaliters, and the volume available to a user is subject to availability of supply. The tenure of entitlements is not permanent, as it is with property rights to land or chattels.[2] Establishing permits to use, rather than rights of ownership, and vesting greater regulatory power in state governments provided important distinctions between the Australian legal doctrine on the one hand, and those of England and the western United States on the other. In essence, those formulating the 'Australian' doctrine rejected the riparian and prior appropriation doctrines, and gave the water authority a greater degree of control over water property rights than these traditional water doctrines, in order to avoid the legal disputes which have occurred in other countries (Davis 1968).

9.2 Defining the Use of Water in Australia

The use of water resources in Australia began, as it did in England, under riparian common law.[3] Under riparian law, landholders abutting rivers and streams can utilize the water provided that sufficient instream water remains in the streams to meet the demands of other riparian landholders. Such an informal system of water allocation works well provided there is an abundance of water available; however, it was soon realized that water is a scarce resource in Australia's dry environment.[4] The informal system of water allocation led to many conflicts and water allocation became a critical issue for Australia's development. As well, the riparian philosophy was found wanting because it excluded non-riparian water use throughout the nineteenth century, including the gold mining ventures of 1850s.

The Victorian colonial government established a Royal Commission in 1884 to examine alternative water doctrines (Clark 1970, 1982). Alfred Deakin, a

[1] Davis (1968) provides a (somewhat dated) comparison of American and Australian water doctrines.

[2] There are a number of different forms of water permits of varying duration. The majority of water entitlements for irrigation are usually issued for 10–15 years.

[3] Instream water that flows through more than one person's land could be deemed as common property and consequently subject to common law of ownership. States have legislated that the right to use and control surface water be vested with specific government agencies.

[4] The vast majority of Australia is defined by Davidson (1969) as arid or semi-arid with variable precipitation. The population is concentrated along the coastlines, with over 70 % living in urban centers scattered along the eastern coastline. The more arid inland regions have tended to be dominated by rural communities (including irrigation) which consume 82 % of total water used in Australia and depend upon less reliable inland streams.

strong advocate of irrigation and later Prime Minister of Australia, was appointed chairman. As part of his responsibility, Deakin traveled overseas extensively.[5] Deakin observed that water rights policies in Colorado were causing conflicts between holders of riparian and prior appropriation rights, and leading to costly and time-consuming disputes and legal battles. California at that time was divided into warring factions resulting in messy confrontations (Davis 1968). To avoid conflicts over ownership and priority of use of the water resources of Victoria, the Deakin committee (Deakin 1881) recommended that:

(a) the riparian rights system in Victoria should be abolished and the State Government be given supreme control over all water resources of the State;
(b) riparian users be issued with a license to divert water for domestic and stock use,
(c) a standard unit of measure be adopted for water consumption,
(d) a formal system of licensing be established, and
(e) non-riparian users be given a statutory right of easement to the rivers and streams.

The ideas developed by the Deakin Commission culminated in the Irrigation Act 1886 (Vic.). It nationalized the right to use water, gave easement of aqueduct to non-riparian farmers and formalized a licensing system. The idea of nationalization of water resources was not new. It had been preceded by the Spanish Law of Water (1866) and Northern India Canal and Drainage Act (1873) (Davis 1968). The Victorian Act however represented a radical departure from existing common law notions of riparian rights which existed in Australia at that time. The Act effectively abolished any new riparian rights, and replaced existing rights with statutory rights in order to assert State authority. These rights were later to be become known as statutory riparian rights. The government reserved public lands abutting watercourses, thereby ensuring that no new riparian rights could be established. Deakin believed that all river banks in Victoria should be owned by the State and that all riparian rights should be abolished (Deakin 1881, pp. 440–441). In fact, very few riparian rights existed in Victoria prior to the Irrigation Act 1886 (Vic.). The government gained control over riparian land by making purchases whenever this land was offered for sale. Vesting the right of control of the water resources and riparian land with the State was criticized by some as socialism in action (Davis 1968).

The right to the "use and flow and to the control" of all water in watercourses was vested with the government. "The purpose of the 'vesting provision' was to establish the legislative basis for the State to act as the grantor of rights and thereby ensure that water could be allocated fairly between all users" (Department of Water Resources, Victoria (DWRV) 1986, p. 21).

[5]Many of the basic elements of Australian water allocation law stem from Deakin's assessment of irrigation systems in the western United States, particularly California and Colorado, and in India, Egypt and Italy.

A similar Royal Commission to that held in Victoria was appointed in New South Wales. Known as the 'Lyne' Royal Commission 1884–1887, it came to the same conclusions as Deakin: nationalize the water resource by vesting the ownership of the water resources of the State with the government so as to avoid the conflicts that were resulting from individual rights of ownership in California.

Following the recommendations of these two Royal Commissions, state legislation was enacted throughout Australia that vested the ownership of water resources with respective state governments. The acts were seen as a "landmark for Australian water legislation" as they "avoided a number of complex problems of riparian use encountered overseas and greatly facilitated planning and development procedures and policies related to allocation of the resource" (Schegen and Donohue 1983, p. 927). In Australia there was no longer an individual right to water, but rather an entitlement granted at the discretion of the government. The philosophies of a non-priority permit doctrine of entitlements, rather than rights of ownership, had been established.

Other states soon followed the lead of Victoria and New South Wales with their own legislation. In Queensland (Qld) the Rights in Water Conservation and Utilization Act, 1910 (Qld)[6] vested the control of all natural waters with the Queensland Government. The ownership of the beds and banks of watercourses was vested with the Government that removed, theoretically, riparian rights in regulated river basins. One major problem with the 1910 Act was that it did not give the Government any administrative infrastructure to regulate water allocation, and hence did not produce any dramatic changes in the use of the resource.

In 1926 the Water Act 1926 (Qld) was proclaimed, giving the Government powers over the allocation of the State's surface water resources. It went further by providing aid for landholders to establish water facilities, and by conferring control over sub-artesian water. In addition, the Act (sub-section ll(a)) introduced rights for non-riparian owners of land to obtain licenses, and broke the nexus between water and its riparian use. Section 4(1) of the Water Act 1926 (Qld) gave the Government agency[7] "the right to the use and flow and the control of the water at any time in ... all watercourses which flow through the land of two or more occupiers and all lakes and springs that are situated within the land of two or more occupiers."

The ownership of riverbanks and the effect on the rights of riparian landowners was not seriously contested in Queensland until 1983. In the case Nalder v Commissioner for Railways [1983] Queensland Law Report (1) 620 it was held that the Water Act 1926 (Qld) did not remove the riparian common law right of supply.[8] It was further held that vesting the control and allocation of water with

---

[6]Water legislation commenced in Queensland with *The Water Authorities Act 1891* and *The Irrigation Act 1891*, both of which provided for the construction and maintenance of dams and weirs.

[7]Other examples of the Governments' control over surface water include the *Water Act 1912 (NSW)*, Sect. 4A (1); *Water Resources Act 1976 (S.A.)*, s.6; *Water Act 1958 (Vic)*, s. 4.

[8]These issues are discussed further in *Travis v Vanderloos* (1984) 54 L.G.R.A. 268; and *Reid v Chapman* (1984) 37 S.A.S.R. 117.

the Queensland Government did not confer or imply ownership by the Government under the Water Act 1926–1987 (Qld). This distinguished Queensland legislation from that in Victoria and New South Wales.

The Queensland legislation that has followed the 1926 Act has been developed on a needs basis without any apparent overall direction. As a result, numerous amendment Acts have appeared, as well as a number of issue-specific Acts.[9] The result was that Queensland's water resources management had become unwieldy and in need of rationalization.

This historical background sets the scene for understanding the institutional changes to water management which have occurred in Australia in recent times. While each State developed its own legislation, the demands for institutional change were common to all as there were problems associated with the very nature of Australia's water economy.

The legislation formalizing the transfer of water entitlements was introduced in New South Wales in 1987 and in 1989 in Queensland and Victoria.

9.2.1 Water Markets in New South Wales

In the 1983/1984 water year a trial water transfer scheme was introduced in New South Wales under an amendment to the Water Act 1912 (NSW). At its inception these annual transfers were restricted to supplementing short-term water shortages, rather than encouraging long-term efficiency, and were small in number relative to the size of the potential market.

In December, 1986, NSW introduced permanent transfer arrangements. The Water Act 1912 (NSW) was amended in 1987. Division 4C of Sect. 20AH, of the amended 1912 Act allowed that the holder of an entitlement (transferor) may, with the approval of the Ministerial Corporation, transfer the whole or part of the water allocation for the entitlement to the holder of another entitlement (transferee). The transfer could be for a limited period of time or permanent (s.20AH(2)). There was also the possibility for transfer between different private schemes (s.20AH(3)) and long-term intersectoral transfers. Recognizing the potential problems, permanent transfers are subject to environmental assessment and public enquires, and the conditions of the transferring license are subject to change by the water authority. The possibility of permanent transfers opened the way for long-term adjustments to promote economic growth while considering both the social and economic consequences of the trade (Water Act 1912 (NSW), s. 20AI (6)).

---

[9]The most significant of these are the *Water Resources Administration Act 1978–1984;* the *Water Act 1926–1987;* the *River Improvement Trust Act 1940–1985;* the *Farm Water Supplies Assistance Act 1958–1984;* the *Irrigation Areas (Land Settlement) Act 1962–1972;* and the *New South Wales-Queensland Border Rivers Agreement Act 1946–1968* (QWRC 1987).

Therefore the effects upon regional economies and the equity of distribution could come into consideration in New South Wales but is still subject to testing in the Courts.

9.2.2 Water Markets in Queensland

To meet demands for institutional change, the Water Act 1989 (Qld) introduced transferable water entitlements and highlighted a need for the environmental requirements of rivers and streams. After a period of testing the concept of transferable water entitlements in the Border Rivers region, the then Queensland Water Resources Commission (QWRC) introduced transferability across the State in 1989 with the Water Resources Act 1989 (Qld). Under Sect. 10.17 of the Act an irrigator could transfer all or part of his or her allocated water to another irrigator within the same water area. (Water Resources Act 1989 (Qld), s.10.17(1)).

The legislation specified conditions for the transfer in terms of approval by the Commissioner, ownership of the license, duration of the transfer, and burden of the administrative costs associated with the transfer. The Act did not provide for intersectoral transfers and was unclear as to the rights of third parties affected by transfers. The rationale for prohibiting intersectoral transfers was to test the concept of transferability first within the irrigation sector. "Intersectoral transfers are unlikely to become a reality until well into the future" (Fenwick 1990, p. 221). The transfer of water was not even available within the agricultural sector. Transfers were restricted to water entitlement holders only. Such limited reforms seem unjustified provided the water authority had the power to intervene in transfers which have third party affects, or are not promoting efficiency or social equitable, but typified the cautious approach to water trading taken by most states at that time. All transfers were subject to the approval of the Commission (Water Resources Act 1989 (Qld), s. 10.17(5)).

9.2.3 Water Markets in Victoria

The Government of Victoria viewed transferability as removing the nexus between land and water. The major benefit was that irrigators could obtain more water through the market without having to purchase more land. The Rural Water Commission of Victoria (RWCV)[10] regarded the introduction as a logical extension of the current system where irrigators could transfer an entitlement between parcels

[10]The Rural Water Corporation was previously the Rural Water Commission. There was limited change to the role of the water authority as a result of the name change.

of land under common ownership[11]; under transferability, the requirement for common ownership was removed. Transfers prior to 1986 were restricted to seasonal transfers, up to irrigators' licensed quantity, and subject to channel capacity, no detriment to other irrigators, and no adverse drainage and salinity consequences (RWCV 1986). The quantity for saleable transfer had been restricted to maintain a minimum of 30 % of the total water right or original allocation. The acquired quantity, nonetheless, was subject to the discretion of the Rural Water Commission of Victoria (RWCV 1986).

Following a trial period and the introduction of new legislation, the transfer of licenses could be permanent or temporary (Sect. 62(3)). In considering a transfer, the Minister had regard to matters in Sect. 53, which made reference to the initial conditions for issuing a license, and Sect. 56(1) which outlined the conditions of licenses. Section 53 also made reference to Sect. 40; this means that applications for transfer were effectively subject to the same hydrological and environmental conditions that were applied to a new license application.

Bulk allocations could be transferred on either a permanent or temporary basis to another government authority. This could result in substantial changes in the distribution of income derived from the resource throughout society. The transfer of bulk allocations, like any other transfer, is subject to objection and public hearings under Sect. 40 of the Act. There was pressure to divert water away from agricultural use to urban and industrial use (IC 1992) as was by this time common in the U.S.. Yet it was not clear how Sect. 40(j) was to be interpreted in response to an application for transfer across sectors of the economy, i.e. between agricultural, urban and industrial uses.

In concert with the development of water markets was recognition that riverine ecosystems were in decline. Further blue-green algal outbreaks and the associated media coverage place pressure on governments. Regulation of flows had implications for the emerging water markets.

9.3 Environmental Consideration

Concern for the preservation of riverine ecosystems was becoming an important component of water management in Australia as in other parts of the world (Thompson 1991, p. 155).[12] During the 1980s, community concern for the environment in Australia came to the fore at the same time that water markets were emerging and the links between trade and environment were being explored.

Society was now concerned with environmental allocations as well as promoting an efficient use of the water in extractive enterprises. It was hoped that as the

[11]Common ownership in this context is ownership by deed. In other words, the same person or pastoral company can transfer quota between their properties.

[12]For example, in Canada, the "most serious water problems are not related to inadequate supply at all, but to degraded water quality and to disrupted flow regimes" (Pearce et al. 1985, p. 48).

process of reform developed, a new social contract for the care and use of water resources had been or would soon be embraced by water authorities in Australia. Milner and Knights (1986), among others, argued that for such an ethic to develop, policy makers needed to recognize allocating water for environmental purposes as important. They argued that unless provisions for the water requirements of the riverine ecosystem were placed in the context of overall water allocation; it would continue to be regarded as a residual use of water.

Any allocation or allowance for the riverine and riparian ecosystem was seen as ineffective without a systematic framework for environmental decision making and a formal recognition of water allocations for environmental use. Unless environmental allocations were recognized, any decision would lack standing and certainty of supply for the riverine ecosystem (DWRV 1986, p. 19). Water managers' approaches to meeting environmental water requirements were a mixture of restrictions by command on trade that affect the flow of rivers and streams and the levels of water tables, as well as direct allocation to the environment by licensing. Each State has developed its own legislation and approached the issue of the environment differently.

9.3.1 Water Legislation for the Environment in Queensland

The Water Act 1989 (Qld) failed to provide explicitly for environmental flows. The role of the DPIWR was to co-ordinate plans for the conservation of the waters of Queensland (Sect. 3.11(g)(ii)), yet how these plans were to be implemented was not clear. In considering an application for a license to extract water, for example, the Commission could inquire into availability and sufficiency of water to supply the requirements of riparian owners, licensees, permittees, the applicant and the water requirements of other government authorities. No direct consideration was given for environmental requirements. Section 4.18 outlined the procedure of inquiry by the Commissioner to grant or refuse an application for a license. Section 4.18(1)(a)(e) made reference to persons specified in Sect. 2.2(a). Section 2.2(a) referred to restrictions on the rights in water vested with the Crown in terms of the rights of other authorities conferred by the Water Act 1989 (Qld) or any other Act. Unless environmental considerations were specific in the procedural implementation of the Act, then environmental requirements could only be met if they happen to coincide with extractive objectives.

9.3.2 Water Legislation for the Environment in New South Wales

The two main pieces of water legislation in New South Wales were the Water Administration Act 1986 (NSW) and the Water Act 1912 (NSW). The Water Administration Act 1986, Sect. 4, established the objectives of the Department of

Water Resources, which were to promote the commercial benefits of development consistent with environmental requirements. The only other specific references to the environment in the Water Administration Act 1986 (NSW) were made in relation to the functions of the Water Administration Ministerial Corporation. Section 4(j) allowed the Corporation to "integrate the management of water resources with the management of other natural resources" and Sect. 12(3)(i) provides for such measures as the Ministerial Corporation thinks fit for environmental protection. The Water Administration Act 1986 (NSW) allowed regulations to be developed to make allocations for environment explicit. For example, the NSW Government had made an annual allocation of 50,000 ML of water from regulated flows for the conservation of wildlife in the Macquarie Marshes.

The Water Act 1912 (NSW) also explicitly allowed for restrictions to be imposed upon licenses to protect environmental flows and associated riverine ecosystems. The Ministerial Corporation also had the right to enter the market and purchase entitlements for any public purpose (Water Act 1912 (NSW), Division 4C, s. 20AL), which could include water for environmental flows. This opened the debate on the role of state and federal agencies as custodians of water for the environment and was arguably the forerunner to the Federal environmental water holder that now exists and can trade water in the Murray Darling basin.

9.3.3 Water Legislation for the Environment in Victoria

Legislation in Victoria led the way in explicitly recognizing the environment as a legitimate user of water resources (which would be a necessary prerequisite for formal environmental water holdings), with direct water allocation being made for the environment under the Water Act 1989 (Vic.). "Environmental problems attributable to rising water tables and consequential salinity are particularly acute across northern Victoria" (IC 1992, p. 171). Division 2 of the Water Act 1989 (Vic.) established the Water Resource Assessment program to monitor the condition of Victoria's water resource.[13] The program gave the Minister the power to do anything necessary to conduct monitoring work (Sect. 23(1)). Section 20 of the Act gave guidelines as to matters to be taken into consideration in determining whether a flow is reasonable. The guidelines included matters to determine whether or not the flow is likely to damage any waterway, wetland or aquifer (Sect. 20(i)), and whether or not a development takes account of the likely impact of works and activities,[14] given the availability of data at the time the work is established.

---

[13]The collection, collation, analysis and publication of water data has been formalized in the *Water Resources Assessment Program* which was established under Division 2 of the Act.

[14]"Works and activities" in this context is any development activity related directed to taking water from the river. This could include, for example, the development of pumping systems.

The Water Act 1989 (Vic.) also provided a formal means of protection and enhancement of the environmental qualities of waterways and instream uses.[15] Under this Act water allocations could be made to a number of instream water uses, such as maintenance of aquatic, riparian, floodplain and wetland ecosystems; maintenance of aesthetic, scientific and cultural values; water-based recreation; commercial fishing; water quality and navigation.[16] Water could be allocated to the environment as a bulk entitlement, known as an order granting entitlement, or as an instream license.

The order granting entitlement was a means of quantifying the amount of water, specified in terms of volume or the level of flow past a given point or by reference to a shared flow or capacity storage (Sects. 43(a) and 43(b)). These entitlements could be transferable and are used for the protection of the environment, the conservation policy of the government and the water returning to the discretion of the water authority (Sect. 43(i)).

A license for the instream use of water under Sect. 52 was defined in a similar fashion to order granting entitlements; that is, it was defined in terms of the location and rate or level of flow at specified times (Sect. 52(2)(b)). An application for an instream license was subject to the same assessment as for extractive use licenses, and such inquires as may arise under Sect. 40 (b to m).

All future water resource developments are subject to environmental consideration. When bulk allocation and individual licenses are granted for extractive use, conditions can be attached to the allocation to protect the environment (Sects. 43 and 56 respectively). In considering an application for bulk entitlements and licenses, the Minister must consider the impact of the allocation on the waterways, riverine and riparian environment, as well as water quality.

The Minister also has the discretion to attach any conditions to an entitlement or bulk allocation deemed necessary to protect the waterway environment. The Act allows any water authority, the Minister for conservation and environment and the Minister for planning and urban growth to request the Minister for water resources to declare an environmental or recreational area if the area is owned by a government authority. The Department of Conservation and Environment has guidelines for incorporating environmental water requirements which need to be followed in the planning of a new water project. Sections 46 and 62 of the Act allow the transfer of bulk entitlements and individual licenses between different users, including transfers to and from the environment according to need. Where bulk entitlements already exist, licenses may be issued to ensure a bulk entitlement is maintained at a specific level for a specific purpose.

Under Sects. 36 and 52 of the Water Act 1989 (Vic.), a government authority may apply to the Minister for an instream water entitlement or license. The

[15]While other Acts also covered environmental issues, such as the *Environmental Protection Act 1987*, the legislation did not overlap nor cause fragmentation of responsibilities, so that in most cases the water authority needed to only consult one Act.

[16]This interpretation was taken from the definition of "instream" in the Act.

Department of Conservation and Environment has taken the role of custodian of water allocations for the environment. Water can be allocated for extractive use as a bulk entitlement or an individual license (Water Act 1989 (Vic.), Sects. 36 and 52). Bulk allocations can take the form of a volume, a level of flow or a share of flow or storage (Sect. 43). Capacity sharing[17] gives management the flexibility to use storage capacity and planning beyond a water year. Surface licenses may be allocated to ensure that, for example, a storage's capacity is managed to maintain the volume and timing of flow required for downstream environmental purposes such as wetland management (Dept. of Conservation and Environment 1990).

Formally defined water allocations for the environment can be established by the issuing of new allocations or the purchase or recoupment of existing allocations. While the legislation exists to allocate water for the environment, no water allocations were made, primarily because there was little unallocated water available for many years. During that time unallocated water became part of the environmental flow by default. Environmental allocations are subject to the same conditions under the Act as any other application for an entitlement. One of the conditions is an assurance that the certainty of tenure of existing licensees is protected. The issuance of a new allocation in most river systems in Victoria at that time would, depending upon its size, seriously reduce the availability of water for existing extractive uses. In fact, for many years no new licenses were issued once the 1989 Act was proclaimed. If an allocation was made to the environment, the existing extractive users would have to be compensated (Sect. 56(1)(a)(x)). Such compensation and protection of existing entitlements may suggest that the above provisions for the environment are little more than good intent.

Alternatives to directly allocating water for the environment include managing the market to produce a flow regime more akin to the needs of the riverine ecosystem. In approving a transfer of a license under Sect. 62(6)(b) the Minister may amend the conditions of the license in accordance with Sect. 56(1), which specifies the conditions of a license when it is originally issued. The means that the Minister could limit trade to those transactions which favor conservation of the riverine ecosystem.

Other alternatives include the purchase of existing extractive allocations for environmental purposes in the market. Such an option would have to be financed. Ryan (1991) suggested imposing a tax upon market transfers, the revenue generated from which could be used to fund the purchase of entitlements. An alternative option is to recoup a proportion of water traded at the time of transfer for environmental flows. Eventually, a federally funded buyback scheme formed the basis of water for environmental use.

---

[17] *Capacity Sharing* involves property rights to water defined in terms of a share of the capacity of river storages and their inflows rather than their contents (Dudley and Musgrave 1988, p. 649). This form of property right has been introduced in Victoria. Modeling trade of capacity sharing rights is beyond the scope of this study, but is recognized as an area for further research and application of the approach used in this study.

While water entitlements for environmental use were being established and restrictions to trade could be made on the basis of environmental damage, full integration of extractive and instream demand for water in the emerging market environment was yet to be fully realized.

9.3.4 Improvements to State Water Legislation

It would appear that, of the State water legislation considered, the reforms in Victoria provided the clearest direction for the water authority in terms of transferability and the allocation of water for environmental use. Even then, the Water Act 1989 (Vic.) appeared to contain limited reform of intersectoral water transfers. While bulk allocations could be moved between sectors of the economy, there are still potential benefits which could be derived from the retail trade of individual water entitlements between sectors of the economy. Efficiency and social equity are not mentioned explicitly in the legislation governing intersectoral transfers. Section 40 of the Water Act 1989 (Vic.) needs to be modified to give greater power to the water authority to prevent trade which does not promote efficiency or social equity.

The Water Act 1989 (Qld) could be improved by more clearly defining the role of the water authority as a social policy maker. It is evident from Court proceedings that this Act did not give clear guidance on the social role of the water authority. Section 3.11 needs to define what is meant by "the best advantage of the public interest" because this could potentially be interpreted as encompassing social policy or more narrowly as solely hydrological issues. In considering an application for transfer the legislation gives the water authority the broadest brief possible, including as it does any "other matters" the Commissioner considers important. This broad nature could, however be the downfall of the legislation as undefined bounds are vague and open to dispute. Section 10.17(5) could be rewritten to define more precisely the protection of and process for assessing, third party and environmental effects of trade. Furthermore, there appears no rational reason to inhibit intersectoral trade. If the public interests were well defined and protected, Sect. 10.17 could be expanded to include the transfer of water to industrial, urban and environmental uses.

The water acts of all States collectively could be improved by clearly defining the meaning of an "equitable distribution" and a "beneficial distribution" of water, particularly between current and future users. The judiciary in New South Wales has interpreted equity in terms of a utilitarian viewpoint. This may be the view the government wishes to take. If not, it would be advisable to define equity more clearly under the Act. Like the Queensland Act, the Water Act 1912 (NSW) gives the water authority, in this case the Ministerial Corporation, an open slate to consider such matters as it thinks fit. Such a broad brief is, however, likely to lead to dispute, and it may be advisable for the legislation to define more clearly the parameters for consideration. In terms of trade, markets were fragmented throughout the landscape and as a result thin. Coordination of trade across the landscape and associated state boundaries became the responsibility of the federal government.

9.4 National Initiatives

National approaches to the issue of water management in Australia and water trading in particular have focused on the Murray Darling Basin. The Basin spans the east coast of Australia from south east Queensland to its mouth in South Australia, over 1,000 km in length and across five states and territories. Management of the basin has a long history. The first management agreement, the River Murray Water Agreement, was implemented in 1915 and stayed in place until the Murray Darling Basin agreement in 1993 which came into effect under the Murray-Darling Basin Act, 1993.

The Murray Darling Basin agreement set in place the notions of coordination with specific management targets. As with catchments throughout Australia, water management across the catchment evolved from the expansionary phase where water was seen as a limitless resource available for economic development of rural communities, to the mature phase where water was fully (and in some cases over) allocated and environmental decline was becoming evident with blue green algal outbreaks. Realizing the impact of over allocation on both extractive users and the environmental condition of the catchment a number of initiatives were activated which had potential flow on effects on water markets.

9.5 Basin Wide Water Initiatives

Given state jurisdictions and an associated raft of different water extraction entitlements reducing the level of basin wide extraction was difficult without the support of the States. The main body for State, Territory and Federal discussion is the Council of Australian Governments (COAG). COAG consists of the state and territories Premiers and Chief Ministers (similar to U.S. State Governors) and the Prime Minister. The Council of Australian Governments, realizing the extent of over allocation of water in the basin have imposed a number of supply constraints initiatives and developed basin wide water markets.

The Water Act 2007 (Cwlth) changed the nature of water management in the basin. It effectively established the Murray Darling Basin Authority with the power to develop and then enforce a plan for the basin. In terms of water trading, the Act provides for conversion of water entitlements into a single tradable water access right, the establishment of a national water market and an environmental water holder who could trade environmental water entitlements – the building blocks of a national water market.

The Act established the notion of a basin wide market for trading water, subject to hydrological and environmental constraints. The objectives in terms of trading arrangements for the Murray-Darling Basin are to facilitate efficient water markets within the basin, minimize transaction costs, create a suit of tradable water products, recognize the needs of the environment; and protect third party interests (Water Act 2007, S.3(3)).

The Act also provided a platform for the development of an overall basin plan. In 2011 a Murray Darling Basin plan was developed under amendments to the Water Act 2007 (Cwlth), subparagraph 44(2)(c)(ii). The plan focuses specifically on reducing extraction levels to restore environmental flows in the system to sustainable levels, effectively reducing the aggregate levels of extraction to 10,873 GL/year (historic extraction is approximately 26,000 GL per year on average). Reducing aggregate extraction was seen as achievable through sustainable diversion limits. Water trading under the plan allows free trade in surface water and groundwater access entitlements subject to physical or environmental reasons. Such restrictions include channel capacity and transmission losses in the case of surface water or hydraulic connectivity in the case of groundwater. In both cases potential impacts on third parties are explicitly considered. Finally, the impact on the needs of the environment is also explicitly included.

A cornerstone of the plan is to buyback water for environmental use. A new form of market is emerging in which industry (irrigators, rural town and water using industry) are now effectively competing with a federally funded tender market for water supply. Sections of the catchment have specific buyback targets. The impact of these buybacks will be to reduce supply to both the permanent and temporary water markets in the basin. The exact impact will be realized over the coming decade.

The plan establishes and operationalizes a formal environmental water holder. Over time the Commonwealth entered into buy back arrangements with irrigators at the point that "[a]s at 30 April 2013, the Commonwealth environmental water holdings totaled 1,582,826 ML of registered entitlements" (http://www.environment.gov.au/ewater/about/index.html). The use of the environmental water is aimed at restoring flow regimes to 1992 levels as defined in the Murray-Darling Basin Plan. The buyback scheme comes in direct competition with the permanent and to some degree the temporary water markets in the basin. Commonwealth environmental water holdings are tradable water rights and managed under the same trading and carryover rules, and charged the same fees, as equivalent entitlements.

The impact of having an environmental water holder in the market is yet to be fully realized. The buyback schemes to acquire water for environmental use has effectively reduced supply to water markets. In the future when water is plentiful it is possible that the environmental water holder will increase supply and conversely become a major buyer in direr period. The impact of a large seasonally dependent trader in the market is likely to make prices more volatile and subject to the requirements of the water holder.

9.6 Conclusion

Water management in Australia has developed from riparian rights to entitlements shares of the available water resource. Water emerged into a mature phase with high and conflicting demands for the available resource and with it changes in water law. Trade in the earlier years was the informal outcome of land transfer. Legislation was required to break the nexus between land and water and breaking that nexus opened the way for water to be viewed as a chattel which could be traded.

Trade began with poorly defined entitlements and was spatially explicit. Regions developed informal markets of bilateral trade. Over time more formal markets emerged, the most successful being the Goulburn Murray Water exchange. Through an evolution of water laws and government agreements institutional barriers to trade have been removed. The array of state and local water entitlement schemes have been rationalized so that trade can occur in a common currency.

Trade emerged in concert with issues of hydrological uncertainties, the over allocation of available water supplies and a declining riverine environment. The interrelated nature of these issues meant that as various government agencies and legislators grappled with supply and environmental issues, they directly impacted on the evolution of water markets. Placing a cap on aggregate water extraction increased the water demand for water allocations. The rationalization of entitlements in the Murray Darling Basin led to a more tractable and tradable water right which in principle extended the opportunity for water trading throughout the basin.

Finally recognition of environmental needs in legalization ensures to some degree that sustainable take limits conforms to notions of sustainable development. As demand for water continues to increase in a world of climate change and the maturity of water management evolves further, so the legislation underpinning the use of water will also need to evolve.

References

Clark C (1970) The economics of irrigation, 2nd edn. Pergamon Press, Oxford

Clark S (1982) Water law and scope for changes. In: Irrigation water: policies for its allocation in Australia. Australian Rural Adjustment Unit, Armidale

Davis PN (1968) Australian and American water allocation systems compared. Boston Coll Ind Commer Law Rev 9(3):647–676

Davidson BR (1969) Australia wet or dry? The physical and economic limits to the expansion of irrigation. Melbourne University Press, Carlton

DCE (Department of Conservation and Environment) (1990) Water for the environment: information bulletin on the water act 1989 and protection of the environment, Melbourne

Deakin A (1881) Progress report of the Royal Commission on water supply, irrigation in Western America, 2 Vict. Parl. Papers 731

Dudley N, Musgrave W (1988) Capasity sharing of water resources. Water Resour Res 24(5):649–658

DWRV (Department of Water Resources, Victoria) (1986) Water law review. Discussion paper. Government Printer, Melbourne

Fenwick TD (1990) Water allocation, and transfer practices in Queensland. In: Proceedings of an international seminar and workshop on transferability of water entitlements, CWPR, Armidale

IC (Industry Commission) (1992) Water resources and waste water disposal. Draft report, Canberra

Milner H, Knights P (1986) Hydrologic analysis – a de facto means for evaluating surface needs. In: Proceedings of the specialist workshop on surface needs and water uses, Canberra

Pearce PH, Bertrand F, Maclaren JW (1985) The framework of water rights legislation in Canada. Canadian Institute of Resources Law, University of Calgary, Calgary

QWRC (Queensland Water resources Commission) (1987) Green paper, water resources bill, Brisbane
RWCV (Rural Water Resources Commission of Victoria) (1986) Transferable water rights. Discussion paper, Melbourne
Ryan N (1991) Management to supply water requirements for conservation purposes now and in the future. In: Dendy T, Coombe M (eds) Conservation in management of the river murray system – making conservation count, proceedings of the third Fenner conference on the environment, department of environment and planning, Adelaide
Schegen FJ, Donohue EP (1983) Water allocation: the Queensland approach. In: Fifth Afro-Asian regional conference, Australia
Thompson A (1991) Water allocation for the environment – the Canadian experience. In: Proceedings of water allocation for the environment: an international seminar and workshop, CWPR, Armidale

Court Cases Cited

Reid v Chapman (1984) 37 S.A.S.R. 117.
Travis v Vanderloos (1984) 54 L.G.R.A. 268

Legislation Cited

Spanish Law of Water (1866)
Northern India Canal and Drainage Act (1873)
Water Act 1912 (NSW)
Water Act 1989 (Qld)
Water Act 2007 (Cwlth)

Chapter 10
Water Trading in Australia: Tracing its' Development and Impact Over the Past Three Decades

Sarah Wheeler, Henning Bjornlund, and Adam Loch

Abstract This chapter describes why and how water markets have evolved in Australia. The various changes that have occurred in Australia's water markets from their early inception to their current relative maturity are canvassed throughout. It outlines how various groups—mainly irrigators and governments—have used water markets, as well as some of the general benefits (and costs) that are associated with water markets, including environmental costs.

Keywords Water markets • Murray-Darling Basin • Farmer adaptation • Economic analysis • Water policy

10.1 Introduction

Markets have arisen worldwide in response to the global challenge of sharing water among competing users. Policies that promoted unsustainable water consumption resulted in rivers running dry and water quality deterioration (Hearne and Easter 1995). In Australia, examples of deteriorating water quality arose from incidences of

---

S. Wheeler (✉) • A. Loch
School of Commerce|UniSA Business School, Center for Regulation and Market Analysis, University of South Australia, Room WL3.51 Way Lee Building, 37 – 44 North Terrace, 5001 Adelaide, SA, Australia
e-mail: sarah.wheeler@unisa.edu.au; adam.loch@unisa.edu.au

H. Bjornlund
School of Commerce|UniSA Business School, Center for Regulation and Market Analysis, University of South Australia, Room WL3.51 Way Lee Building, 37 – 44 North Terrace, 5001 Adelaide, SA, Australia

Economics Department, University of Lethbridge, C576 University Hall, 4401 University Drive, Lethbridge, AB T1K 3M4, Canada
e-mail: Henning.Bjornlund@unisa.edu.au; henning.bjornlund@uleth.ca

K.W. Easter and Q. Huang (eds.), *Water Markets for the 21st Century: What Have We Learned?*, Global Issues in Water Policy 11, DOI 10.1007/978-94-017-9081-9_10,

widespread algal-bloom outbreaks, stressed rivers and wetlands, changes to riverine integrity, soil erosion and declining native fish species. In response, governments stopped issuing new entitlements to water, and sought to reduce existing levels of consumptive use extraction to improve ecosystems and water quality. The 1992 Earth Summit in Rio de Janeiro represented a milestone in how to address water problems (Sitarz 1993). In the Rio Declaration and Agenda 21 the international community stated there should be a focus on managing demand for water, reallocating water from existing to new users, and recognizing the needs of the environment.

A common approach to reallocation has been the adoption of market-based instruments, especially for natural resources such as water (Grafton et al. 2011). The challenge for water markets is to manage the increasing demand for water among competing users and to ensure that it is used efficiently for desired ends, while promoting environmental sustainability. Markets can reflect the scarcity value of water in a way that administrators of water systems have trouble doing. Further, markets do not allocate water in inflexible ways based on political concerns or custom (NWC 2011a).

The main economic benefit of water markets is that they should enable limited resources to be put to their most productive uses by distributing them to those that value them most highly both in the short- and long-term. This reallocation benefit results in three forms of economic efficiency: (i) *allocative efficiency*: changes in water resource demanded or used motivated by seasonal conditions, commodity price adjustments, cropping choices and other short-term decision-making requirements; (ii) *dynamic efficiency*: changes in water resource demanded or used that stem from structural alterations such as new investment opportunities, regulatory shifts in access arrangements (e.g. extraction limits or embargos) or personal strategic choices (e.g. retirement); and (iii) *productive efficiency:* changes in water prices offer incentives for the efficient use of water resources as either an investment or input for productive outcomes.

10.2 Australian Water Markets

Australia represents a country that has adopted water markets on a large-scale, particularly in the Murray-Darling Basin (MDB). In the Australian context, most water access rights in the past century were issued when the country was in a 50-year wet period (Fig. 10.1). This contributed substantially to the over-allocation of access rights and a variety of environmental problems, low flows and growing scarcity (Davidson 1969).

Generally speaking, two water markets exist in the MDB. First, the entitlement market in which water access entitlements are traded (otherwise known as permanent trading). These provide exclusive access to a share of the water resources within a water resource plan area (predominantly in perpetuity). These entitlements yield seasonal volumetric water allocations which can be extracted from the water source during that season and put to beneficial use by holders of a water access right. These allocations are announced at the beginning of each season as a percentage the

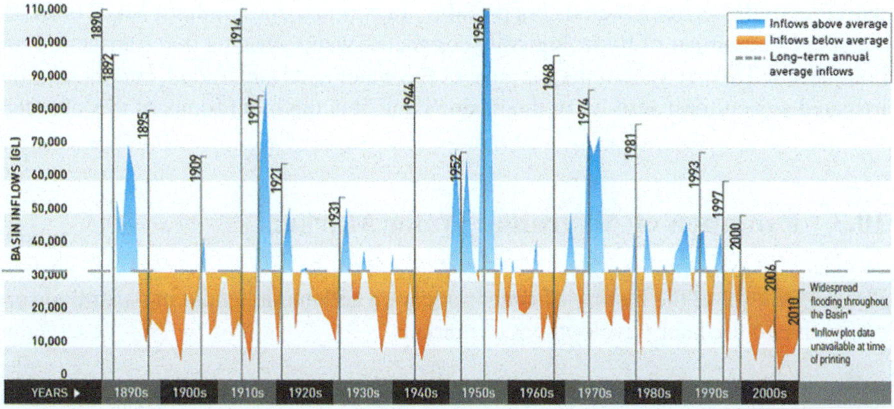

Fig. 10.1 Murray-Darling Basin flows 1890–2000s (Source: MDBA 2012a)

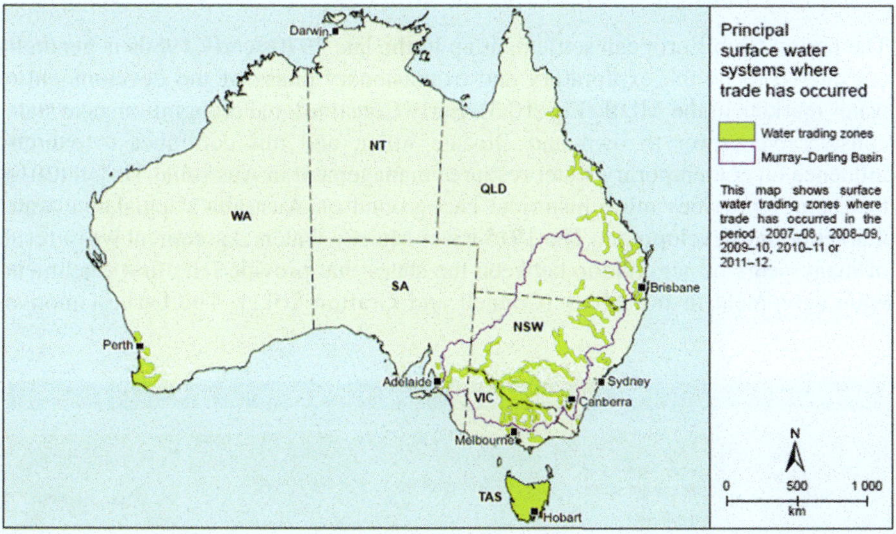

Fig. 10.2 Surface water trade areas in Australia (Source: National Water Commission (NWC) 2013: 7)

total access right depending on seasonal conditions. Second, the allocation market is where seasonal allocations are traded (otherwise known as temporary trading). Figure 10.2 illustrates the principal surface water trading areas in Australia, where some trade had occurred by the start of 2013 (NWC 2013).

The MDB (indicated in Fig. 10.2) comprised 94 % of the volume of entitlements and allocations traded across Australia in 2010–2011 and 2011–2012 (NWC 2011b, 2013). It consists of the southern MDB (83 % of volume traded across Australia in 2010–2011) and the northern MDB (11 % of volume traded across Australia in 2010–2011). The northern and southern MDB are not hydrologically linked and therefore trade between them is not possible. In contrast, the southern MDB (sMDB) consists of a number of hydrologically connected water systems that transcend state boundaries. This area accounts for most of the water used and traded, and most of the irrigated agricultural activity in the basin. Thus, it is the main focus of this chapter.

10.3 Evolution of Australian Water Markets

Several features of the history of water allocation and management in Australia were conducive to the later development of markets.

10.3.1 Exploratory and Expansionary/Development Phases in Water Market Policy

The period from European settlement up to the late 1970s/early 1980s is generally considered to be the exploratory and expansionary phase of the development of water markets in the MDB (Fig. 10.3). Early Constitutional recognition gave states considerable power to own and allocate water, and this continues to exercise influence on contemporary water resource management in Australia. Tisdell (2014) in Chap. 9 provides more historical background on Australia's legislative water market policy development. The 1915 River Murray Waters Agreement was a result of many years of negotiation between the states that provided the first catchment-wide agreement in the MDB (Connell and Grafton 2011). The earliest motives

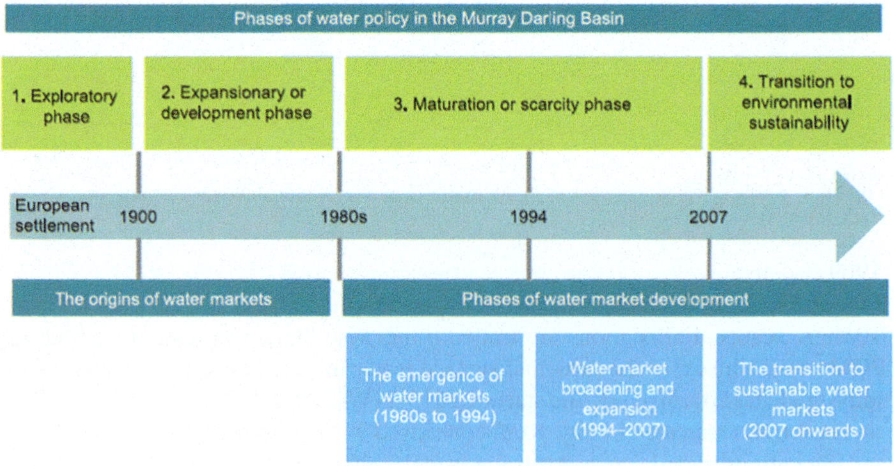

Fig. 10.3 The evolution of Australian water markets (Source: NWC 2011d)

for Australian water trade often stemmed from water scarcity as a consequence of drought or regulatory pressures. There is anecdotal evidence that unofficial temporary water trades occurred during the 1940s drought in Australia.

Temporary transfers of water rights were also permitted in New South Wales (NSW) during the droughts of 1966–1967 and 1972–1973, in Victoria during 1966–1967, and in a restricted version over the period from 1982 to 1983 to not long before its more general introduction in 1986–1987. These limited examples of water trading were one time responses to isolated and temporary water shortages, but they became the precursor to more formal water markets (NWC 2011d).

10.3.2 Maturation or Scarcity Phase

Aside from periodic drought restrictions, by the late 1960s there was an increasing general awareness of the negative environmental consequences from over-allocating sMDB water resources (in other words, too many water access rights, relative to volumes normally available, had been granted by the states). As a consequence, the mandate of the River Murray Commission was expanded from managing water demand to also include water quality issues (MDBC 2007). In the interim, evidence of environmental degradation including toxic blue-green algal-bloom events and irrigation-induced land salinization drew wider public attention to the negative consequences of unsustainable water use. The 1980s saw the real recognition of the need to manage these issues, and the River Murray Commission powers were extended again to cover salinity control, among other environmental and management issues (MDBC 2007). As a result, the second phase of water market development in Australia was characterized by an increasing emphasis on sustainability rather than regional development and engineering solutions (NWC 2011d). Crase (2008) characterizes this phase of market extension and maturity as being initiated by two impulses: first a greater individualism among traditional water users (for example, irrigators became less concerned about the negative community impacts of trade and began recognizing the business and personal benefits to be gained from transferring water entitlements) and second the transition from developmentalism to sustainability. In response to sustainability concerns the South Australia (SA) government recognized potential negative impacts of water over-allocation, placing a moratorium on the issuing of new water entitlements in 1969 (Loch et al. 2013). This approach was followed in 1979 with the introduction of volumetric allocations for each water access entitlement—resulting in a general 10 % reduction in the total volume of entitlements (Bjornlund and O'Callaghan 2003). Regulatory restrictions such as these, combined with significant demand pressures from the growing horticultural industries, provided an additional early driver for the formation of water markets in SA.

The introduction of water as an economic commodity, however, was typically met with resistance. Steps toward market development were incremental and initially trade was only allowed in the allocation market. SA was the first state in

Australia to offer provisions for formal water trading of both water access entitlements and allocations in 1983, and trade within irrigation districts began in 1989. Not until 1995 was trade between private diverters and irrigators within irrigation districts allowed, following clear legislative backing in 1994 (as highlighted in Fig. 10.3) (NWC 2011d).

In NSW, legislation permitting the transfer of water allocations was enacted in 1983, with water access entitlement trade legalized in 1989 between individual right holders. The first transfer in the Murray Region was registered in 1992 (Bjornlund and McKay 2000). Water access entitlement trade involving irrigation districts was not possible until individual districts were privatized; a process which commenced in 1991 and was completed for the five largest districts by 1995 (NSW Office of Water 2011). There was an increasing push in the 1980s to privatize irrigation districts that had traditionally been under government management. For example, in 1987 the Government Highland Irrigated Districts in South Australia were transferred to eight self-managed irrigation trusts (McKay 2001). In Victoria, water allocation and access entitlement trade was formally included in the new *Water Act* in 1989. Trading in water access entitlements did not take place until required regulations setting out the rules of such trade were passed in September 1991. The first Victorian water access entitlement transfers were therefore not formally registered until January 1992. Trade was made increasingly more flexible with successive regulatory changes during the 1990s; e.g. trade between district irrigators and private irrigators were allowed in 1994 (Bjornlund and McKay 2003). This again highlights why 1994 is of such importance in Fig. 10.3, as it represents the first trans-boundary water transfers.

It became clear in the 1990s that policy and institutional changes were needed to allow markets to operate more effectively, and to meet their outcomes of promoting efficient and sustainable use of water by directing it to high-value uses. By the 1990s broadening of water reform and trade intensified. A substantial body of evidence documented how the states were unable to manage MDB water resources effectively without federal coordination and investment. This was partly due to developmental and salinity pressures of the 1970s and 1980s, which had led to the breakdown of the 1915 River Murray Waters Agreement (Connell and Grafton 2011). A new MDB Agreement was crafted in 1992 following federal-state negotiation. It provided the basis for measures to advance management of land, water and environmental resources on a basin-wide scale (Papas 2007) via market-based instruments. The momentum of this agreement was maintained by the creation of the Council of Australian Governments (COAG) in 1992. COAG facilitated state and federal collaboration and the establishment of the Working Group on Water Resources Policy, which included senior civil servants' input from all the states. The Working Group on Water Resources Policy's (1994) report resulted in the 1994 COAG National Water Reform Agenda which, as part of a wider National Competition Policy, was the focus of subsequent COAG meeting agendas. For example, it was not until 1994/1995 that water trade activity increased dramatically when inter-district trade was allowed.

10.3.3 Transition to Environmental Sustainability Phase in Water Market Policy

The third stage (Fig. 10.3) of water market development in Australia was therefore characterized by major reforms to ensure that water markets addressed scarcity issues and sustainability requirements. This has been classified as the year 2007, which saw the introduction of a Commonwealth Water Act (2007) (Australian Parliament 2007), as discussed later in the chapter. These reforms were widely adopted by the states and water users. This process required massive legislative, regulatory and stakeholder reform in line with the elements for efficient water markets detailed below. Further reforms followed with the unbundling of land and water, removal of trade restrictions and the use of markets for environmental purposes.

10.4 Water Market Reforms

In outlining the formation and ongoing evolution of Australian water markets two elements need to be considered. One is the role of particular market institutions and reforms that shape the way markets operate. This includes factors that underpin successful market operation, such as: (i) defining and securing property (access) rights; (ii) making changes that allow markets to function more effectively, such as developing carry over provisions that allow unused water at the end of the season to be carried forward to the next season; and (iii) removing barriers to trade out of irrigation districts and between regions. This latter kind of activity requires federal-state coordination. The second relevant element is the increasing role of the federal government in providing coordinated and coherent management of the MDB, directed particularly to its sustainability. In part, this sustainability is underpinned by having efficient and effective markets.

In line with reforms detailed above, the Commonwealth has assumed an increasingly active role in MDB water policy to ensure state collaboration and coordination. This is signified by the establishment of the Murray Darling Basin Commission in 1988 and the National Water Commission (NWC) in 2004 (NWC 2011c). Following these institutional reforms, between 1992 and 2004 a number of federal initiatives aided the extension and maturity of water markets. First, the 1992 Murray-Darling Basin Agreement, which not only created several new organizations but also formalized institutional coordination to improve the health of the Murray River. Second, the promotion of water trade, including trade between the states, was formalized as one way of achieving more collaborative management of the MDB. Subsequently, between 1994 and 2008 there were a series of nationally agreed reforms (such as: secure property rights; access caps; the National Water Initiative (NWI); unbundling of water access rights; carry-over provisions; and other regulatory instruments) that allowed for the consolidation and extension of water markets.

10.4.1 Secure Property Rights

The assignment of secure property rights is a prerequisite for efficient market exchange, as well as for avoiding commons dilemmas in natural resources use (Smith 1981). Secure property rights are crucial in encouraging investment in water use and productive outputs (Bjornlund and O'Callaghan 2003). Secure property rights are also essential for successful market reallocation of natural resources such as water (Loch et al. 2012). Australian water markets have developed to recognize *de jure* property rights, supported by formal legal instruments which, if challenged judicially or administratively, would most likely be upheld (Schlager and Ostrom 1992).

10.4.2 Caps on Resource Use and Unbundling

A nationally consistent scheme of entitlement and allocation trading is also essential for providing certainty to both consumptive users and the environment. In Australia this scheme relied on defining the water resource pool available for consumption. In essence, it was one method of moving toward defined property rights. In addition, rules for minimizing transaction costs were outlined (NWC 2011a). As discussed, COAG established a Working Group on water resources policy to examine, among other things, barriers to the effective transfer of water between competing uses. This led to COAG initiatives for the unbundling of land and water rights, and principles for water trading arrangements (ARMCANZ 1996). Each of the states subsequently worked to legally separate land and water assets, establish new water property rights, and develop the means of transferring those rights between users (Hamstead et al. 2008). Transfers of water remained limited during these early phases. Bundling access to water with land was a barrier to easy movement of water between users, but had been axiomatic to water policy throughout Australia's history. There were, and are, pervasive fears that severing the connection would privatize water, facilitate the emergence of water barons, cause communities to lose water, and strand delivery assets (NWC 2011d). Previously, an access right entitled the holder to irrigate an area of land. Over time this was changed by the states into volumetric water access entitlements, and the COAG and NWI reforms enabled separation of these rights from land.

The separation of land and water was necessary for water markets to work effectively and efficiently. Under the NWI, further unbundling took place into four components (Fig. 10.4): (i) a water access entitlement which gives the holder a right to receive seasonal allocations depending on availability; (ii) a water allocation account to which the seasonal allocations are credited; (iii) a delivery share which ensures that the holder has the right to get allocation water delivered to their property; and (iv) a water use license that enables the water access entitlement holder

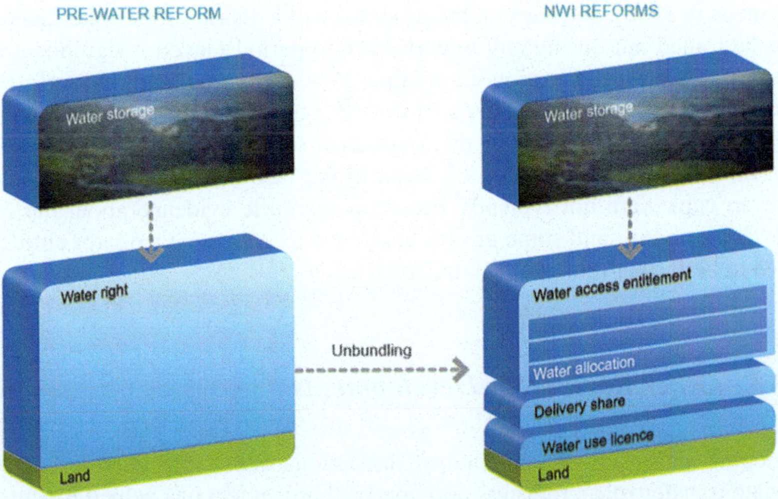

Fig. 10.4 NWI unbundling of property rights process in Australia (Source: NWC 2011d: 83)

to use water for beneficial gain subject to proven best practice use requirements. To irrigate, a farmer needs the last three rights but does not necessarily need a water access entitlement, as allocation water can be credited to their account either in the form of yields from a water access entitlement held by the account holder or by purchasing allocation water from the holder of another water allocation account. The first three of these rights can be traded separately while the water use license is attached to the land on which it is approved.

This unbundling has had four important impacts on the promotion of water trading. First, trade in both water access entitlements and allocations can take place with minimum oversight as they do not grant any right to use the water. Hence, there are no concerns over water use impacts, since this is addressed via the process of issuing a water use license. Second, it is possible to sell a water access entitlement without terminating the delivery or water use rights and/or the allocation account (ACCC 2010). Third, farmers can sell their water access entitlements, but have to continue paying any associated fixed water charges unless they can find a buyer for their delivery share. This can resolve concerns over asset stranding and cost-sharing implications for remaining irrigators. Finally, irrigators can sell their water access entitlement and continue to irrigate, as this does not preclude the purchase of water allocations, or stop them from repurchasing water access entitlements at a later date. Consequently, in the MDB well-defined, secure and unbundled water property rights increasingly enable irrigators to transfer water access entitlement and allocation assets between one-another in response to risk attitudes, seasonal conditions or strategic planning (Loch et al. 2013).

General moratoriums on the issuing of new entitlements within all states, a cap on total extraction of water from the MDB introduced in 1997 (interim cap announced in 1995), and caps on the proportion of a districts total water license that could be traded out during any one given time period played a significant role in encouraging water reallocation via trading. While it is true that these reforms also activated previously unused (sleeper) or under-used (dozer) water entitlements—increasing overall water use in the short-term (Crase 2008)—they represented a major advance in state cooperation, even in the face of competing state interests. However, caps were not typically based on scientific evidence about the MDB's water requirements, and some groups continued to assert that irrigator entitlements needed to be further reduced (MDBC 2007).

10.4.3 Other Regulatory Developments

Developing consistent water trading institutions (i.e. carry-over arrangements, interstate transfer rules, registers and approval processes) has helped to strengthen water market activity and efficiency, partly by reducing transaction costs. Such reforms can aid the development of innovative water use, management and transfers between consumptive users, as well as the creation of sustainable water use systems in environmental sectors (Loch et al. 2013). Regulations, for instance, were developed which ensured that trades were consistent with the hydrological features of particular areas and thereby minimized negative impacts on river water quality. This included the specifying of trading zones and the documentation of rules that indicated whether trades within and between zones would be approved or associated with penalties. In a related sense, there have also been specifications about how transmission losses were to be accounted for and what the potential water trade impacts were on such losses.

10.5 Adoption of Water Markets by Users

In response to these reforms, escalating water scarcity and government intervention in entitlement markets to purchase water for the environment, the adoption and use of water markets in Australia has increased rapidly over the past three decades (Fig. 10.5). Initially trading in water allocations grew rapidly in response to the 1995 interim cap and drought, with entitlement trading increasing slowly until the large-scale government market entry in 2007–2008.

Many economic studies have investigated what influences farmers' adoption of technology innovation. Overall, they generally conclude that early adopters of innovation tend to: be younger; be more educated; have higher incomes; have larger farm operations; be leaders; and be more reliant on primary sources of information (e.g. results from scientific experiments and agricultural trials) than later adopters. The adoption of water trading, however, does not exactly parallel the adoption of other innovations which frequently simply involve buying and installing a physical

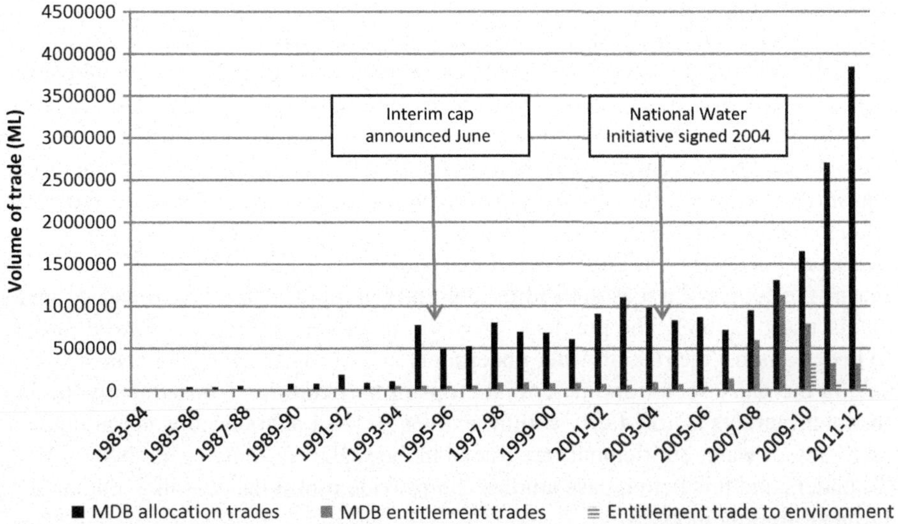

Fig. 10.5 Water allocation and entitlement trade in the sMDB (Source: Adapted from data in NWC 2011c, 2013)

input or output. Water trading is an innovation that involves the farmer acquiring new skills or rearranging their farm management or decision-making processes. This can mean that the adoption of water trading can be more difficult to predict than simple innovation theory might suggest. When comparing the adoption of water trading in the Goulburn Murray Irrigated District (GMID – Australia's largest irrigation district) over time, Wheeler et al. (2009) found that the adoption of water allocation trading broadly conformed to the general model of agricultural innovation adoption. Specifically, early adopters were more likely to: be more educated and have a farm plan; have higher incomes and have worked less years on the farm; and be female. Other variables, however, indicated that influences (such as region, water allocations, farming technology, and farm type) were also important for water trade adoption and, specifically, whether the farmer bought or sold water allocations. Further, other economic studies that model participation in water trade over time highlight dynamic factors such as water scarcity, climate, water allocations and water/commodity input and output prices as important influences (Wheeler et al. 2008, 2010, 2012).

In 2010–2011, it was found that 47 % of the irrigators had traded in the water allocation market in the past 3 years, with slightly higher participation by those in horticulture (NWC 2012). Participation in water entitlement trading in Australia has never been as widespread as water allocation trading. However, it too has shown a steady recent increase, with a major spur taking place in 2008 when the federal government entered the market under its' *Restoring the Balance* program to recover entitlement water for environmental use (discussed later).

Farmers use a combination of participation in the entitlement and allocation markets to help them optimize the management of their businesses. For instance, they may choose to reduce their reliance on entitlements and enter the allocation market to buy or sell water opportunistically. It often makes economic sense to sell a portion of entitlements and buy water allocations as needed (NWC 2012). On the one hand, water entitlement trading was often considered a tool that facilitated transition toward a different farming system over a few years (change in farm size, change in farm activities, exit from the farm sector, etc.). On the other hand, trading of water allocations was mostly described by farmers as providing greater flexibility during the course of a season to adjust their mix of inputs (such as in the dairy sector as discussed below). The purchase of water allocations have also allowed farmers to hedge against the risk of water uncertainty and losing long-lived assets (e.g. herd size in the dairy sector and perennial crops in horticulture). Consequently, trading in water markets is now fairly common: over 30 % of announced water allocations and 10 % of water entitlements are traded in the sMDB in some years (NWC 2012). To understand this better, case studies can provide useful insights in understanding farmers' trading motives.

10.5.1 Dairy

Australian farmers have become sophisticated in the way they use water markets. Dairy farmers, for instance, dominated early entitlement markets in Victoria, buying water allocations to maintain feed production during droughts. Thus, in the late 1990s they paid prices for water allocations far in excess of the profit they could generate from using the water to produce fodder to protect their stock, and banks were willing to fund this practice (Bjornlund 2003). However as scarcity persisted and demand from horticultural growers increased prices further they learned how to determine the point at which they should sell water and buy substitute feed. Between 2007–2008 and 2008–2009, demand from downstream horticulturalists and high prices for most water trade products allowed dairy farmers to sell water allocations, buy feed for their herds and maintain production (NWC 2012).

At the beginning of 2009, drought conditions and a fall in milk prices increased the debt of many dairy farmers in the Sunraysia region of Victoria, and they began selling entitlements. This trend has diminished with better rainfall since late 2009–2010. While milk production in the area fell at the beginning of 2005–2006, this was due more to drought than to water trading. Given drought, farm restructuring, low milk prices and the water buyback (Sect. 10.6 provides more details on the buyback) it is estimated that while there were 2,000 dairy farms in the Sunraysia region in 2006–2007, this number had reduced to 1,400 by 2010–2011. The sale of water allowed exiting farmers to exit having paid down their debts and, possibly, with some capital reserves (NWC 2012).

10.5.2 Horticulture

The dominant horticultural industries in the sMDB are wine grapes and almonds. Grapes are grown in the Sunraysia area (mainly in NSW), the NSW Riverina and the SA Riverland. Wine production grew steadily throughout the 1990s, but has slowed down since the mid-2000s. The gross value was about AUD$1 billion in 2007–2008 and AUD$600 million in 2008–2009 (NWC 2012). In contrast, almond production has grown dramatically in Australia. Almond production is currently worth AUD$250 million/year, but this is predicted to increase to AUD$600 million/year by 2016.

Water trading has increased significantly among horticulturalists to secure water supply that ensures the survival of capital-intensive crops, and facilitates the planting of new crops in new locations. For instance, in the 1990s farmers in the Red Cliffs, Merebein and Mildura regions of Victoria shifted from producing dried vine fruit to wine grape production. In the face of historically low allocation levels in 2006–2008, horticulturalists purchased large volumes of water allocations (and paid very high prices) from rice growers, dairy farmers and mixed farmers typically located in NSW. Many horticulturalists stated that they had learned hard lessons from this period, and changed their strategy for the future (Loch et al. 2012).

In the Sunraysia region (both Victoria and New South Wales), restructuring of the industry has been widespread given the small block sizes, drought and water prices. Many horticulturalists sold water entitlements and exited the industry, while others used the proceeds to retire or reduce debt (NWC 2012). Trade has therefore helped horticulturalists reduce farm debt and aided structural adjustment in relation to plummeting commodity prices, particularly among grape growers (NWC 2012; Wheeler and Cheesman 2013).

10.5.3 Annual Crops

Annual crop (broadacre) irrigators have considerable flexibility in adapting to water scarcity, since in some cases they can switch to dryland production (or not put a crop in at all). They may also choose to switch from water-intensive to less water-intensive crops. In the rice and cotton industries, however, selling water has appeared to be in some years a less risky strategy than growing crops, since water sales can provide certain cash income. For example, rice and cotton producers may, instead of purchasing water, sell water allocations and reduce the area under production (Loch et al. 2012; NWC 2011b). Thus like horticultural water users discussed above, water markets have helped sMDB broadacre farmers find willing buyers or sellers to meet their water management requirements under uncertain seasonal conditions.

Despite much argument in Australia about the fallacy of one of the driest countries in the world growing rice and other high water intensive crops (Australian

Parliament 2011), we believe it is essential that both permanent high value users and more flexible annual lower value users co-exist. If all water went to high-value perennial users and annual cropping ceased, the economic consequences of drought would be worse as no temporary supply would be available from farmers growing annual crops. Conversely, during wet years perennial users would have excess supply which would not be put to economic use. Only via the coexistence of permanent and annual crops with effective and adaptable market mechanisms put in place, can economic benefits be maximized under any given climatic condition.

10.5.4 Urban Water Users

Purchasing water on markets for urban use has proved to be a cost-effective method of augmenting urban water supplies in periods of drought in Australia. Water suppliers in Adelaide, Canberra, local councils in southern NSW and several regional water authorities in Victoria have purchased allocations and/or entitlements in the past. For example, in 2008–2009, SA Water bought 160,000 ML of water, and between 2006 and 2009 Coliban Water (Victoria) bought 28,000 ML to alleviate water restrictions in Bendigo. The wetter conditions since 2010 have seen many authorities reducing their purchases, or not purchasing at all (NWC 2012).

10.6 Increasing Federal Water Market Intervention

Since the implementation of the NWI there have been a range of policy and legislative changes that have attempted to ensure increased state collaboration and more coordinated management of the MDB. From 2004 onward several state- and federal-based recovery programs operated to recover environmental water in the MDB. The first was the 2004 inter-governmental agreement on addressing over-allocation and achieving environmental objectives in the MDB which assigned AUD$500 million to secure 500 GL of water for key environmental assets in the sMDB.

Following an extended period of drought and low water supply, the Howard government released its *National Plan for Water Security* in early 2007 (Howard 2007). A new Commonwealth *Water Act* (2007) was also introduced, establishing and enforcing environmentally sustainable limits on water extraction, and creating the impetus for a Basin-wide environmental watering plan. New administrative bodies such as the Murray-Darling Basin Authority (MDBA) replaced the MDB Commission while the Commonwealth Environmental Water Holder was also created to manage acquired consumptive water and thereby increase environmental flows (Australian Parliament 2007).

The *Water Act* 2007 gathered disparate state water planning and trade mechanisms into a single over-arching framework requiring Commonwealth approval and deferral where conflict arose. In March 2008 the Commonwealth announced a new policy called *Water for the Future* (WFF). WFF encompassed an AUD$12.9

billion investment over 10 years to 2018–2019, and placed significant emphasis on the role of water markets to recover water for the environment. Beginning in 2007–2008, the Commonwealth (as part of WFF) directly entered water entitlement markets to secure environmental water from willing sellers under its AUD$3.1 billion *Restoring the Balance* (RtB) program, in order to meet MDB plan targets (Crase and O'Keefe 2009). Additional funding of up to AUD$310 million per annum for 2014–2015 has been provided to bridge any remaining gap between the level of water returned to the MDB under existing WFF initiatives and the level required to be returned under the final MDB Plan. Table 10.1 illustrates the water buybacks that have taken place in Australia since the early 2000s. These buybacks represent the largest purchase of entitlements in the world.

In November 2012, the federal Parliament passed the MDB Plan into law confirming a target for sustainable diversion that limited recovery volumes to 2,750GL. This outcome is to be achieved by 2019 with a performance (and target) review process scheduled for 2015 (MDBA 2012b). As of December 31 2013, the RtB program had recovered water entitlements with a long term average yield of 1,138GL for environmental sites in the MDB (DSEWPaC 2013a) (see Table 10.1). The recovery effort will be assisted by the provision of an additional AUD$1.77 billion to achieve strategic buyback within the MDB and to overcome environmental water delivery constraints along the river systems out to 2024 (Wroe 2012). Importantly, the Basin Plan also provides significant opportunity for improving water trade rules, operations and integration in the sMDB. Specifically, new requirements on water trading rules in the plan seek to enhance trade in groundwater entitlements, trade in delivery rights, clearer definitions of water user/service provider rights, delivery and reporting obligations, and allocation announcements. See DSEWPaC (2013b) for more detail on the range of trade-based rules and changes.

10.7 What Have Been the Benefits and Costs of Water Markets?

As a result of this lengthy process of developing, implementing and improving water markets in Australia one might well ask has the significant time, trouble and expense been worthwhile? To answer that, we now examine the benefits and costs of water markets in Australia.

10.7.1 Economic Water Trading Impacts

Young (2013) argues that the greatest cost of water markets in Australia has been the failure of governments to establish clear entitlements and ownership before the development of water markets. As a result, the establishment of water markets in

Table 10.1 Water buy-backs in Australia (as at December 2013)

Program	Year	Purchaser	Purchase mechanism	Water acquired (LTAAY GL[a])	Nominal cost (AUD)
Commonwealth					
The living Murray: pilot environ. water	2004–2009	MDBA	Competitive EOI tender	13.3	\$21.1 M
Restoring the balance	2007–present	Cwlth	Competitive EOI tenders	1,119.2	\$3.1B over 10 years
The living Murray: MDBA purchases	2009	MDBA	Single round EOI tender	18.6	\$50 M budgeted
State governments					
The living Murray: NSW market purchase measure	2004–2009	NSW RiverBank	Competitive EOI[b] yearly tenders	115.3	\$192 M
NSW wetland recovery project	2005–2006	NSW RiverBank	Negotiated on-market purchases	9.4	\$26.8 M
Rivers environmental restoration	2005–present	NSW RiverBank	Competitive EOI yearly tenders and market purchases	113.2	\$181 M
The living Murray: SA water purchases	2006–2009	SA Water, (agent for SA Government)	Purchases from willing sellers (no tender process)	36	\$49.5 M
Victorian stream flow tender	2007	VIC Govt.	Tender for license surrender, reduced license volume, or altered seasonal access	55ML surrendered; 664ML to stream flow mgt plans	No data available

The Snowy river inquiry outcomes implementation deed 2002	2007–2008	Water for rivers	On-market purchases	84	$35.3 M
Narran lakes environmental water purchase	2008	MDBC (NSW DECCW and QLD DERM)	One-time purchase negotiated with a willing seller	10.4	$1.88 M
NGO					
Healthy rivers Australia		Healthy Rivers Australia	On-market allocation purchases and donations	0.5 ML permanent rights; temp. purchases	No data available
Total				**1,519.8**	**$3.6B**[c]

Source: Adapted from Lane-Miller et al. (2013)

[a]The Long Term Average Annual Yield (LTAAY) is the estimated average allocation that would have accrued to a given water entitlement based on historical climate and inflow data from 1891 to 2003

[b]Expressions of interest

[c]This total should be interpreted with care because these values include administrative costs for programs, and are not a complete representation of the cost of all water buybacks. What is most notable here is that the planned expenditure for RtB dwarfs any of the other water buybacks

Australia activated un- and/or under-used water access entitlements, which led to greater water use and the need to spend billions to recover water for the environment (Table 10.1).

Qureshi et al. (2009) argue that water trading will increase economic efficiency because market prices signal the opportunity cost of water explicitly. They also provide incentives to adopt water-saving technologies and reduce inefficient uses of water. In support of this, Peterson et al. (2004) estimated the gains from trade in a dry year at AUD$495 million, while the NWC (2011b) suggested that water trading in the sMDB increased Australia's gross domestic product by AUD$220 million in 2008–2009. Qureshi et al. (2009) similarly found that a reduction in water market barriers in the sMDB would increase annual net returns significantly.

The positive impacts of water trading on the value of agricultural production are difficult to estimate with precision. However, between 2005–2006 and 2008–2009 (drought years) while water use across the MDB fell by 53 %, the gross value of irrigated agricultural production (GVIAP) fell by only 27 %; suggesting that the capacity to trade water supported agricultural production. Computable general equilibrium modeling suggests that in the sMDB intra- and inter-regional trading contributed between AUD$270 and AUD$370 million to agricultural production during the dry years of 2007–2008 and 2008–2009 (Wittwer and Griffith 2011). Trade of water into SA directed water to higher value users, generating a net benefit of AUD$35 million and ameliorating the general economic impacts of low water allocations. Production in the sMDB was estimated to be AUD$845 million higher than it would have been in the absence of water trading (NWC 2012).

10.7.2 Social Water Trading Impacts

There remains much debate about the negative social impacts from water trade. Concerns include: vulnerability from water transfers away from a regional area; reduced community spending and reinvestment; population losses as farmers and/or younger people elect to seek employment elsewhere; consequent employment opportunity reductions; declining taxation base, loss of local services and businesses, regional production changes (e.g. shifts to dry-land farming); and legacy issues for the remaining farmers (e.g. higher variable farm operating costs, stranded asset problems and/or pressure to rationalize marginal operations) (Australian Parliament 2011). However, many such social issues in the MDB predate the introduction of water markets, and are predominantly associated with ongoing rural structural change. Research has suggested that the rural communities that are most vulnerable are smaller irrigation-dependent towns (EBC et al. 2011). Several studies have attempted to identify links between water trading and observed community changes (e.g. Fenton 2006; Edwards et al. 2008). However, while they all identified the impacts mentioned above, they failed to separate the impact of trade from the impact of other structural change. In contrast, it has been argued that trading has allowed irrigators and communities to better deal with the consequences of prolonged drought and policy pressures (Bjornlund et al. 2011).

10.7.3 Environmental Costs of Water Trading

Water trade is often argued to have environmental costs; including changes to the location and timing of water use which can affect river hydrology and environmental outcomes (NWC 2012). The recent rapid increase in water trade activity has caused some to question the impacts of large resource reallocation on sMDB environmental conditions and outcomes. Since the late 1960s, it has become apparent that the health of the MDB has been seriously threatened by over-allocation. As mentioned previously, during the implementation of the MDB cap, many water users with un- or under-used water access entitlements were motivated to trade them, and this seriously compromised storage management leading to the continual over-allocation of water in the MDB (Young 2013).

Aside from this historical impact, on balance the evidence suggests that water markets have reallocated resources without significant environmental consequences from increased water use (NWC 2012). Modeling suggests that the hydrological and environmental impact of water trade between 1998–1999 and 2010–2011 was relatively small and positive in terms of their environmental impacts. This was largely due to water moving downstream during the drought, reducing summer flow stress, and creating no change to winter flow patterns. Negative outcomes from trade may eventuate in cases where water trade results in a detrimental change to the volume, location and/or timing of water use. But modeling suggests environmental benefits from water trade are likely to be greater in drier rather than wetter periods (NWC 2012). That is, if predictions of the effects of climate change are accurate then increased river flows in the sMDB from water trading in the future will most likely benefit the environment.

To date, most of the policy focus on water recovery has concentrated on volumes of water access entitlements as the sole environmental metric. However, such a focus on entitlement volume ignores the nuances involved with environmental water management and the non-linearity between volume and environmental outcomes (Crase et al. 2011). For example, water management is not just dependent upon volume; it is also dependent upon timing (i.e. across seasons and across years). In addition, there has been a call for increased use of alternative derivative water products (such as water options) to be used in buyback to increase the flexibility and cost-effectiveness of the program (Productivity Commission 2010; Wheeler et al. 2013). Efficiency increases also contribute to over-allocation because they decrease returns to rivers (Adamson and Loch 2014). Young (2013) also argues that as efficiency improves, allocations should be commensurately reduced.

Concerns have been expressed that while Commonwealth purchases may look impressive on paper, the volume of water delivered to the environment is, or is likely to be, relatively small. The NWC reports that less than 50 % of water available to the Commonwealth Environmental Water Holder was delivered in the sMDB (NWC 2012). The Productivity Commission (2010) found that buyback may not help sites where additional flows are needed (NWC 2011a). Not being able to supply environmental water to the places that need it, at the times it is needed, may provide motives for increased water trade by the CEWH, and the introduction of more sophisticated derivative products providing more flexible ways for irrigators and the CEWH to manage the allocation of annual supply between environmental and productive uses (Loch et al. 2013).

10.7.4 Further Lessons for Water Trading

Australia has made a choice that, generally, irrigators' rights to water are paramount. Historically, this has meant that sleeper and dozer rights were recognized when the Cap was introduced in the 1990s and irrigators who had never (or rarely) used their water could sell their rights on the market. These irrigator rights continued with large-scale water buyback starting from 2007 to 2008, and where voluntary acquisition was the government's objective, prices above the market were paid. At the same time, billions of dollars have been directed toward irrigation infrastructure upgrades. The acquisition of these water access entitlements also continues to ignore the fact that irrigators historically do not use all their water they are allocated, and that some of this unused water was supplementing annual river flows anyway (which is the case for South Australia, but for NSW and Victorian irrigators' unused or unallocated water remains in storage). There is also the issue, which needs further research, that groundwater use can be a substitute for some surface water use. Given that there historically has been a lack of regulation over groundwater use in the MDB, many irrigators have sold their surface water entitlements and increasingly turned to using groundwater as a substitute.

As Australian water markets mature, expansion of existing modes and products for trade will eventuate similar to other market arrangements. As trade assists water users to manage risk one such expansion may see intertemporal trade of water increase, similar to that in pollution permit markets (Yates and Cronshaw 2001). Where water users can bank (carry-over) water for future use, intertemporal trade may allow that water to be transferred between users across water years, subject to accounting requirements (e.g. a deduction for inter-period evaporation). There is also increasing pressure in Australia to allow water users to trade in derivative products (e.g. forward contracts), where uncertainties about intertemporal allocation and supply can be hedged via contracting between parties ahead of those events (Leroux and Crase 2010). Since recent regulatory changes have provided only limited capacity to write intertemporal trade agreements in Australian water markets, the benefits of such arrangements are currently being investigated.

Finally, as evidenced in this chapter, there has been considerable reform in water policy in Australia. It is obvious that Australia, and in particular the pace of reform water policy in the MDB, has left many policy makers and water managers in a period of fatigue. Considerable change is still needed to just implement the water market trading changes of the Basin Plan, let alone consider further changes. It is likely that to enable further state, industry and federal agreement on water issues will most likely require the pressure of another severe drought. MDB history shows that droughts are invariably the trigger for new water policy reform.

10.8 Conclusion

Since the introduction of water markets in Australia over three decades ago, water access entitlement and allocation trade have been increasingly adopted by private individuals, organizations and governments. Irrigators use water markets to manage water supply scarcity risks and demand requirements, and to facilitate farm restructuring and exit. Governments use water markets to correct the consequences of past over-allocation decisions to provide and/or maintain key strategic environmental outcomes in the MDB. It is expected that further adoption of water markets, as well as improved water management and use, will be essential for coping with future climate change impacts.

Although the initial introduction of water markets in Australia activated water licenses that had never been used—and there is an argument that if policy had set the consumptive limits correctly in the beginning many costly policy initiatives such as *Water for the Future* would not have been required—the economic benefits of water trade are quite clear, offering significant potential for assisting water users to adapt to climate change and ongoing water scarcity. That said, there is scope for further water market improvement in the areas of trade rules; trade processing timeframes; trade product expansion into derivatives; improvements in the depth and access of market information; and reductions in transaction costs associated with water trade activity. Further, community concerns remain about the impacts of water trade, although it is hard to disentangle many of these impacts from the ongoing general structural change in Australian agriculture.

The ongoing challenge of climate change and water scarcity management means that there needs to be continuing adjustments to policy, institutional and governance arrangements in the water sector to deal with such issues. Further flexibility and adaption will allow water users to continue to cope with the ongoing challenges of farming in Australia.

References

ACCC (2010) Water trading rules: final advice. Australian Competition and Consumer Commission, Canberra

Adamson D, Loch A (2014) Possible negative feedbacks from 'gold-plating' irrigation infrastructure. Agric Water Manage. doi:10.1016/j.agwat.2013.09.022

ARMCANZ (1996) National principles for the provision of water for ecosystems. Occasional paper SWR No. 3, Department of Environment, Sport and Territories, Canberra

Australian Parliament (2007) Water act 2007. Commonwealth of Australia, Canberra
Australian Parliament (2011) Of drought and flooding rains: inquiry into the impact of the guide to the Murray-Darling Basin plan. House of Representatives Standing Committee on Regional Australia, Canberra
Bjornlund H (2003) Efficient water market mechanisms to cope with water scarcity. Int J Water Resour Dev 19:553–567
Bjornlund H, McKay J (2000) Are water markets achieving a more sustainable water use. In: Proceedings from the 10th world water congress, Melbourne, 12–17 March
Bjornlund H, McKay J (2003) Elements of an institutional framework for management of water for poverty alleviation in developing countries. In: Ünver O, Gupta R, Kibaroğlu A (eds) Water development for poverty reduction. Kluwer Academic Publishers, Boston
Bjornlund H, O'Callaghan B (2003) Property implications of the separation of land and water rights, 9th Annual Pacific-Rim Real Estate Society conference, Brisbane, 19–22 January
Bjornlund H, Wheeler S, Cheesman J (2011) Irrigators, water trading, the environment, and debt: perspectives and realities of buying water entitlements for the environment. In: Grafton Q, Connell D (eds) Basin futures: water reform in the Murray-Darling Basin. ANU Press, Canberra
Connell D, Grafton Q (2011) Water reform in the Murray-Darling basin. Water Resour Res 47(W00G03). http://dx.doi.org/10.1029/2010WR009820
Crase L (2008) An introduction to Australian water policy. In: Crase L (ed) Water policy in Australia: the impact of change and uncertainty. Resources for the Future, Washington, DC
Crase L, O'Keefe S (2009) The paradox of national water savings: a critique of 'water for the future'. Agenda 16:45–60
Crase L, O'Keefe S, Dollery B (2011) Presumptions of linearity and faith in the power of centralised decision-making: two challenges to the efficient management of environmental water in Australia, Western Economic Association international conference, Brisbane, 26–29 April
Davidson B (1969) Australia wet or dry? The physical and economic limits to expansion of irrigation. Melbourne University Press, Victoria
DSEWPaC (2013a) Progress of water recovery against proposed 2750 GL SDL reduction [Online], Canberra, Department of sustainability, water, environment, population and communities. Available at: http://www.environment.gov.au/water/policy-programs/entitlement-purchasing/progress.html. Accessed 16 Apr 2013
DSEWPaC (2013b) Department of sustainability, environment, water, population and communities. Water charge and water market rules. www.environment.gov.au/water/australia/water-act/rules.html. Accessed 28 July 2013
EBC, RMCG, Marsden Jacob Associates, Econsearch, McLeod G, Cummins T, Guy R, Cornish D (2011) Community impacts of the guide to the proposed Murray-Darling basin plan: report to the Murray-Darling Basin Authority. Murray-Darling Basin Authority, Canberra
Edwards J, Cheers B, Bjornlund H (2008) Social, economic and community impacts of water markets in Australia's Murray-Darling Basin region. Int J Interdiscip Soc Sci 2:1–10
Fenton M (2006) The social implications of permanent water trading in the Loddon-Campaspe irrigation region of northern Victoria. North Central Catchment Management Authority, Huntly
Grafton RQ, Libecap G, McGlennon S, Landry C, O'Brien B (2011) An integrated assessment of water markets: a cross-country comparison. Rev Environ Econ Policy 5:219–239
Hamstead M, Baldwin C, O'Keefe V (2008) Water allocation planning in Australia – current practices and lessons learned, waterlines occasional paper no 6. National Water Commission, Canberra
Hearne R, Easter W (1995) Water allocation and water markets: an analysis of gains-from-trade in Chile, World Bank technical paper 315, Washington DC
Howard J (2007) The national plan for water security. Prime Minister & Cabinet, Canberra
Lane-Miller C, Wheeler S, Bjornlund H, Connor J (2013) Acquiring water for the environment: lessons from natural resources management. J Environ Policy Plan 15:513–532
Leroux A, Crase L (2010) Advancing water trade: a preliminary investigation of urban irrigation options contracts in the Ovens Basin, Victoria, Australia. Econ Pap 29:251–266

Loch A, Bjornlund H, Wheeler S, Connor J (2012) Allocation trade in Australia: a qualitative understanding of irrigator motives and behaviour. Aust J Agr Resour Econ 56:42–60
Loch A, Wheeler S, Bjornlund H, Beecham S, Edwards J, Zuo A, Shanahan M (2013) The role of water markets in climate change adaptation. National Climate Change Adaptation Research Facility, Gold Coast
McKay J (2001) The modern Australian history of corporatisation and privatisation of water supplies and irrigation – lessons for regulators. In: van der Wel B (ed) Water history: lessons for the future. Hydrological Society of South Australia, Adelaide, pp 87–96
MDBA (2012a) Water inflows and significant flooding/drought events. Murray-Darling Basin Authority, Canberra
MDBA (2012b) Water act 2007 – the Basin Plan. Commonwealth of Australia, Canberra
MDBC (2007) A brief history of the Murray-Darling Basin agreement. Murray-Darling Basin Commission, Canberra
NSW Office of Water (2011) Water reform in the NSW Murray-Darling Basin – summary of regional water reform and environmental water recovery in NSW 1996–2011. NSW Department of Primary Industries – Office of Water, Sydney
NWC (2011a) Australian water markets: trends and drivers, 2007–08 to 2009–10. National Water Commission, Canberra
NWC (2011b) Australian water markets: trends and drivers, 2007–08 to 2010–11. National Water Commission, Canberra
NWC (2011c) Strengthening Australia's water markets. National Water Commission, Canberra
NWC (2011d) Water markets in Australia: a short history. National Water Commission, Canberra
NWC (2012) Impacts of water trading in the southern Murray-Darling Basin between 2006–07 and 2010–11. National Water Commission, Canberra
NWC (2013) Australian water markets: trends and drivers 2007–08 to 2011–12. National Water Commission, Canberra
Papas M (2007) The proposed governance framework for the Murray-Darling Basin. Macquarie J Int Comp Environ Law 4:77–90
Peterson D, Dwyer G, Appels D, Fry J (2004) Modelling water trade in the southern Murray-Darling basin, productivity commission staff working paper, productivity commission, Melbourne
Productivity Commission (2010) Market mechanisms for recovering water in the Murray-Darling Basin. Productivity Commission, Canberra
Qureshi ME, Shi T, Qureshi SE, Proctor W (2009) Removing barriers to facilitate efficient water markets in the Murray-Darling Basin of Australia. Agric Water Manage 96:1641–1651
Schlager E, Ostrom E (1992) Property-rights regimes and natural resources: a conceptual analysis. Land Econ 68:249–262
Sitarz D (1993) Agenda 21: the Earth summit strategy to save our planet. Earth Press, Boulder
Smith R (1981) Resolving the tragedy of the commons by creating private property rights in wildlife. Cato J 1:439–468
Tisdell J (2014) A short history of the evolution of water management in Australia. In: Easter KW, Huang Q (eds) Water markets for the 21st century: what we have learned. Westview press
Wheeler S, Cheesman J (2013) Key findings from a survey of sellers to the restoring the balance programme. Econ Pap 32:340–352
Wheeler S, Bjornlund H, Shanahan M, Zuo A (2008) Price elasticity of water allocation demand in the Goulburn-Murray irrigation district. Aust J Agric Resour Econ 52:37–55
Wheeler S, Bjornlund H, Shanahan M, Zuo A (2009) Who trades water allocations? Evidence of the characteristics of early adopters in the Goulburn-Murray irrigation district, Australia, 1998–99. Agric Econ 40:631–643
Wheeler S, Bjornlund H, Zuo A, Shanahan M (2010) The changing profile of water traders in the Goulburn-Murray Irrigation district, Australia. Agric Water Manage 97:1333–1343
Wheeler S, Zuo A, Bjornlund H, Lane-Miller C (2012) Selling the farm silver? Understanding water sales to the Australian government. Environ Resour Econ 52:133–154

Wheeler S, Garrick D, Loch A, Bjornlund H (2013) Evaluating water market products to acquire water for the environment in Australia. Land Use Policy 30:427–436

Wittwer G, Griffith M (2011) Modelling drought and recovery in the southern Murray-Darling basin. Aust J Agric Resour Econ 55:342–359

Working Group on Water Resources Policy (1994) Report of the working group on water resources policy to the Council of Australian Governments, Working Group on Water Resource Policy, Canberra

Wroe D (2012) Gillard pumps $1.7 billion into Murray-Darling [Online], Canberra, Fairfax Media Group. Available: http://www.yasstribune.com.au/story/423400/gillard-pumps-17b-into-murray-darling/. Accessed 30 Oct 2012

Yates A, Cronshaw M (2001) Pollution permit markets with intertemporal trading and asymmetric information. J Environ Econ Manage 42:104–118

Young M (2013) Trading into and out of trouble: Australia's water allocation and trading experience. In: Maestu J (ed) Water trading & global water scarcity: international perspectives. RFF Press/Taylor & Francis/Routledge, New York, pp 94–110

Chapter 11
Trading into Trouble? Lessons from Australia's Mistakes in Water Policy Reform Sequencing

Mike Young

Abstract The changes in land-use practice and investment that flow from the modification of an abstraction regime to allow water trading can bring significant economic gains. If these gains from trade are to be unequivocally beneficial to all members of society and to the environment simultaneous reform of the abstraction regime may be necessary. In particular, it is critical to understand how trading will affect return flows, the capture of overland flows and abstraction from connected water resources.

Failure to attend to the sequence of reforms needed to establish a robust abstraction regime capable of sustaining the pressures from trade can be very expensive. In retrospect, it can be argued that Australia got its water reform sequence wrong. As a result and unnecessarily, she had to spend billions of dollars restoring balance to the Murray Darling Basin. The cost to society of restoring balance to abstraction arrangements appears to be greater than the benefits that flowed from the rapid development of water trading. Those who recommend a transformational change to a policy regime have a responsibility to consider the system-wide consequences of adopting the change they recommend.

Keywords Water trading • Sequencing • Abstraction regime • Cost benefit • Policy reform • Transformational

---

M. Young (✉)
Water and Environmental Policy, Global Food Studies Program, The University of Adelaide, Adelaide, SA 5005, Australia

Gough Whitlam and Malcolm Fraser Chair in Australian Studies, Harvard University, Cambridge, MA, USA
e-mail: mike.young@adelaide.edu.au; myoung@seas.harvard.edu

K.W. Easter and Q. Huang (eds.), *Water Markets for the 21st Century: What Have We Learned?*, Global Issues in Water Policy 11, DOI 10.1007/978-94-017-9081-9_11,

11.1 Introduction

Many years ago Australia was described by the poet Dorothea Mackellar as a "land of droughts and flooding rains".[1] In response, the country has developed a vibrant irrigation industry. One of the most commonly stated reasons for this vibrancy is the way that Australia has allowed its water sharing agreements to evolve, reformed its entitlement and allocation regimes and allowed water users to trade water with one another. Australia's most cost-effective water sharing and associated trading regime can be found in its Southern Connected River Murray system.

Transition from a traditional regulatory water-abstraction regime[2] to one that attempts to take full advantage of bottom-up market process has not been easy. Three parallel processes have been critical to the transformational policy changes that have occurred in Australia. The first was the development of a robust water sharing regime at the Basin level. The second has been the conversion of licenses to take water into entitlements and the separation of these licenses from land titles. The third has been the development of water accounting, entitlement registration and administrative protocols that have made low transaction cost water trading possible.

The reform process began in the 1960s with a few cautious trades that tested the feasibility of moving the location of water use without harming third parties. Water trading expanded gradually during the 1970s and 1980s as irrigators and administrators began to realize that that there was an economic case for allowing users to trade water with one another. Trading took off in 1996 following development of a National Competition Policy and introduction of a "cap" on surface water use in the Murray Darling Basin (MDB). The so-called Murray Darling "cap" placed a limit on the maximum volume of surface water that could be diverted from any part of the system in any 1 year. The National Competition Policy required all Australian governments to make it possible for anyone to invest in water and in effect be able to buy and hold a water entitlement without having any intention to use it. This, it was argued and subsequently demonstrated, would drive innovation and encourage the transfer of water to places where it would make the greatest contribution to the economy (Young 2012).

With the onset of a long dry spell at the start of this century and collective agreement to a National Water Initiative in 2004, water trading in the Murray Darling Basin became the norm. This was followed, in 2007, by the development of national legislation that enabled Australia's Federal Government to establish an independent Murray Darling Basin Authority and, in 2012, use a Basin Plan to set a suite of sustainable diversion limits (Australia 2007, 2012). Those interested in the

---

[1] See http://www.dorotheamackellar.com.au/archive/mycountry.htm

[2] Throughout the world the terms used to describe the regimes used to manage water vary considerably. In this paper, the term "abstraction regime" is used to describe the full suite of institutional arrangements used to control the amount of water that may be taken from a water body and how it may be used.

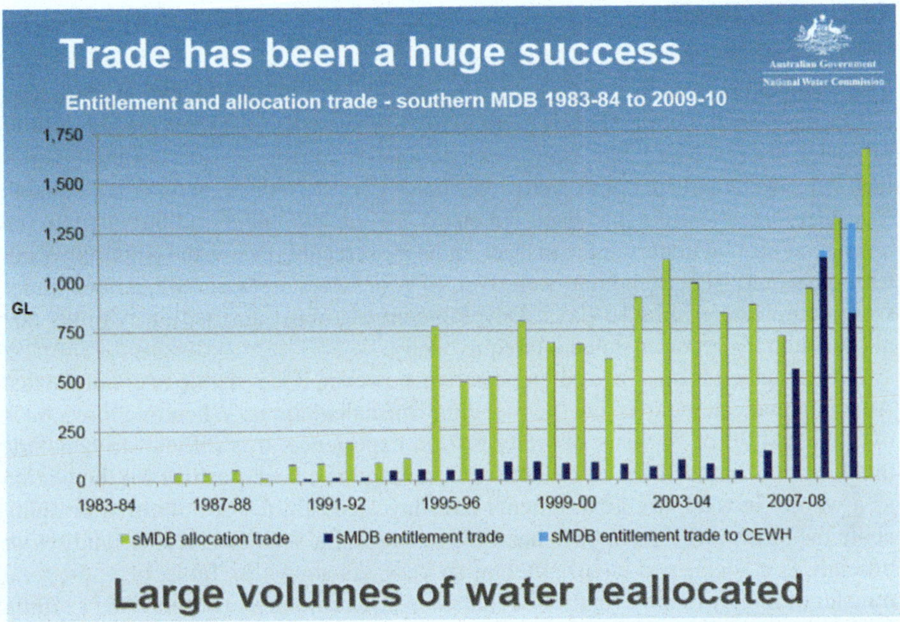

Fig. 11.1 National Water Commission slide presented to Australian Competition and Consumer Commission conference, Brisbane, July 2011 (Available at http://www.accc.gov.au/system/files/Presentation%20by%20Chloe%20Munro.pdf

detail should read "A short history of the development of water trading in Australia" (NWC 2012).[3] Skinner and Langford (2013) provide a detailed description of the reform journey taken in the Murray Darling Basin.

At conferences throughout the world Australian officials tend to present water trading, especially in the Nation's Southern Connected River Murray System, as an "outstanding success!" (see Fig. 11.1). Economists too have been keen to present the advantages that trading has brought to Australia. For example, Grafton et al. (2012) have drawn attention to the fact that water trading enabled the impact of the drought to be much less than it otherwise would have been. The late Jim McColl and I have made similar observations (see, for example, Young and McColl 2005). Peterson et al. (2004) have estimated that the gains from water trading in a dry year to be in the vicinity of $495 million. Water use per hectare has been halved and there has been much innovation (Young 2012). The internal rate of return on holding a water entitlement over the first decade after trading was opened up was well over 15 % per annum (Bjornlund and Rossini 2007). In dry years, as much as 40 % of the water allocated to irrigators is traded (NWC 2008, 2009, 2010, 2012).

[3]Available at http://www.nwc.gov.au/__data/assets/pdf_file/0004/18958/Water-markets-in-Australia-a-short-history.pdf

11.2 The Reality

The argument presented in this chapter is different. Yes, the development of trading, per se, has been a success. This is particularly true when viewed from the perspective of water users. There has been considerable innovation, water use efficiency has improved markedly and there has been considerable structural adjustment. Considerable attention has been given to the development of the institutional arrangements that underpin the market. In its most recent review, the National Water Commission (2012) found that well over 90 % of interstate water allocations trades were completed within 10 days. Development of robust abstraction regimes that allow trading is recommended – unequivocally.

When viewed from a broader perspective, however, it is possible to conclude that Australia made some massive "sequencing" miscalculations. When one looks back over the last 20 or so years of water reform experience, it is difficult to conclude that the net benefits of the introduction of trading have been positive for the Nation as a whole. In total, the governments have invested billions of dollars in bringing about the reforms that have been heralded by Australia's political leaders and senior officials as a social and environmental success. As shown in Table 11.1, the total transfer of money to the irrigation sector since 2012 amounts to around A$11 billion in both nominal and real terms or around A$3,500 per ML.

When distributed among the 14,340 surface and groundwater irrigators in the Murray Darling Basin in 2012, the government investment required to restore health to the Murray Darling Basin has required a transfer payment from taxpayers to the irrigation sector of around A$750,000 per irrigator for an asset which, at the time the reform process began in 1993/1994 was issued only for a short period of time and definitely was not a perpetual right[4] to a volume of water.[5] From this perspective, the picture looks a little different. Yes, the introduction of water trading has been to the massive benefit of irrigators and irrigation communities but when one looks at the cost of the investment required to establish the administrative regime one must ask whether or not there was a better reform pathway and if that pathway could have been followed.

The argument put forward in this chapter is that those advocating water reform need to consider the benefits of first putting in place a robust abstraction regime. Robust abstraction regimes withstand the test of time. In particular, they are designed to work elegantly when a system is under stress such as during a severe

---

[4]The long-term component of many water licenses were converted into perpetual shares after states agreed to the National Water Initiative in 2004 (COAG 2004).

[5]Number of farmers irrigating in the MDB in 2011. At the time of writing, one Australian dollar is roughly equivalent to one US Dollar. Conversion from nominal to 2012 dollars is approximate as details on actual expenditure is only available for a range of years. Discounting for future years based on an assumption of 2 % per annum inflation. Entitlements were converted into entitlements in perpetuity following completion of the National Water Initiative in 2004 (COAG 2004).

Table 11.1 MDB programs to improve the environment by reducing allocations to the irrigation sector via entitlement purchase, infrastructure savings and regime reform (Nominal $ and $2012 assuming 2 % per annum inflation)

Year	Program	Water returning to the environment (GL)	Transfers to irrigation sector (Millions Australian $)	
			Nominal $	$2012
1987	Victorian Murray Wetlands Environmental Water Agreement	27.6	–[a]	–[a]
1993	Barmah-Millewa Forest Environmental Water Agreement	100–150	–[a]	–[a]
1998	Murrumbidgee Environmental Contingency Allowance	25–200	–[a]	–[a]
2000	Murray Additional Environmental Water Allowance	5.4	–[a]	–[a]
2000	NSW Murray Wetlands Environmental Water Agreement	30	–[a]	–[a]
2002	2002 Lower Darling River Environmental Contingency Allowance	30	–[a]	–[a]
2004–2009	The Living Murray Initiative	487	$500 + $150[b]	$756[b]
2004–2009	Cap to NSW Water Sharing Plans	206	–[a]	–[a]
2004–2009	Other State Recovery Programs	77	–[a]	–[a]
2004–2012	Water for Rivers	55	–[a]	–[a]
2009	Water gifted from Queensland to the Commonwealth	11	–[a]	–[a]
2009	NSW Wetlands Recovery Program	4	–[a]	–[a]
2009–2012	NSW Riverbank Program	41	$105	$108
2009–2012	NSW Rivers Environmental Recovery Program	14	–[a]	–[a]
2013–2019	Sustainable Rural Water Use and Infrastructure Program	75	–[a]	–[a]
2013–2019	Restoring the Balance as at 30 Sept 2012	1,094 + 1264	$3,000 + $5,900[c]	$8,675[c]

(continued)

Table 11.1 (continued)

Year	Program	Water returning to the environment (GL)	Transfers to irrigation sector (Millions Australian $)	
			Nominal $	$2012
2019–2024	Restoring the Balance extension announced 26 October 2012	450	$1,700	$1,411
Total		6,118	A$11,455	A$10,590

Source: Volumetric data adapted from Skinner and Langford (2013) and updated. Program expenditure information from government statements

[a]Agreement or change to an administrative arrangement. No direct transfer of money to the irrigation sector involved unless one concludes that the irrigation industry should pay for these improvements

[b]A$500m for the core program plus an additional A$150m for a complementary environmental works and measures program

[c]A$3000m for the purchase of entitlements and A$5,900m for investment in "savings" projects which through improvements in water use efficiency enable the transfer of water entitlements to the environment without any adverse effect on productivity

drought (Young and McColl, 2005). When the reform process began, irrigators held entitlements only to the amount of water that could be taken sustainably from the river and the groundwater systems associated with it. They had no guarantee that the volumes of water they had become accustomed to taking would continue to be available and they had no entitlement to compensation if the amounts they were allowed to take was reduced (Fisher 2006).

A full cost benefit analysis of the celebrated Australian reforms has yet to be conducted. As indicated by the above observations, however, the result of a full cost-benefit analysis would most likely not be a positive one. Yes, the value of irrigation entitlements has risen and trading has brought significant increases in regional income but it required a massive transfer of capital from the rest of the nation. Examining the benefits to the nation of fixing the problem and including an estimate of the value of the resultant environmental improvement, Grafton et al. (2011) observe that "from 2001 to 2009 a water allocation that would have given less to irrigated agriculture and more to environmental flows would have generated between half a billion and over 3 billion U.S dollars in overall economic benefits.

Two very important caveats need to be made. The lessons that have emerged from the Australian experience in water reform are proving to be extremely valuable to other nations. Moreover, it is possible that no cheaper reform pathway may have been politically feasible. As Marshall et al. (2013) observe, Australia's Federal Governments used the very large transfer payments to State Governments and to irrigators to keep them from obstructing the reform process.

11.3 Sequencing the Reform Process

When trading in the Southern Connected Murray Darling Basin took off in the 1995/1996 irrigation system (see Fig. 11.1) the cap was deemed to be an "interim" cap. Those responsible for administration of water use in the Murray Darling Basin knew that the system was already over-allocated. Those advocating the expansion of trading were either a) not worried about the declining environmental health of this internationally significant river system or b) not concerned about the adverse effects that the introduction of water trading would impose on its environment.

In retrospect, this failure to think through the implications of introducing a trading without simultaneously fixing flaws in the water allocation regime is surprising. The clues were there but economists and policy makers seemed to be blind to them. "In the Darling River in 1991, a toxic bloom of blue-green algae occurred over a distance of 1,000 km and caused the New South Wales government to declare a state of emergency."[6] Moreover, in July 1998, when governments agreed to finalize the Cap, they set it at the "volume that would have been diverted under 1993–1994 levels of development."[7] Once again, the language is clear. Administrators were worried about the implications of allowing "further development."

In November 2002 – before the long drought emerged – governments put first one and then two and then three dredges in the mouth of the River Murray. Despite all the agreements and plans being made the River Murray had stopped flowing to the sea! Murray Darling Basin Officials were warning that:

- by 2020, unless significant action is taken, it is expected that River Murray salinity at Morgan will fail to meet the World Health Organization's desirable drinking water standards over 50 % of the time (MDBC 1999); and
- between 20 and 40 % of irrigation water needs to be returned to the stem of the River Murray so that it can be restored to a healthy working river (Murray-Darling Basin Ministerial Council 2002).

Around 2000, concerned ecologists were attempting to work out how to fix the problem and soon came to the conclusion that 1,500 GL of water entitlements being used by irrigators had to be returned to the Southern Connected River Murray System. Officials, however, were unable to agree on this action and, in 2003, they agreed to a Living Murray program that would recover 500 GL of water as a first step to restoring this system to health (Skinner and Langford 2013).[8]

[6] See http://www.mdba.gov.au/river-data/water-quality/bga

[7] See http://www.mdba.gov.au/sites/default/files/archived/cap/Striking_the_Balance_Report_97_98.pdf

[8] The Murray Darling Basin Plan has since identified a need to reduce entitlements by a further 3,200 GL.

11.4 Water-Use Efficiency and Return Flows

Water trading is a pretty simple concept. Every irrigator is given a water allocation and told that if they don't want to use this allocation they can sell "their" allocation to someone else. With the introduction of trading, water soon became valuable and it did not take long for irrigators to decide to become more efficient and to seek new sources of water. As a result, in the 5 years following the opening up of the water market in 1995/1996 groundwater and surface water use went up by 25 % (Bryan and Marvanek 2004). This increase was not due to a breach of the cap. It was due to the ways that the introduction of water trading changed land-use practice and investment in irrigation technology.

First and foremost, irrigators quickly began to invest in more efficient irrigation practices. The problem was that, as these irrigators became more efficient, less water flowed back to the river and, as a result, less water is available for others to use, for conveyance and for the environment. Return flows, as they are called, went down and use went up.

This reduction in return flows would have had no adverse impact on river and aquifer health if the amount of water allocated to irrigators was reduced as fast as the efficiency of water use increased. But, in the Murray Darling Basin no such policy regime existed. A double counting problem existed (Young and McColl 2003, 2008; Qureshi et al. 2010). In fact, to this day this flaw in the allocation regime remains. The current Murray Darling Basin Plan still assumes that return flows will not decline as the technical efficiency of irrigated water use increases.

The new Murray-Darling Basin Plan (Australia 2012) does, however, require the management of the second source of flow reduction. If you are a flood irrigator and you start investing in recycling systems, the first thing you do is to decide that none of "your" water should drain or flow back to the river. The next step is to make sure that none of the rain that falls onto your property runs into the river. This is achieved by building structures that divert as much water as possible into on-farm storages. Development of ways to account for the capture of overland flows in a robust manner is one of the frontiers in Australian water management.

To make matters worse for the river, irrigators also began looking for new cheaper sources of water. One of the obvious new cheaper sources was to begin using groundwater that they had a license to access but had not yet decided to use. Those that could access groundwater began to do so and sell off the surface water that they had been using. The problem was that this groundwater used to flow into the river and it was not long before groundwater contributions to river flow started to decline.

11.5 Groundwater Connectivity

Not putting a cap on groundwater at the same time that surface water was capped is another policy failure that could have been averted. Especially as groundwater trading was introduced at the same time as surface water trading. Given the opportunity to sell surface water, a significant number of irrigators already licensed to take groundwater began to develop this resource that they had been keeping in reserve. From a system perspective, however, this meant that less groundwater flowed to the river. Early estimates of the size of this impact were kept confidential but were thought to be between 4 and 7 % of water use in the Southern Connected Murray Darling System (Young and McColl 2003).[9] Subsequent estimates placed the Basin-wide estimate at 670 GL (Walker et al. 2009) which is very close to the upper 7 % estimate that experts said was likely to be the size of the effect. Once again the problem could have been fixed but it wasn't and warnings from those advocating trade in the 1990s – including this author – were non-existent. A cap on groundwater was needed and, as with surface water, there was and still is a need for an arrangement that ensures that as groundwater efficiency increases, the groundwater cap has to come down. The good news is that the new Murray Darling Basin Plan includes a cap on groundwater as well as surface water use.[10] Plans are being put in place to reduce the extent of double accounting in the arrangements used to control how much water is used. Arrangements to account for changes in flows between ground and surface water resources still have to be put in place.

11.6 Optimizing Storage

One would have thought that all these policy sequencing errors would have been sufficient for alarm bells to be going off everywhere but this is only part of the story. It took 10 years for an economist – Donna Brennan – to point out that the way trading rules were specified meant that too much water was being used in dry times and not enough was being saved for use at a later time. From 1995 until 2007 it was not possible to carry forward unused allocations from 1 year to the next. The rule was "use it or sell it because you could not save it." All allocations had to be used within a season. With the exception of general-security irrigators in the Murrumbidgee System, carry-forward of water from 1 year to the next was not allowed. In systems with large dams, optimizing storage between seasons is as important as optimizing water use within a season.

Brennan (2008) found that when the benefits from trading were considered from a multi-seasonal perspective, the losses that resulted from storage mis-management were greater than the benefits from trading! Brennan found that trading had increased water use in dry times so much so that it deepened the economic impact of the drought and that the cost to the Australian economy of trading, without the opportunity to carry forward unused water from year to year was greater than the within-season benefits of trading. Almost as soon as Brennan's research results

---

[9]Now available at http://www.mdba.gov.au/sites/default/files/archived/mdbc-GW-reports/2178_Projections_of_GW_extraction_rates_and_the_CAP.pdf

[10]The science around the limits that have been set remain controversial.

came out, all Governments involved in the allocation of water in the Southern Connected River Murray System changed allocation policies so that unused water could be carried forward. The damage to the economy of 11 years of market-based mis-allocation had been done! Optimization of storage management and water use for both commercial and environmental purposes is critical (Grafton et al. 2011).

Unfortunately, the story gets worse than this. Driven by the market and rising water prices, many irrigators sold water entitlements and used the money they received to install very expensive and very efficient irrigation systems. These irrigators and the local town people who sold equipment to them liked all this new investment. They thought that this new business opportunity would be sustainable. The problem was that no-one told them that water trading systems and water allocation systems they were relying upon were seriously flawed. In fact, much of the growth was being achieved unsustainably by taking water needed to keep the river flowing and keep its ecosystems healthy. As a result, when the severity of all the above and other problems was revealed, communities became very angry – even though the government promised to fix this problem using tax revenues collected primarily from other people. In response, Australian governments have decided to restore balance to the Murray Darling Basin's abstraction regime by spending massive amount of money on investments in infrastructure savings and purchasing entitlements for the environment.

The final question that needs to be asked is one of whether or not the government was wise in deciding to fix the mistakes made. Cost-benefit analyses such as those undertaken by Grafton (2011), Morrison and Hatton MacDonald (2010) and the CIE (2011) suggest that the answer this question is yes as the value of the non-market benefits from the purchase of water for the environment and other non-consumptive uses like recreation appears to be greater than its value to irrigation. Grafton (2011) and the Productivity Commission (2010) have also been quick to point out that the approach taken to begin restoring health to the Murray Darling Basin ecosystems could have been much more cost-effective. In particular, much more emphasis could have been given to purchasing entitlements and much less emphasis given to the attainment of savings via investment in infrastructure.

11.7 Learning from Australia's Sequencing Mistakes

When mistakes of the magnitude described above are made, developed country governments need to act and fix the problems they create. The good news is that Australia is trying to do this and after a very difficult period has now put a suite of new administrative arrangements in place and they are much more robust. What can be learned from this experience?

The first and most important lesson for economists is that they need to be careful when making recommendations designed to increase the efficiency of resource use and encourage innovation. Markets have little respect for bio-physical conditions. They do, however, respond well to feedback loops. Robust abstraction regimes are needed. In particular, these need to signal increases in resource scarcity and the importance of changes in non-market conditions that lie outside the market.

When advocating a change in the way rights are specified and the opportunities associated with trading them, economists and policy makers need to pay attention to the consequences of doing so. Simplistic recommendations need to be replaced with ones that take account of biophysical realities. Attention also needs to be given to the importance of sequencing policy reforms. The pathway chosen should be one that results in continuous improvement not one that allows environmental or any other form of decline.

With hindsight it is clear that Australian policy reformers got the reform sequence wrong. If it had fixed the Murray Darling Basin's water allocation regime at the same time as it introduced trading or better still before trading was introduced then it could have saved well over ten billion dollars. Looking forward, all water users can expect to benefit from the mistakes that Australia has made. Water trading makes economic and social sense. It makes total sense when the entitlements and allocations being traded derive from abstraction regimes that have hydrological integrity.

References

Australia (2007) Water Act 2007. Parliament of Australia, Canberra

Australia (2012) Water Act 2007: Basin Plan. Commonwealth Parliament of Australia, Canberra

Bjornlund H, Rossini P (2007) An analysis of the returns from an investment in water entitlements in Australia. Pacific Rim Prop Res J 13(3):344–60

Brennan D (2008) Missing markets for storage and the potential economic cost of expanding the spatial scope of water trade. Aust J Agric Resour Econ 52(4):471–485

Bryan B, Marvanek S (2004) Quantifying and valuing land use change for integrated catchment management evaluation in the Murray-Darling Basin 1996/97-2000/01. Stage 2 report to the Murray-Darling Basin Commission. Available at www.clw.csiro.au/research/society/peru/publications.html

CIE (Centre for International Economics) (2011) Economic benefits and costs of the proposed Basin Plan: discussion and issues. The Centre for International Economics, Canberra

COAG (Council of Australian Governments) (2004) Intergovernmental agreement on a national water initiative between the Commonwealth of Australia and the Governments of New South Wales, Victoria, Queensland, South Australia, the Australian Capital Territory and the Northern Territory

Fisher DE (2006) Water law and policy in Australia: an overview. Environ Law Report 36(4):10264–76

Grafton RQ (2011) Economic costs and benefits of the proposed Basin Plan. In: Connell D, Grafton RQ (eds) Basin futures: water reform in the Murray Darling Basin. ANU E Press, Canberra

Grafton RQ, Chu HL, Stewardson M, Kompas T (2011) Optimal dynamic water allocation. Water Resour Res 47(6):W00G08-13. Reprinted in Grafton RQ, Pittock J, Tait M, White C (eds) (2013) Water, security, economics and governance. Tilde University Press, Prahran

Grafton RQ, Libecap G, Edwards EC, O'Brien R, Landry C (2012) Comparative assessment of water markets: insights from the Murray-Darling Basin, Australia and the western USA. Water Policy 14:175–193

Marshall GR, Connell D, Taylor BM (2013) Australia's Murray Darling Basin: a century of policy centric experiments in cross-border integration of water resources management. Int J Water Gov 1:197–218

Morrison M, Hatton MacDonald D (2010) Economic evaluation of environmental benefits in the Murray-Darling Basin, Report prepared for the MDBA

Murray-Darling Basin Commission (1999) The salinity audit of the Murray-darling basin: a 100-year perspective. Murray-Darling Basin Commission, Canberra

Murray-Darling Basin Ministerial Council (2002) The living murray: A discussion paper on restoring the health of the River Murray. Murray-Darling Basin Commission, Canberra

NWC (National Water Commission) (2008) Australian water markets report 2007–08. National Water Commission, Canberra

NWC (National Water Commission) (2009) Australian water markets report 2008–09. National Water Commission, Canberra

NWC (National Water Commission) (2010) Australian water markets report 2009–10. National Water Commission, Canberra

NWC (National Water Commission) (2012) Australian water markets: trends and drivers 2007–08 to 2011–12. National Water Commission, Canberra

Peterson DC, Dwyer G, Appels D, Fry J (2004) Modelling water trade in the Southern Murray Darling Basin. Staff Working Papers 31925. Productivity Commission, Melbourne

Productivity Commission (2010) Market mechanisms for recovering water in the Murray-Darling Basin, Final Report, March

Qureshi ME, Schwabe K, Connor J, Kirby M (2010) Environmental water incentive policy and return flows. Water Resour Res 46(4):1–12

Skinner D, Langford J (2013) Legislating for sustainable basin management: the story of Australia's Water Act (2007). Water Policy 15:871–894

Walker G, Evans R, Barnett B, Daamen, C, Parsons S (2009) The implications of surface water groundwater interaction on water resources of the Murray Darling Basin. CSIRO, Canberra. Available at http://irrigation.org.au/documents/publications-resources/conference_papers_2009/201009_Session12_RichardEvans.pdf

Young MD (2012) Trading into and out of trouble. Australia's Water allocation and trading experience. In: Maestu J (ed) Water trading and global water scarcity. Resources for the Future Press Water Policy Series, Washington, DC

Young MD, McColl JC (2003) Robust reform. The case for a new water entitlement system for Australia. Aust Econ Rev 36(2):225–34

Young MD, McColl JC (2005) Defining tradable water entitlements and allocations: a robust system. Can Water Resour J 30(1):65–72

Young MD, McColl JC (2008) Double trouble: the importance of accounting for and defining water entitlements consistent with hydrological realities. Aust J Agric Resour Econ 53:19–35

Chapter 12
Exploring the Reluctance to Embrace Water Markets in Alberta, Canada

Henning Bjornlund, Alec Zuo, Sarah Wheeler, and Wei Xu

Abstract Politically Alberta has acknowledged the need to reallocate existing water allocations to meet future demand using voluntary water transfers. However, support for water markets among irrigators has been slow to emerge, laws do not allow private entities to buy and hold water to meet in-stream needs and among the general public there has been a high level of skepticism and opposition in response to early attempts to reallocate or share water held by irrigation district. This threatens the ability of Alberta's Water for Life strategy to achieve its objective of meeting future supply in a sustainable manner. This chapter explores the reasons for this reluctance across southern Alberta using a series of surveys of irrigators and other residents conducted over a 9 year period and identifies issues of concern to stake-holder groups which need to be taken into account when marketing water sharing policies. We find distinct variation in policy preferences for water sharing across

H. Bjornlund (✉)
School of Commerce|UniSA Business School, Center for Regulation and Market Analysis, University of South Australia, Room WL3.51 Way Lee Building, 37-44 North Terrace, 5001 Adelaide, SA, Australia

Economics Department, University of Lethbridge, C576 University Hall, 4401 University Drive, Lethbridge, AB T1K 3M4, Canada
e-mail: Henning.Bjornlund@unisa.edu.au; henning.bjornlund@uleth.ca

A. Zuo • S. Wheeler
School of Commerce|UniSA Business School, Center for Regulation and Market Analysis, University of South Australia, Room WL3.51 Way Lee Building, 37-44 North Terrace, 5001 Adelaide, SA, Australia
e-mail: alec.zuo@unisa.edu.au; sarah.wheeler@unisa.edu.au

W. Xu
Department of Geography, University of Lethbridge, University Hall, 4401 University Drive, Lethbridge, AB T1K 3M4, Canada
e-mail: wei.xu@uleth.ca

K.W. Easter and Q. Huang (eds.), *Water Markets for the 21st Century: What Have We Learned?*, Global Issues in Water Policy 11, DOI 10.1007/978-94-017-9081-9_12,

the region depending on the level of water scarcity, environmental degradation and economic dependence on water experienced across space. Irrigators' willingness to share water with other sectors of the economy depends on context and purpose.

Keywords Reallocation • Water markets • Water sharing policies • Alberta • Stakeholder surveys

12.1 Introduction

Water scarcity is an emerging problem in the southern part of Alberta, Canada, due to increased demand and the environmental impact of current extraction levels. The provincial government is proposing voluntary reallocation as a key instrument to meet new demand and conservation objectives. Despite this, water trading has not been widely adopted by irrigators. Attempts to reallocate, or share, the water currently held by irrigation districts with other sectors of the economy and the environment have been met with opposition from all sectors of the community within southern Alberta, including irrigators. Consequently, proposals for water sharing have faced lengthy and costly processes that have frequently ended up in the courts. It is therefore important to better understand the reasons for this opposition and slow adoption and how different sectors of society believe that new demand and water conservation objectives should be met. This chapter investigates these issues based on previously published findings from extensive surveys of people living across the South Saskatchewan River Basin (SSRB) in Southern Alberta between 2003 and 2012. The aims of the surveys were two-fold: (i) to investigate actual and intended market adoption by irrigators and explore the reasons for their reluctance to adopt water markets, and (ii) to investigate how peoples' perception about water sharing differed across space and regions experiencing different levels of water scarcity and environmental degradation.

12.2 Approaches to Water Sharing

Reallocation or reductions of existing entitlements to extract water from water resources can be implemented in two different ways. The first is a command and control approach, where governments administratively reallocate existing entitlements to extract water or reduce their current entitlements with or without compensation. The second is a market based approach, where current entitlement holders are granted the right to trade their entitlements (in part or whole; permanently or for a specified term) to new entities needing water, or to other existing entitlement holders seeking to expand. This approach can also be used by governments, non-government organizations or individuals to buy extractive entitlements from existing users and leaving them in the water source to reduce overall diversions, thereby increasing in-stream flows (Wheeler et al. 2013a; Lane-Miller et al. 2013). Governments have generally been reluctant to adopt the first approach as it might require legislative changes and because it is politically sensitive and therefore associated with high political costs. Increasingly, governments and international institutions have therefore promoted market based approaches to achieve these outcomes and to help minimize the socio-economic consequences of sharing water. Concerns have centered on the perception that reducing irrigators' right to water extraction may reduce their ability to generate income, thereby reducing the value of their assets. This is perceived to then reduce economic activity within associated local communities, resulting in loss of employment opportunities, a declining population base and an associated decline in the local revenue base. In short, rural communities fear that sharing water via markets will threaten the viability of irrigation communities (Bjornlund et al. 2013c). These concerns put an end to the process of administrative reductions in extraction rights in New South Wales in Australia in 2006 and caused the Federal government to focus only on a program of buying water back from irrigators and upgrading irrigation infrastructure to secure environmental outcomes (Loch et al. 2011). However, while some Australian rural communities have experienced some adverse effects caused by reduced water availability since trading was introduced, it should be noted that the operation of water markets expanded during a period of severe and prolonged drought. It is therefore not possible to disentangle the effects of market operation from the effects of drought (NWC 2011). Further, it is difficult to isolate the impacts of water markets from the structural adjustment pressures that rural communities have encountered in the past several decades (NWC 2011). Finally, it has been found that trading has assisted irrigators and irrigation communities in Australia to adjust to increased water scarcity (Bjornlund 2004).

The magnitude of such impacts will, however, depend on how strategies to reduce access to water are implemented and how irrigators respond to such reductions. Irrigators could respond by improving their irrigation efficiency (i.e. increasing the proportion of diverted water taken up by the plant), changing to crops or varieties that need less water for the same yield, or using crops that produce a higher return per unit of water used. These responses would likely see the value of production remain relatively stable, minimizing the perceived negative impacts. If reductions in diversions are achieved by governments buying water from irrigators, proceeds could be used to finance an alternative way of living, reduce debt, reinvest in more efficient irrigation technology, or change to more water efficient crops. In some instances no production losses will be experienced as irrigators own surplus water (e.g. Wheeler et al. 2013b). If more productive or efficient irrigators buy water entitlements from unproductive or inefficient irrigators, the overall productivity and the value of production will increase.

If the policy objective is to facilitate a reduction in water diversions to increase in-stream flows to improve water quality, then a note of caution is needed, at least in jurisdictions such as Alberta, where irrigators have the consumptive right to all the water they are entitled to extract. In such circumstances, improving on and off-farm irrigation efficiency might actually result in reducing in-stream flows in two ways. First, where on-farm savings are made, irrigators may trade the saved water or increase their irrigation. Second, water saved through off-farm infrastructure investment to reduce leaks may not really be saved because the original water lost was contributing to in-stream flows via return flows. In effect, only the water previously lost to evaporation or seepage into very deep or very saline aquifers represents true water savings. This issue has caused significant debate and concern in Australia with its focus on upgrading irrigation infrastructure (Adamson and Loch in press).

The administrative and the market approaches reflect different perceptions of the type of property right irrigator have to extract and use water. In the context of land owners' agri-environmental property rights, Vera-Toscano et al. (2008) identified two competing property right theories which could be applied to water. First, the theory of absolute property rights posits that owners of the extractive right to water have the right to use it as they see fit; any reallocation of water to the environment, or sharing of existing rights, therefore has to be voluntary decisions and must be compensated. The second theory is based on a premise that resources such as water are communal property and individuals do not have absolute water rights. Groups such as irrigators are trustees only and are granted the right to use the water in pursuit of the interest and values of the society to which the water belong. Hence water should be transferred to the environment or shared with other users (without a guarantee of compensation) if that better reflects the changing social values. Vera-Toscano et al. found that the majority of citizens believe farmers do not have absolute property rights, but they do believe some public compensation or subsidization is needed to offset the impact on farmers.

This discussion raises at least four questions related to water sharing: (i) Should it be compulsory for water right holders to adopt water saving practices and share water?; (ii) Should such actions be subsidized by the taxpayers, or funded by the resource users?; (iii) What should the relative roles of government and markets be in water allocations? and (iv) What should happen to the water saved? In section five we report research that gauges irrigators as well as rural and urban residents' preferences about policies addressing these questions in the context of current debates in Alberta.

12.3 Water Sharing in the Alberta Context

In Canada, control over water resources and their management is largely vested in the provinces. While Canada has a National Water Policy it has had little effect on the way water is managed across Canada and there are renewed calls for a new and more comprehensive National Water Policy (de Loe 2009). Under British Colonial rule, access to water extraction followed the British common law doctrine of riparian rights under which only landowners with land abutting a water source had right to its extraction and use, under a complex set of rules. In semiarid regions, such as the Canadian prairies and Australia, this doctrine limited development because water was scarce and the number of rivers few and far between. This limited the area of land that could be irrigated and thereby restricted closer settlement of these areas. Victoria (a state in southeastern Australia) therefore abandoned the riparian doctrine as early as 1886 and vested ownership of water in the crown which issued licenses to water users.

Water availability in Canada varies significantly. It ranges from ample to plentiful in most parts of the Eastern and Maritime Provinces, as well as British Colombia (BC) in the west, while being scarce across the Prairie Provinces of Alberta, Saskatchewan and Manitoba, but also in pockets of BC and Ontario. In the Prairie Provinces, availability varies from plentiful in the thinly populated north to scarce in the more densely populated south. This is particularly the case in Alberta where in the northern part supply is plentiful, but population pressure and economic activity is low. Water scarcity is not a problem in this region but water quality problems are emerging because of the impacts of the mining sector, especially the oil sands (Kelly et al. 2010). On the other hand, the southern part has a diverse economy, is densely populated, experiences one of the highest rates of population and economic growth in North America, but it has a limited water supply. This is especially true in the SSRB, which contains 65 % of all irrigated land in Canada.

12.3.1 Water Allocation in Alberta

Following the Victorian (Australia) lead, in 1894 the Northwest Territory (of which Alberta then was part) introduced a new Irrigation Act, largely modeled on the Victorian Act of 1886, abandoning the riparian doctrine. It also vested ownership of water in the crown and issued licensed allocations to individuals wanting to extract water. These licenses were, however, issued under the prior allocation system, which borrowed from the prior appropriation system used in California and other parts of the US west. Under this system, access to extract water is accorded in order of priority of the license. During periods of droughts, where everyone cannot get all the water to which they are licensed, senior license holders are satisfied in full before junior license holders get access to water and some junior holders might not receive any part of their licensed allocation. The prior allocation doctrine has been retained through a series of Water Acts, most recently in the Water Act, 1999. It has been argued that this might not continue to be the most beneficial way of allocating Alberta's water and in 2009 the Minister commenced a process of developing a new water allocation and management framework (Bjornlund 2010). However, nothing has happened and a new consultation process is currently under way in Alberta.

12.3.2 The Emergence of the Need for Water Sharing in Alberta

The Alberta Government first responded to the signs of strained water resources in 1991 when it limited water extraction within the SSRB by introducing guidelines capping the amount of water that can be allocated for irrigation. In 2001/02, the province experienced a severe drought and irrigators in the southern part of the SSRB were subject to cuts to their allocations. In response, they took collective action to share their water on a one time basis and trading in water assignments took place (Nicol and Klein 2006). Following this drought, the Alberta government placed a moratorium on the issuing of new licensed allocations within the southern tributaries of the Oldman River (AE 2003a). Following the 1999 Water Act and the Irrigation Districts Act of 2000 and spurred on by the 2001/2 drought, the government commenced a process of public consultation to develop a new water management strategy, resulting in the 2003 *Water for Life* strategy (AE 2003b). This stated that water within the SSRB is already fully or overcommitted and that demand for water is likely to increase due to population and economic growth, at the same time that there is increased demand for in-stream flows.

As a result of these developments, a water management planning process commenced within all basins. In 2005 the Draft Water Management Plan for the SSRB was released (AE 2005). This plan clearly documented that the SSRB was fully, or over, allocated and many river reaches suffered negative environmental impacts as a result of current diversion levels. In response, the Alberta Government decreed that no new applications would be accepted for licensed water allocations within the SSRB, except for the Red Dear River (AE 2005). The draft plan also documented a number of factors which will exacerbate the current problem of water allocation in the future: (i) water demand from the non-irrigation sector could increase by 35–67 % by 2021 and by 52–136 % by 2046; (ii) irrigation has the potential to expand by up to 10 % and 20 % using water from the Oldman and Bow Rivers respectively and, (iii) the 1996 population of 1.3 million may increase to over two million by 2021 and more than three million by 2046. In addition, current climate change predictions suggest that the region is likely to face a change in the pattern and type of precipitation, further increasing pressures on water resources (Byrne et al. 2011).

While water in the SSRB is fully or over-allocated, the full impact of this has not yet been felt as most license holders only use a fraction of their allocations. The fear is that water sharing could activate this unused water and that improving irrigation efficiency could increase net use of water, increasing environmental stress within the rivers, as has been the case in Australia (e.g. Young 2013).

The role of the irrigation sector cannot be overestimated in discussing the need for water sharing and the need to meet increased in-stream requirements, because it controls 87 % of all licensed water allocations and most of the senior licenses (AE 2003c). The sector is therefore by far the biggest potential source for water to meet new demands. There are two types of irrigators in Alberta: private irrigators who hold their own licensed allocations under the 1999 Water Act and provide their own infrastructure to divert the water to their fields and district irrigators who are members of one of Alberta's 13 irrigation districts. District irrigators do not hold their own licenses but are registered on the district assessment roll as owners of a number of irrigated acres, which guarantees them access to a share of their district's licensed allocation, relative to their number of irrigated acres. The irrigator's share of the water allocation is delivered to their fields by the district infrastructure and they pay a fee for maintenance and administration of this infrastructure and a minimum of 25 % of the cost of its improvement. The Provincial government pays for head works and in-stream structures to deliver the water to the point where the district diverts it, as well as providing a maximum of 75 % (Government pays 75 % up to a maximum annual limit) of the cost of infrastructure improvements.

12.3.3 Water Trading in Alberta

Water trading under the Water Act (1999) can only take place subject to an approved water management plan and the approval of the Director, under the Administrative Guidelines for Transferring Water Allocations (AE 2003c). License holders have several ways in which they can share water. They can buy, sell, or lease their licensed allocations, under this system the license itself with all its rights and obligations is sold or leased. License holders can also assign the right to extract water under their license to another license holder during a given season. Under that system the control of the license remain with the license holder for the duration. However, assignments cannot be made to a water user who does not already hold a license and cannot therefore be used to assist the establishment of new water users. Also, existing water users can only purchase assignement up to their licensed allocation, that is to replace any cuts in allocation due to scarcity. Hence, assignment can not be used a mean to increase production due to changes in market demand, only to maintain production during scarcity.

District irrigators' ability to trade water is set out in the Irrigation Districts Act, 2000 (IDA). Trade between irrigators within the same district is relatively unrestricted. However, trading water to an entity outside their district needs the approval of a majority of irrigators via a plebiscite. Irrigation districts have another option to share their water to meet outside needs. Under the IDA, they can enter into a water supply contract without selling part of their licensed allocations. Such contract can be for longer or shorter terms. To do this, their license has to make explicit provision for this and specify how much water can be used for non-irrigation purposes.

Under the 1999 Water Act, licensed water allocations can only be granted for extractive purposes. As a consequence, environmental organizations, semi-public institutions and individuals cannot vote with their money by purchasing water rights and leaving the water in the river to meet their environmental or recreational objectives, as has occurred in places such as Australia and the US (e.g. Wheeler et al. 2014).

12.3.4 What Are the Policy Objectives?

One of the main objectives of the *Water for Life* strategy is to secure the supply of sufficient water of adequate quality to meet Alberta's future human and economic needs. Recognizing that the water resources of the SSRB are already over-allocated, it posits that a 30 % increase in water use efficiency and productivity over 2005 levels will be the main source of meeting new demand. The strategy also emphasizes that existing licenses will be respected and therefore sees voluntary and mutually beneficial transfers of water between existing and new users as the means by which water will move to meet new demands. However, it also stresses that economic instruments will be used as necessary to achieve these outcomes. The government has been vocal in promoting water sharing and especially in arguing the need for sharing by the irrigation sector. As we will see below, this explicitly expressed opinion was identified as one of the main factors driving irrigators to vote yes to a water transfer out of a district (Lafreniere et al. 2013a) and has caused the Alberta Irrigation Project Association, the organization representing the interest of the 13 irrigation districts in Alberta, to publicly announce its 'people first policy' stating that during drought the irrigation districts will assure that water is supplied to meet human needs before the needs of irrigation. While this remains the case, the government has yielded to public protests from a range of stakeholders against voluntary reallocations which has caused delays to a number of attempts which otherwise are clear examples of exactly what the government is promoting (see Sect. 12.4.3).

12.4 Perceptions of Water Sharing in Alberta

This section is based on surveys conducted with irrigators within the St. Mary Irrigation District (SMRID) in 2003 (Nicol and Klein 2006), irrigation district managers and board members in 2005 (Bjornlund et al. 2007, 2008a), district irrigators in 2006 (Bjornlund et al. 2008b, 2009), private irrigators in 2007 (Bjornlund et al. 2008b; Nicol et al. 2010), a cross section of both district and private irrigators in 2012 (Hall et al 2012) and a separate 2012 survey of district irrigators (Lafreniere et al. in press, 2013a, b).

12.4.1 What Is the Opinion of the Irrigation Industry

The irrigation industry, as represented by the managers and board members of the 13 irrigation district, were initially reluctant to participate in water sharing, because they believed that their water rights are secure and cannot be challenged and that they have the right to use their water as they wish (under the condition of their license) or not to use it. A 2005 survey, to which 60 % of all managers and directors responded, indicated only a 28 % agreement with the *Water for Life* strategy's plan to use economic instruments as necessary to meet conservation and productivity objectives (Bjornlund et al. 2008a). Only 24 % agreed with the use of any kind of economic instrument, with only 5 % agreeing with trading in licensed allocation and

15 % with lease or assignments of a license. The highest level of agreement with any one instrument was the use of subsidies to improve irrigation equipment, but even for that measure only 21 % were in favor (Bjornlund et al. 2007). Similarly, only 26 % considered the objective of a 30 % improvement in water use efficiency and productivity achievable. Only 26 % agreed that any water saved through improved water use efficiency should go to the environment, while 36 % indicated that it should go to support local industry and 44 % suggested it be used to meet the growing needs of cities and municipalities (Bjornlund et al. 2008a).

These early results did not indicate a strong willingness by irrigators to participate in water sharing. Given the ongoing debate and continued political and public pressure to share water, the sector is starting to understand that they need to be proactive and that the long-term control of their water licenses is not as secure as they once thought. This has caused a shift in official statements from the sector. One example of that is the announcement of the People First policy. The districts have started two processes to protect their licenses. First, they have applied for amendments to their licenses so that they can supply more of their water for non-irrigation purposes, which will enable them to share their water without selling their licensed allocation. Second, they are going through a process of plebiscites to approve expansion of their irrigated area so they will use more of their licensed allocations. Both of these attempts have been meet by opposition as they will increase overall diversion and worsen water quality in already degraded rivers (discussed later in this chapter).

12.4.2 Irrigator Participation in Water Sharing: Private and District Irrigators

As previously discussed, two types of irrigators have very different opportunities to trade under the *Water Act, 1999* and the *Irrigation District Act, 2000*. This has influenced their level and type of market participation. The two groups also differ significantly in farm type, production and irrigation intensity. District Irrigators have most of their land under irrigation, produce more cereal and specialty crops, have more efficient irrigation technologies and management practices, depends more on off-farm work and have a lower level of expectation of family succession. On the other hand, private irrigators have much larger properties, most of their land is in dry land farming (mainly for grazing for cow-calf operations) and most of their irrigated land is for fodder production. Both activities are in support of feedlots, either selling to feedlots or supplying their own. Also, many of the private irrigators are located in the foot hills of the Rocky Mountains where the environmental conditions and the scenic beauty differs from the flat and intensely irrigated central SSRB, which houses most of the irrigation districts. Generally, private irrigators have irrigation as a small part of their overall business, directed mainly to producing input for their final product. They are therefore less dependent on irrigation as most of what they produce can be substituted by buying in fodder; they are therefore less concerned about water and also more willing to consider trading their water allocations (Bjornlund et al. 2008b).

The first activity in water markets took place during the drought of 2001/02 when seasonal allocations were reduced to near 50 % of regular allocations. The trading that took place was mainly of assignments. The records of the St. Mary River Irrigation District show that 12.3 % of irrigators participated in trading during the season, trading 3.5 % of the water allocated during that year. Nicol and Klein's survey, representing 40 % of all the irrigators that traded, found that water moved out of lower value crops (26% of water sold caused reductions in some commodities: 21 % in wheat production, 21 % in Barley, and 34 % in forage) and into specialty crop production (43 % of water bought went into potato production, 43 % to other specialty crops and 24 % to forage). The water also moved from irrigators with less efficient technology or irrigators currently not irrigating the land (flood irrigation 12 %, wheel move sprinklers[1] 48 %, and not irrigating the land 18 %) to irrigators with more efficient technology (Pivot 75 %). About 60 % of the sellers reported that they did not need the water for their own crops (reflecting the fact that most irrigators use only 60 % of their allocations in normal years) while 69 % found that selling their water provided a good income opportunity and 42 % considered selling their water more profitable than applying it to their own crop (Nicol and Klein 2006). The overall assessment was that during this season the market played a critical role in allowing specialty crop producers to maintain production and fulfill their contractual obligations to processing plants, which in turn could continue to provide jobs in the community.

In the first 5 years after trading was introduced in 1999, 23 applications for the transfer of licensed allocations were filed. However, only six of those could be considered market transactions, the remainder represented irrigators using the opportunity to transfer all or part of their extraction right from one point to another to make their operations more efficient. By 2008, only three of the six transfers had been approved and three more applications had been filed. The six first transfers suggest that water moved to more efficient and higher value users, but that sellers mainly sold unused water. This indicates a very low level of activity. A primary reason for this is probably that generally irrigators hold licenses in excess of what they need, so that expansion can take place without the use of markets. However, there is also evidence that lack of market information, price uncertainty, cumbersome and lengthy approval processes and the potential for a 10 % holdback of the transferred water under the Water Act 1999 are major factors impeding trade (Nicol et al. 2008).

The surveys conducted in 2006 and 2007 of district and private irrigators respectively asked them about their intention to trade water over the next 5 year, but did not ask about actual market participation due to the very low level of market activity

[1]Wheel move sprinkler systems are also known as wheel-line, sideroll, or lateral-roll irrigation machines (see http://extension.usu.edu/files/publications/publication/engr_bie_wm_08.pdf).

Table 12.1 Actual and intended water trading reported in 2006, 2007 and 2012 surveys

	Next 5 years[a] 2006(7) – 2011(12)		Last 5 years[b] 2007–2011		Next 5 years[b] 2012–16	
Trading type:	District (%)	Private (%)	District (%)	Private (%)	District (%)	Private (%)
Sold/sell leased water	37	37	3	2	8	15
Bought/buy leased water	NA	NA	17	8	38	30
Sold/sell licensed allocation	7	22	2	1	4	16
Bought/buy licensed allocation	61	42	28	8	43	30

[a]2006 and 2007 surveys: respondents were asked if they were willing to sell lease water if a buyer was available and the price made sense; for licensed allocation, respondents were asked if they were willing to sell if the price was right and convert to dry land farming and if they were willing to buy licensed allocation if they wanted to expand their irrigated area or needed more water to maintain it (Bjornlund et al. 2008a)

[b]2012 survey: Respondents were asked about the likelihood of them undertaking each of the four trading activities on a 1 (highly unlikely) to 7 (highly likely) scale. The percentages reported here are those rating each trading activity 5–7 (Hall et al. 2012)

(Table 12.1). Approximately 37 % of both private and district irrigators indicated they would lease out their water allocation for a year if somebody offered to buy it and the offered price was financially desirable. When it came to the intention to sell licensed allocation, 22 % of private irrigators were willing to consider this and convert to dry land farming compared to only 7 % for district irrigators. Finally, irrigators were asked whether they would consider buying additional licensed water allocation if they wanted to expand or needed it to maintain their irrigated area. A much higher percentage was willing to consider buying, but the outcome was reversed: 42 % of private irrigators were willing to consider a purchase, compared to 61 % of district irrigators. Relative to the actual activities prior to 2007, these figures are very high.

A survey of 319 private and district irrigators across the SSRB in January 2012 asked irrigators about both their actual participation in water trading in the last 5 years and their likely participation over the next 5 years (Table 12.1). Only 1–2 % had sold licensed allocations over the last 5 years from 2007 to 2011 (which is 7–12 years after trading was introduced) (Hall et al. 2012). There was a higher level of participation in buying licensed water allocations with 28 % of district irrigators and 8 % of private irrigators participating. The high level of buying among district irrigators is a result of the recent plebiscites to expand the irrigated area within some districts. Hence most of these 'purchases' represent purchases of the right to expand their irrigated area on the district's assessment roll. They therefore do not represent water transfers between individual irrigators. This is a significant increase in market activity compared to the mere nine arms-length transfers filed by 2007 (Nicole et al. 2008).

There has also been quite some activity in the use of short-term leasing of water, especially among district irrigators; 17 % having bought leased water and 3 % sold it. Private irrigators have also been active in this market, with 8 % having bought leased water and 2 % sold it (Table 12.1). This suggests that some private irrigators are venturing into more intensive and high tech production of high value crops, which is supported by anecdotal evidence. Irrigators' intentions to participate in water trading in the next 5 years from 2012 to 2016 indicate that market activities are likely to continue to increase significantly for both buying and selling licensed allocations and short term leases and both among district and private irrigators. Again, the intention to buy is much higher than to sell and the intention to buy is much higher among district irrigators, while the intension to sell is much higher among private irrigators.

The patterns in the relative trading behavior among private and district irrigators are consistent with the expectations previously discussed and reflect the different nature of the two groups. Private irrigators are far more willing to consider selling as they are better able to manage with less water, are less dependent on irrigation, and have more direct powers to make decisions to trade. District irrigators are more willing to buy, which reflects their higher level of dependence on irrigation. However, it is interesting to note the high level of private irrigators' willingness to consider buying. Again, this might reflect the emergence of a new group of private irrigators venturing into specialty crop production and investing in the most sophisticated irrigation technology and management practices. There is clear anecdotal evidence of the emergence of this group of private diverters.

The above findings suggest that both actual and intended trading activities are increasing. This is consistent with experiences in other jurisdictions, such as Australia, albeit at a slower pace (Bjornlund 2006). There are likely to be at least four reasons for these increases: (i) water trading is a new agricultural management innovation, and any adoption of a new innovation will increase slowly in the beginning, and increase at a more rapid rate later (Wheeler et al. 2009); (ii) water trading is increasingly discussed in policy debates and irrigators are starting to realize its potential as means of expanding, consolidating or adjusting their operations to become more stable and financially viable; (iii) several districts are planning to hold a plebiscite to expand their irrigated area allowing, individual irrigators to expand their operations to be more financially viable and, (iv) irrigators start to realize that their water represents significant value – both as a short-term tradable consumptive commodity and as a long-term financial asset (Bjornlund et al. 2013c).

12.4.3 The Specific Role and Opportunities for Districts

As previously noted, the irrigation districts hold some of the largest and most senior licenses and have traditionally only used about 60 % of their licensed allocation during normal years. While their conveyance infrastructure has been undergoing constant refurbishment over many years, there are still opportunities to improve these systems and thereby 'save' substantial volumes of water (please note the previously discussed significant problems with this). The irrigation districts are therefore considered the most likely source of supply to meet future demand. There is evidence of irrigation districts being willing to secure water for new processing industries moving into the area and providing markets for their produce and to meet the needs of local small towns. Also during the drought in 2001/02, when junior licenses were likely to be cut off, the districts agreed to temporarily abandon the seniority principle and accept proportional sharing. This agreement was encouraged by the fact that some junior license holders were processing industries and if they were cut off the irrigators would lose lucrative markets. However, the action was hailed as a major success and an example of the irrigation sector's willingness to share their water.

The first major attempt to share a district's water came in 2006/07 when a developer wanted to build a new shopping center, casino and race course. The municipality of Rocky View did not have water to supply the development. It was just outside the city boundaries of Calgary and its pipelines would be the cheapest and most efficient way to supply the water and Calgary controls water licenses to supply three times as many people as it currently has. However, Calgary demanded that Rocky View surrender the control of the land to Calgary in order to supply the new development. The municipality refused to do this and the developer had to find a different source of supply. The second option was to convey the water from the nearby Red Dear River where new licenses could still be issued. However, this would be an inter basin transfer which requires the approval by the Minister. This was not forthcoming, due to the issue's political sensitivity and pressure from environmental groups. Hence, the developer commenced negotiations with the Western Irrigation District (WID) to purchase part of their license. This would be by far the most expensive and least efficient solution as it would require the construction of a pipeline basically duplicating the existing infrastructure controlled by the City of Calgary (Bjornlund 2010).

The management of the WID came to an agreement with the developer to sell 2,500 dam^3 to the Municipality of Rocky View at a price of C$6,000 per dam^3 (One Dam^3 = 1,000 m^3); a record breaking price in Alberta and possibly most other jurisdictions, such as Australia and the US (Bjornlund et al. 2013c; Saliba and Bush 1987). The WID suggested using the proceeds to replace an old leaking canal with a new pipeline, 'saving' more water than they sold. The transfer should be a clear example of a voluntary transfer, a win-win proposal and exactly what the *Water for Life* strategy calls for. However, the transfer required the approval of the irrigators in a plebiscite. Prior to the plebiscite, the management argued in letters and at meetings that the sale would be beneficial as more water would be saved and the district and its irrigators would be better off. Also, the district management felt the pressure from government to share their water and argued that failing to do so might result in the government simply taking the water away. The debate during town hall meetings became extremely heated and came close to physical confrontations between opponents and supporters of the transfer. The plebiscite eventually narrowly approved the transfer but it caused a serious split among the irrigators. The approval of the transfer by the Alberta Government was also controversial, several stakeholders protested and tried to prevent the approval and a Calgary Water Authority tried to block it through the courts. However, the case was eventually dismissed and the transfer approved.

A study by Lafreniere et al. (2013a) looked into the reasons for the narrow acceptance and found that the district management had failed to consider the opinions of its members and address their concerns. The irrigators are generally opposed to surrendering any part of their licensed allocation as they find it too final and fear for the long-term security of their supply. The study identified four distinct groups of irrigators, each basing their vote on different perceptions and interests. One group voted yes as they feared that the provincial government would simply take their water if they voted no. A second group voted yes because the money would be spent in such a way that more water would be saved than sold. A third group voted no because this solution was the least efficient way of supplying the new development, which was only made necessary because Calgary tried to use its control of water to extend its domain. The fourth group voted no because they did not agree with the way the proceeds were spent as it did not benefit them directly or they did not feel that this particular pipe, and the irrigators supplied by it, needed it.

When district irrigators across the SSRB were asked how they intended to vote if a plebiscite was held to transfer water out of their district to meet the need of a non-agricultural user (without being provided with information about any other attributes of the transfer) only 5 % gave some indication of considering voting yes, and nobody said that they for sure would do so. This suggests a very low willingness to share the district's water with non-agricultural users (Hall et al. 2012). In the same study, private irrigators were asked how likely they would be to sell some or all of their licensed allocations to a non-agricultural user; 16 % gave some indication that they might be willing to do so. Comparing this finding with those in Table 12.1, suggests that private irrigators' willingness to sell their water is not influenced much by the type of buyer.

Considering the importance to Alberta of sharing the districts' water, and the low level of willingness to do so, it is important to understand irrigators' motivations for voting yes or no in a plebiscite. Lafrenier et al. (2012) tried to better understand how the attributes of a transfer influence the intended vote, how big a proportion of irrigators were influenced by which attribute and the relative importance of each attribute. Based on Lafreniere et al. (2013a), six attributes were tested: (i) the purpose of the agreement (supply irrigation, a municipality or the environment); (ii) the price paid (at market value, above market value or a market record); (iii) personal proximity to benefits (proceeds used to upgrade infrastructure on which the respondents were dependent or not); (iv) water saving (all the water sold would be saved by reducing water losses or half the water sold would be saved); (v) Government intervention (the government wants the deal approved or the government does not care) and, (vi) environmental efficiency (the district was the closest supplier or the district was not). Respondents were asked to rate their likely voting intention under a number of different scenarios, representing different levels of the six attributes. Conjoint analysis were applied to establish the relative importance that irrigators gave each attribute, while cluster analysis were used to group irrigators according to their preferences for the attributes (Lafreniere et al. in press).

A group of 17 % voted no to all scenarios regardless of the attributes of the transfer. These respondents were not included in the conjoint or cluster analysis. The order of importance of the attributes is: (i) the purpose of the agreement; (ii) the price paid; (iii) water savings; (iv) environmental efficiency; (v) personal proximity to benefits and (vi) government intervention. As could be expected, the higher the price, the more water saved, being the closest provider, and government support resulted in a higher likelihood of voting yes. Interestingly, the most important influence was purpose and transfers for municipal and environmental purposes were more likely to attract a yes vote than if the purpose was for irrigation. It was also surprising that local community benefits (price, water saving and environmental efficiency) were considered more important than personal benefit and the government's opinion. That the government's opinion was the least important is surprising considering the high level of importance attached to it identified by Lafreniere et al. (2013a).

Five segments were identified, depending on which attributes irrigators were most influenced by: (i) water savers (25 %), were most likely to vote yes if the proceeds were spent on saving all the water sold; (ii) the Greenies (17 %), were most likely to vote yes if the water was going to the environment, but also ranked as second in importance was the district being the closest provider of the water; (iii) the Municipal Friends (17 %), were most likely to vote yes if the transfer was for municipal uses, but were also the most likely to vote yes if the transfer was for irrigation; (iv) Personal Gainers (14 %), were most likely to vote yes if they personally gained from the way the proceeds were invested but were also most influenced by the price being at a market record and, (v) Efficiency Savers (8 %), were most likely to vote yes if the proceeds from the transfer were invested in improving system efficiency in such a way that the volume of water saved by the improvements equaled the volume of water sold. No segment was formed around the government's opinion and the five groups were not influenced significantly different by it. This indicates that the opinion of the government is not considered as important by the broader irrigator group as it was among the WID irrigators (Lafreniere et al. 2013a).

Reflecting the low level of support for water transfers out of districts, district managements have pursued the option of entering into agreements to deliver water for non-irrigation purposes because they do not include the sale of part of their license. To do this many districts need a license amendment. In 2003, the St Mary Irrigation Districts got an amendment of one of its licenses allowing them to divert 270,000 dam^3 for other purposes. It has been argued that this undermines the purposes of the Water Act, 1999, which was designed to give the public a strong interest in water (Bankes and Kwasniak 2005). In 2007 a number of districts applied for similar license amendments. This was strongly opposed, mainly by environmental NGOs, because, they argued, that such changes would allow irrigation district to act like water brokers selling water to whoever they want to, at whatever price they can agree on. Further, they argued that the process of supply agreement circumvents the requirement for public scrutiny and the environmental assessment required for the transfer of licensed allocations (Christensen and Droitsch 2008). In response, the government announced that it would cease to process applications for license amendments while it reviews current policies. Amendments were finally approved in 2011 as Alberta Environment and the Environmental Appeal Board denied public interest groups standing to challenge the decision because only people directly affected by an amendment can appeal. A number of environmental groups sought judicial review by the Queen's Bench of Alberta. This review concluded in January 2013 and ruled that there is no public standing at the Alberta Environmental Appeal Board for those not directly affected (Fluker 2013).

The issue of sharing the districts' water has been controversial both among irrigators and the public. Considering the importance of water sharing for the future of Albertan, this raises a critical question: how do the citizens in Alberta think water should be shared? This is explored in Sect. 12.5.

12.5 Exploring Albertan's Policy Preferences for Water Sharing

This section is based on three surveys of non-irrigator residents of SSRB. One was conducted in 2009 in Lethbridge (the second largest city in the basin), located in the Southern part of the SSRB, and four small towns outside of Lethbridge totally dependent on irrigation for economic activity (Bjornlund et al. 2013a, b). The southern part of the basin has experienced water restrictions in the past and has the most environmentally degraded river reaches. The second survey was conducted in 2010 in Calgary, (Alberta's largest city) and Strathmore, a smaller irrigation town, which is the administrative center of the Western Irrigation District (Russenberger et al. 2011, 2012; Bjornlund et al. 2013a). The third survey was conducted in 2012 with young adults across the SSRB (Bjornlund et al. 2013d).

12.5.1 What Are the Policy Options Explored

Flowing from the discussion in section two, the policy options tested in the survey were aimed at eliciting irrigators' and the public's perception of four questions: (i) should it be compulsory for water right holders to adopt water saving practices and share water?; (ii) should such actions be subsidized or funded by resource users?; (iii) what should the relative roles of government and markets be in water allocations? and (iv) what should happen to the water saved? The policy statements

Table 12.2 Policy options tested in the surveys

GOV_1	The government, rather than market forces, should decide who gets to use Alberta's water
GOV_2	If water is to be traded among irrigation districts and/or municipalities, the government should set the price
GOV_3	If an irrigation district or municipality is not using all of its allocated water, the government should be able to take that water for environmental purposes without compensation
ENV_1	Private individuals and groups should be able to hold water licenses for environmental protection
ENV_2	Public funds should be used to improve irrigation systems only if the water that is saved is left in rivers
ENV_3	The government should buy water from current water license holders, such as irrigation districts, so that more water can be left in the river for the environment
ENV4	Minimum flows of water should be set for all rivers, and only the water above those minimum flows should be available for economic purposes such as irrigation
LIC_1	All water licenses, no matter when they were issued or for what purpose, must be honored
LIC_2	Water that is saved through improved water use efficiency should be used to increase economic activity
LIC 3	Public funds should be used to help larger water users to become more water efficient

tested reflect three policy orientations: (i) give strong powers to the government in water allocation issues (GOV); (ii) protect the environment (ENV) and, (iii) protect existing license holders (LIC) (Table 12.2).

12.5.2 How Do Policy Preferences Vary Across the Urban to Rural Gradient

The four locations tested represent different locations on the rural to urban gradient with Calgary being the most urban center followed by Lethbridge, Strathmore and RTMS (the four small irrigation communities of Raymond, Taber, McGrath and Stirling). The level of agreement and disagreement with the ten policy statements were significantly different across the four locations. The more urban the area, the more likely people are to support policies that give the government a strong role in water allocation (GOV1,2,3), protect the environment (ENV1-4) and support large water users to become more efficient (LIC 3). On the other hand, the more rural the area, the more likely people are to support the right of existing license holders (LIC1,2) (Bjornlund et al. 2013a). Regression analyses suggest that regions with a higher proportion of people employed in agriculture and other resource industries are less supportive of policies granting strong powers to the government and more supportive of policies protecting existing license holders, and regions with a higher population density are more in favor of policies protecting the environment (Bjornlund et al. 2013a).

The study by Bjornlund et al. (2014) included a survey of irrigators across the SSRB and therefore allowed us to expand the analysis reported in Bjornlund et al. (2013a) and for the first time facilitated a comparison of policy preferences between irrigators, those living in irrigation dependent communities (Strathmore and RTMS) and those living in cities (Calgary and Lethbridge). Regression analysis show that support for: (i) a strong role of government is highest among city dwellers, followed by rural residents, with irrigators showing the lowest level of support; (ii) policies protecting the environment follows the same trend; and (iii) policies protecting existing license holders is lowest in cities followed by rural communities with irrigators most in support. These findings are not surprising and are consistent with the extractive commodity theory (Jones et al. 1999). It supports the findings of Bjornlund et al. (2013a), that people living in different settings have different interests in, and experiences of, the environment and its assets, including water.

What might be surprising is that rural non irrigator residents align more closely with city dwellers than with their rural irrigator neighbors, particularly when it comes to policies protecting the right of existing license holders (Bjornlund et al. 2014). These results might reflect other findings in the literature suggesting that the urban-rural gap is narrowing with respect to environmental values (e.g. Huddart-Kennedy et al. 2009; Salka 2001) and that it therefore is not so much place of residence that influence peoples values, actions and policy preferences as it is their level of interaction with resource users (Berenguer et al. 2005). If people have significant interaction with resource users, they are likely to subscribe to the same norms about the environment and, therefore, are likely to express similar policy preferences (Dietz et al. 2005). The survey in Calgary and Strathmore allowed us to explore these issues as it asked questions about respondent's interaction with irrigator families and rural amenities. This study found that closer social ties to irrigation, stronger feeling of community cohesion, and a belief that members of their community support water markets significantly influence the level of support for policies protecting existing license holders (Russenberger et al. 2011).

12.5.3 Does Experience with Water Scarcity and Water Degradation Influence Policy Preferences?

Analyzing the difference in policy preferences between the southern region, (Lethbridge and RTMS) which is most exposed to water scarcity and environmental degradation, with the northern region (Calgary and Strathmore) found significant associations between policy preferences and the region in which the respondents reside. In relation to environmental protection policies (Env1-3), while there were clear urban-rural differences (Lethbridge agreed more than RTMS and Calgary agreed more than Strathmore), the regional difference was clear as demonstrated by stronger agreement in Calgary and Strathmore than in Lethbridge and RTMS. When it comes to government control (GOV2,3), the regional differences again overrode the urban to rural gradient, even though Lethbridge agreed more than RTMS and Calgary agreed more than Starthmore, Lethbridge agreed more than Calgary and RTMS more than Strathmore (Bjornlund et al. 2013a).

The findings regarding policies to protect the environment contradicts the environmental deprivation theory (Lowe and Pinhey 1982) which suggests that people who live under poor environmental conditions have greater environmental concern and therefore should be more supportive of protective policies. The study by Bjornlund et al. (in 2013a), suggests that the most environmentally degraded southern region is least supportive of policies to protect the environment. The explanation for this is likely to be that the southern region is also the most intensively irrigated, most dependent on irrigation, and most exposed to water restrictions. This suggests that within such regions people are more concerned about the viability of their community than they are about the environment. It should also be noted that Lowe and Pinhey's study suggested that urban residents were more likely to live in environments they considered degraded because of phenomenon such as air pollution. They gave little consideration to rural environmental degradation, so their theory may have limited applicability in explaining rural dwellers attitudes regarding environmental degradation. This caveat supports the view that people living in these regions value the environment because of the instrumental uses to which it can be put; their interests are using the environment rather than defending it on the basis of aesthetic or ethical grounds (Berenguer et al. 2005).

12.5.4 Urban Rural Differences Versus Socio-demographic Differences in Policy Preferences

While several studies have identified urban and rural differences in environmental values, behavior and policy preferences (Vera-Toscano, et al. 2008; Graymore and Wallis 2010) others have found that rural and urban areas have different socio-demographic characteristics, for example people living in urban areas tend to be better educated and have higher income than people living in rural areas (Huddart-Kennedy et al. 2009). Some therefore argue that the identified differences are caused by the underlying differences in socio-demographic characteristics rather than rural-urban differences (e.g. Salka 2001).

Bjornlund et al. (2013a), by conducting separate regression models for each of the four regions, is one of the first studies to confirm the influence of socio-demographic characteristics after controlling for location. Bjornlund et al. (2014) controlled for place or residence (or resource dependence) by including a set of dummy variables in their regression analysis of more than 2,000 surveys across the SSRB. One indicated residence within an irrigation community, one indicated a city dweller and the default category was irrigators. They identified several significant socio-demographic characteristics influencing the preference for the three policy orientations after controlling for location: (i) Age; older people are more in support of the government and existing license holders, while younger people are more in support of the environment; (ii) gender; males are more likely to support the government and current license holders but less likely to support the environment; (iii) education; those with more than a high school qualification are more likely to support the government and less likely to support current license holders; (iii) income; those with higher income are more likely to support the government; (iv) rural upbringing; those raised in a rural area are less likely to support the environment and more likely to support current license holders and, (v) involvement in land, water and environmental management; members of Water Planning and Advisory Councils or Land and Water Stewardship Groups are more likely to support existing license holders and less likely to support the government and the environment, while those who are members of conservation groups are more likely to support the government and the environment and less likely to support existing license hollers. The studies by Bjornlund et al. (2013a) and Bjornlund et al. (2013d, 2014) have clearly established that place of residence and socio-demographic characteristics have a significant influence on policy preference for water sharing and markets.

12.6 Conclusions

Water markets in Alberta have evolved very slowly. However, 13 years after the introduction of markets, the level of reported trade over the last 5 years shows increased activity and the level of anticipated trade over the next 5 years is expected to be even higher. Private irrigators indicate a higher level of willingness to consider selling their water based on rational economic arguments, regardless of who the buyer of the water is. Research suggests that information is critical in influencing district irrigators' decisions to sell water. Considering that the irrigation districts hold the largest and most senior licenses and, within normal years, only use about 60 % of the water allocation, it is not surprising that policy makers have great expectation regarding the districts' contribution to meet new demand through voluntary sharing arrangements. It is therefore important that district managers take the attributes of transfers identified in this paper into account when marketing proposals to approve such transfers at a plebiscite.

Acknowledging the high level of concern among irrigators about permanently relinquishing control of any part of a districts licensed allocation, district managers are pursuing an alternative way of sharing their water with other users by entering into supply agreements with non-agricultural users. This option will not require a plebiscite among member irrigators. To facilitate this many districts need an amendment of their license. Applications for such amendments have been met with opposition from a wide range of stakeholder groups. There are significant concerns over social equity and environmental issues over such arrangements as they will not require departmental oversight and will not be subject to the environmental assessments as is required for transfers of licensed allocations. It is also argued that this allows districts to act as water brokers simply trying to sell their water for the highest price, which seems to be contrary to the public interest embedded in the

Water Act. If districts are allowed to supply water under such agreement, they will activate the previously unused proportions of their entitlements which will increase overall diversion and reduce stream flow in the region's rivers, many of which are already suffering environmental degradation from the current level of diversion. The adverse effects of activating 'sleeper' or 'dozer' entitlements have already been a documented concern in Australia.

The research reported in this paper provides significant insight into how various sectors of the community perceive that water should be shared among existing and new users including the environment. There is strong evidence to suggest that the level of resource dependence and the exposure to water scarcity and environmental degradation influence people's policy preference for water sharing. While people living in rural areas align more with their irrigator neighbors than city dwellers when it comes to policy preferences for water sharing, their preferences are more closely aligned with city dwellers than irrigators. This reflects other findings in the literature that the rural-urban gap in environmental values and behavior has been narrowing in recent years and that the level of social interaction with resource users and social norms may be more important.

The overall assessment is that context and geography matter when introducing policies or instruments for water sharing and that policy solutions need to be context specific to ensure their acceptance within the target community. Taking into account stakeholders concerns will reduce conflicts among stakeholders, increase the chance of acceptance and improve the likelihood of achieving policy objectives.

Acknowledgments This chapter builds on several studies conducted by a number of students and funded by a number of agencies. We would like to acknowledge funding from Alberta Innovates: Energy and Environment Solutions, Social Sciences and Humanities Research Council and the Canadian Water Network. We would also like to acknowledge the contribution by PhD student Lorraine Nicol, Masters Students Cameron Parrack, Martin Russenberger, Matthew Hall, Amber Zary and Katie Lafreniere, and postdoc Xinzheng Zhao.

References

Adamson D, Loch A (in press) Possible negative feedbacks from 'gold-plating' irrigation infrastructure. Agric Water Manage. doi:10.1016/j.agwat.2013.09.022

AE, Alberta Environment (2003a) Summary, SSRB background studies. Author, Edmonton

AE, Alberta Environment (2003b) Water for life: Alberta's strategy for sustainability. Author, Edmonton

AE, Alberta Environment (2003c) South Saskatchewan River Basin water allocation. Author, Edmonton

AE, Alberta Environment (2005) Background information for public consultation on the south Saskatchewan River Basin's draft water management. Author, Edmonton

Banks N, Kwasniak A (2005) The St. Mary's Irrigation District License amendment decisions: irrigation district as a law onto themselves. J Environ Law Pract 16(1):1–18

Berenguer J, Corraliza J, Martin R (2005) Rural-urban differences in environmental concern, attitudes, and actions. Eur J Psychol Assess 21(2):128–138

Bjornlund H (2004) Formal and informal water markets – drivers of sustainable rural communities? Water Resour Res 40: W09S07
Bjornlund H (2006) Increased participation in Australian water markets. In: Lorenzini G, Brebbia CA (eds) Sustainable irrigation management, technologies and policies, vol 96, Transactions on ecology and the environment. WIT Press, Southampton, pp 289–302
Bjornlund H (2010) The competition for water: striking a balance among social, environmental and economic needs, Commentary No 302, C.D. Howe Institute, Toronto
Bjornlund H, Nicol L, Klein K (2007) Challenges in implementing economic instruments to manage irrigation water on farms in Southern Alberta. Agric Water Manage 92:131–141
Bjornlund H, Nicol L, Klein K (2008a) Implementing Alberta's water for life strategy: an irrigation industry perspective. Prairie Forum 33(1):167–190
Bjornlund H, Nicol L, Klein K (2008b) Economic instrument and irrigation water management – a comparative study of private and district irrigators in Alberta, Canada. In: Esteve YV, Brebbia CA, Rico CP (eds) Sustainable irrigation management, technologies and policies II, vol 112, Transaction on ecology and the environment. WIT Press, Southampton, pp 3–14
Bjornlund H, Nicol L, Klein K (2009) The adoption of improved irrigation technology and management practices – a study of two irrigation districts in Alberta, Canada. Agric Water Manage 96:121–131
Bjornlund H, Zuo A, Wheeler S, Xu W, Edwards J (2013a) Policy preferences for water sharing in Alberta. Water Resour Econ 1:93–110
Bjornlund H, Parrack C, DeLoe R (2013b) Segmenting residents of southern Alberta for improved understanding of policy preferences for water reallocation. J Soc Nat Resour 26(10–12):1330–1350
Bjornlund H, Wheeler S, Rossini P (2013c) Water markets and their environmental, social and economic impact in Australia. In: Maestu J (ed) Water trading and global water scarcity: international perspectives. Taylor & Francis, Abingdon, Oxfort, UK, pp 68–93
Bjornlund H, Xu W, Zhao X (2013d) Generational differences in policy preferences for water sharing – implications for the future. In: Brebbia C (ed) Water and society II. WIT Press, Southampton, pp 307–332
Bjornlund H, Zhao X, Xu W (2014) Variation in the perspective on sharing water – irrigators, their communities and the wider society. Int J Water Gover 2(1). doi:10.7564/13-IJWG29
Byrne J, Kienzle S, Sauchyn D (2011) Prairie waters and climate change. In: Sauchyn D, Diaz H, Kulshreshtha S (eds) The new normal: the Canadian prairies in a changing climate. Canadian Plains Research Centre Press, Saskatchewan
Christensen R, Droitsch D (2008) Fight to the last drop: a glimpse into Alberta's water future. Ecojustice and Bow Riverkeeper, Vancouver
de Loë R (2009) A Canadian vision and strategy for water in the 21st century. Policy Opt 30(7): 21–24
Dietz T, Fitzgerald A, Shwom R (2005) Environmental values. Annu Rev Environ Resour 30(2005):335–372
Fluker F (2013) No public interest standing at the Alberta Environmental Appeals Board. The University of Calgary Faculty of Law http://ablawg.ca/2013/02/15/no-public-interest-standing-at-the-alberta-environmental-appeals-board/. Accessed on 21 May 2013
Graymore M, Wallis A (2010) Water savings or water efficiency—water-use attitudes and behavior in rural and regional areas. J Sustain Dev World Ecol 17(1):84–93
Hall M, Bjornlund H Wei X (2012) Exploring private and district irrigators' perceptions towards water trading in the south Saskatchewan river basin. SAREC report 2012-1, University of Lethbridge, Alberta, Canada
Huddart-Kennedy E, Beckley T, McFarlane B, Nadeau S (2009) Rural-urban differences in environmental concern in Canada. Rural Sociol 74(3):309–329
Jones R, Fly J, Cordell H (1999) How green is my valley? tracking rural and urban environmentalism in the southern Appalachian Ecoregion. Rural Sociol 64(3):482–499

Kelly E, Schindler D, Hodson P, Short J, Radmanovich R, Nielsen C (2010) Oil sands development contributes elements toxic at low concentrations to the Athabasca river and its tributaries. Proc Natl Acad Sci 107(37):16178–16183

Lafreniere K, Deshpande S, Bjornlund (in press) Using conjoint analysis to assess irrigators' preferences of water transfers in Alberta, Canada. Accepted for publication in J Soc Nat Resour

Lafreniere K, Deshpande S, Bjornlund H, Hunter G (2013a) Extending stakeholder theory to promote resource management initiatives to key stakeholders: a case study of water transfers in Alberta, Canada. J Environ Manage 129:81–91, ERA B

Lafreniere K, Bjornlund H, Deshpande S (2013b) Against the grain: segmenting and profiling irrigators opposed to water transfers in Alberta, Canada. In: Bjornlund H et al (eds) Sustainable irrigation technology and policies IV. WIT Press, Southampton, pp 459–468

Lane-Miller C, Wheeler S, Bjornlund H, Connor J (2013) Acquiring water for the environment: lessons from natural resources management. J Environ Policy Plan. http://dx.doi.org/10.1080/1523908X.2013.807210

Loch A, Bjornlund H, McIver R (2011) An institutional critique of differing allocation models to deliver environmental flows: case studies of Queensland and Victoria in Australia. Environ Plan C 29:745–760

Lowe G, Pinhey T (1982) Rural–urban differences in support for environmental protection. Rural Sociol 47:114–128

Nicol L, Klein K (2006) Water market characteristics: results from a survey of southern Alberta irrigators. Can Water Res J 31(2):91–104

Nicol L, Klein K, Bjornlund H (2008) Permanent transfers of water rights: a study of the southern Alberta market. Prairie Forum 33(2):341–356

Nicol L, Bjornlund H, Klein K (2010) Private irrigators in southern Alberta: a survey of the adoption of improved irrigation technologies and management practices. Can J Water Resour 35(3):339–350

NWC (2011) Water markets in Australia: a short history. National Water Commission, Canberra

Russenberger M, Bjornlund H, Xu W (2011) Exploring the links between values, beliefs, attitudes and social norms and policy preferences for water reallocation in Southern Alberta, Canada. In: Proceedings from the XIV World Water Congress, Recife, Brazil, September

Russenberger M, Bjornlund H, Xu W (2012) Exploring links between policy preferences for water reallocation and beliefs, values, attitudes and social norms in Alberta, Canada. In: Pepper D, Brebbia V (eds) Water and society. WIT Press, Southampton, pp 107–118

Saliba B, Bush D (1987) Water markets in theory and practice: market transfers, water values, and public policy, Studies in water policy and management no. 12. Westview Press, Boulder

Salka W (2001) Urban–rural conflict over environmental policy in the western United States. Am Rev Publ Adm 31(1):33–48

Vera-Toscano E, Gomez-Limon J, Moyano E, Garrido F (2008) Factors determining citizen's attitudes towards agri-environmental property rights. Environ Resour Econ 41:541–561

Wheeler S, Bjornlund H, Shanahan M, Zou A (2009) Who trades water allocations? Evidence of the characteristics of early adopters in the Goulburn-Murray irrigation district, Australia 1998–99. Agric Econ 40:631–643

Wheeler S, Garrick D, Loch A, Bjornlund H (2013a) Evaluating water market products to acquire water for the environment in Australia. Land Use Policy 30:427–436

Wheeler S, Zuo A, Bjornlund H (2013b) Selling water for the environment: how sustainable is it for irrigators. In: Bjornlund H et al (eds) Sustainable irrigation. WIT Press, Southampton, pp 29–40

Wheeler S, Zuo A, Bjornlund H (2014) 'Australian irrigators' recognition of the need for more environmental water flows and their intentions to donate water allocations. J Environ Plann Manage 57(1):104–122

Young M (2013) Trading into and out of trouble. Australian's water allocation and trading experience. In: Maestu J (ed) Water trading and global water scarcity: international perspectives. Taylor & Francis, Abingdon, Oxfort, UK

Chapter 13
Water Markets in India: Extent and Impact

R. Maria Saleth

Abstract Local level water scarcity does motivate farmers not only to improve on-farm water use efficiency but also to evolve new and informal institutional arrangements for inter-farm water sharing. An eminent case here is the spontaneous emergence and growth of water markets (WMs) in many parts of India. Even though Indian WMs are localized, fragmented, and uneven across regions, they are growing in magnitude and gaining in significance. This chapter evaluates the major economic and institutional aspects of Indian WMs based on a critical review and synthesis of available empirical evidence on the subject. Specifically, this chapter attempts to (a) provide an idea about the magnitude and value of water trade, especially at the national level, (b) outline the technical and institutional environment within which Indian WMs are operating, (c) describe their major economic and institutional features, (d) evaluate their efficiency, equity, and sustainability implications, and, (e) suggest the legal and institutional changes needed to make WMs an efficient option for water management in India.

Keywords Groundwater regulations • India • Rental markets • Water rates • Water markets • Irrigation service markets • Water rights • Water trade • Water institutions

R.M. Saleth (✉)
Madras Institute of Development Studies, 79-Second Main Road,
Gandhi Nagar, Chennai 600 020, India
e-mail: r.saleth@mids.ac.in; r.saleth@gmail.com

K.W. Easter and Q. Huang (eds.), *Water Markets for the 21st Century: What Have We Learned?*, Global Issues in Water Policy 11, DOI 10.1007/978-94-017-9081-9_13,

13.1 Introduction

For a monsoon-dependent country such as India, water scarcity[1] has emerged as a major constraint for agricultural production, farm income, and rural employment. How farmers cope with such water scarcity at the field level is as important, if not more, than how policy-makers deal with the same issue at the national level. Local level water scarcity does motivate farmers not only to improve on-farm water use efficiency but also to evolve new and informal institutional arrangements for inter-farm water sharing. An eminent case is the spontaneous emergence and growth of water markets (WMs)[2] in many parts of India. Although the practice of paying for and selling of water is nothing new in the Indian context,[3] the growth of WMs, especially since the expansion of energized pumping technology in agriculture, assumes significance because they occur in an entirely different economic, institutional, and technological environment. While modern water selling practiced by farmers in India has been traced to the 1920s, it has been systematically documented only since the late 1960s (e.g., Patel and Patel 1969; Shah 1985, 1993; Kolvalli and Chicoine 1989; Saleth 1991, 1994; Moitra and Das 2004; Nayak 2007; Mukherji 2008; Manjunatha et al. 2011; Varughese 2012). Furthermore, the recent literature on WMs in India has not only declined but changed its focus. The studies conducted during the 1990s were more detailed with a focus on the internal economic dynamics of WMs, while the recent studies focus more on their institutional features and environmental impacts.

Even though Indian WMs are localized, fragmented, and uneven across regions, they are growing in magnitude and gaining in significance. This chapter evaluates the major economic and institutional aspects of Indian WMs based on a critical review and synthesis of available empirical evidence on the subject. Specifically, this chapter attempts to (a) provide an idea about the magnitude and value of water trade, especially at the national level, (b) outline the technical and institutional environment within which Indian WMs are operating, (c) describe their major economic and institutional features, (d) evaluate their efficiency, equity, and sustainability implications, and, (e) suggest the legal and institutional changes needed to make WMs an efficient option for water management in India.

[1]Water scarcity is defined here in terms of relative physical scarcity of water, i.e., the inadequacy of water supply in relation to water demand in a particular context.

[2]These markets involve essentially groundwater and, to some extent, lift-based surface water, especially in the eastern parts of India. As of now, there are no water markets for flow-based canal or tank water in India.

[3]For instance, Kautilya's *Arthashastra*, dating back 400 BC, specifies the royalty to be paid for water, while south Indian inscriptions dating back 1202 AD notes the practice of selling surface water (see Maloney and Raju 1994: 46).

13.2 Water Marketing: Nature and Magnitude

It is instructive to know at the outset how a typical WM emerges and operates in the Indian context. Due to technical indivisibility and the need to cope with power shortage, Indian farmers often install lift irrigation systems (LIS), i.e., wells, pumps, and water conveyance networks, with a water delivery capacity exceeding their own needs. As LIS involve heavy investment, the excess pumping capacity and surplus water involve an inherent economic loss not only to LIS owners but also to the society as a whole. This excess pumping capacity and surplus water can be used by adjacent farms without LIS or with LIS but insufficient water or costly extraction, thereby, benefiting not just the buyers but also the society at large (Saleth 1998). It is under these conditions that typical WMs emerge in India under a variety of circumstances.

The WMs can occur either in groundwater areas, in canal and tank commands with unreliable water supply, or in areas with lift-based surface water. The wells can be either dugwells or shallow/deep tubewells, the energy source can be either electricity or diesel, and the water conveyance can involve open channels, hoses, or underground pipelines. A few characteristic features of Indian WMs include the following. First, most water sales do not involve any sacrifice of self-irrigation (Varughese 2012). Although Shah (1993) cites cases in Gujarat where farmers, at times, abandon their own cultivation to sell water, they are an exception rather than the rule. As a result, the opportunity cost of water becomes either undefined or when defined (i.e., in terms of the impact of current sale-related water withdrawals on the future pumping costs of both the sellers and those who are the well owning neighbors), it remains outside the economic calculus of sellers under the current systems of water ownership and power pricing. Second, although water sales are essentially for irrigation purpose, sales for non-irrigation uses are not uncommon. These non-irrigation uses include the water sales for brick making and urban domestic use (Shankar 1992; Palanisami 1994; Packialakshmi et al. 2011). Although there have been many active WMs for groundwater in tank-irrigated areas, especially in the southern parts of India, they have declined considerably in recent years due to poor maintenance of tanks and their low recharge potential. Finally, there is also a conceptual issue with considerable legal implications, i.e., whether the sellers are selling the excess pumping capacity (i.e., rental market for LIS), water (i.e., WM), or both (Saleth and Thangaraj 1993; Saleth 2004).

While there is no systematic national-level estimate of the magnitude and value of water trade at present, a few micro studies done in the past do provide some idea especially at the village or regional levels. Of the two West Bengal villages surveyed by Kolavalli and Atheeq (1990), the village in Nadia district showed 85 % of well owners selling water whereas the village in Puralia district reported no water selling at all. In a 7-village study in Vaigai basin, Tamil Nadu, of the well owners studied, only 25 % in wet areas (i.e., canal or tank commands) and 15 % in dry areas (i.e., groundwater region with extreme water scarcity) were selling water (Janakarajan 1994: 52). In the regions adjoining the Union Territory of Pondicherry, only 13 % of the well owners were involved in water sales (Varughese 2012: 58). In contrast, the area under purchased water is projected to be up to 80 % for northern Gujarat (Shah 1993: 205), while a 16-village study in Allahabad district, Uttar Pradesh found it to be about 60 % (Shankar 1992: 12). Yet, in Vaigai basin, Tamil Nadu, the area under purchased water seems to be no more than 30 % of the total irrigated area (Janakarajan 1994). Finally, there are studies which report no water selling at all either in all or part of their sample areas (see Shah 1993: 55; Varughese 2012: 59–60).

Although there is no current national level estimate of area irrigated by WMs, an estimate made in the early 1990s placed the figure at up to 50 % of the total gross irrigated area under private LIS (Shah 1993: 250). This projection neither has any systematic basis nor allows disaggregation either by region, well type, or energy use. While there is no current data for this purpose, the data from the national level sample surveys conducted by the National Sample Survey Organization (NSSO) during 1976–1977 and 1997–1998 are of considerable utility (NSSO 1997, 2000). To better appreciate the implications of this data for water selling, two crucial points are to be noted. First, even though the data are out-dated, they are the only sources of information, providing a basis for a national level estimate of both water trade and its economic worth. Second, while the data for 1976–1977 are in terms of 'pump set rentals', i.e., the proportion of pump sets being rented out, the data for 1997–1998 are in terms of the percentage of irrigated area under five major crops (UFMC) using 'hired irrigation services'.[4] Finally, the data sets cover pump set rentals/hiring of irrigation services occurring both in the canal and non-canal areas.

Even though the data for 1976–1977 pertain to pump set rentals, a closer examination reveals that the phenomenon of pump set rentals inherently involves water selling in the case of all immobile pump sets permanently fitted on dugwells and tubewells or connected with electric supply lines. Pump rentals can only occur independently of water selling in the case of mobile pump sets (e.g., diesel pumps fitted with lift-based surface system or filter points), which can be moved with negligible costs to access water from sources other than those of their owners.[5] The data for 1997–1998 cover the proportion of irrigated area UFMC involving both pure pump set rentals and pump set rentals involving water selling (NSSO 2000: 38). Since the disaggregated information on pump set rentals by water and power sources is provided by the 1976–1977 data, it is possible to isolate pump set rentals involving water sales from those occurring without water sales. This fact can serve as a basis for estimating and projecting the extent of current water trade at the national level.

Table 13.1 provides information on pump set rentals by water and power sources across major states in India. Of the total estimated pump sets of 6.4 million, roughly

[4]The five major crops are in terms of the value and they are not fixed but vary across households and regions (NSSO 2000: S-20).

[5]There is some evidence for such pure rentals. For instance, in Nadia district, West Bengal, farmers rent diesel pump sets to pump water from their own sources on payment of Rs. 1,200–1,500/crop season to pump set owners (Kolavalli and Atheeq 1990: 26). The diesel and maintenance costs are borne by the farmers renting the pumps.

Table 13.1 Pump set rentals and their distribution by water source and energy use: 1976–1977

		Pump set rentals		Pattern of rentals by water source (%)			Pattern of rentals by energy use (%)		
States	Total pump sets reported ('00)	Number ('00)	(%)	Wells	Tube-wells	Others[a]	Electricity	Diesel	Both[b]
Andhra Pradesh	5,971	407	6.82	62.90	19.41	17.69	42.24	57.76	0.00
Assam	2	2	100.00	0.00	0.00	100.00	0.00	100.00	0.00
Bihar	3,547	878	24.76	4.10	70.62	25.28	30.52	58.55	10.93
Gujarat	4,933	116	2.35	93.97	6.03	0.00	84.48	15.52	0.00
Haryana	2,914	138	4.74	2.90	91.30	5.80	51.45	48.55	0.00
Jammu & Kashmir	6	1	16.67	100.00	0.00	0.00	100.00	0.00	0.00
Karnataka	2,669	165	6.18	68.48	0.00	31.52	40.61	57.57	1.82
Kerala	1,038	386	37.18	38.08	0.00	61.92	14.25	84.20	1.55
Madhya Pradesh	3,423	833	24.34	33.49	0.36	66.15	7.08	92.92	0.00
Maharashtra	7,424	1,050	14.14	68.86	0.00	31.14	31.24	67.33	1.43
Orissa	171	69	40.35	5.80	0.00	94.20	5.80	94.20	0.00
Punjab	5,544	170	3.06	4.12	95.88	0.00	25.88	71.77	2.35
Rajasthan	4,228	328	7.76	53.96	5.49	40.55	21.34	77.14	1.52
Tamil Nadu	11,096	267	2.41	87.64	5.62	6.74	78.28	18.72	3.00
Uttar Pradesh	8,722	224	2.57	8.03	89.29	2.68	5.36	94.64	0.00
West Bengal	1,959	974	49.70	0.82	45.28	53.90	4.11	95.07	0.82
Total	63,647	6,008	9.44	35.22	27.83	36.95	24.93	72.66	2.41

Source: NSSO (1984)

[a] It includes lift irrigation from surface water bodies like rivers, ponds, etc.

[b] The category 'both' includes wells having both electric and diesel pumps, the latter used mostly as stand-by

Table 13.2 Extent of hired irrigation services in canal and non-canal areas: 1997–1998

State	Irrigated area (%)	Irrigated area UFMC (%)	Irrigated area UFMC using hired irrigation services		
			Canal area (%)	Non-canal area (%)	All (%)
Andhra Pradesh	77	72	33	35	34
Assam	23	19	*	*	31
Bihar	74	76	60	71	69
Gujarat	70	73	45	35	38
Haryana	93	90	40	37	39
Himachal Pradesh	26	17	*	*	14
Jammu & Kashmir	73	48	3	7	5
Karnataka	46	41	6	23	17
Kerala	40	46	7	7	7
Madhya Pradesh	49	56	18	34	30
Maharashtra	43	44	10	20	18
Orissa	31	30	18	37	27
Rajasthan	68	61	17	30	28
Punjab	95	97	18	21	19
Tamil Nadu	83	84	24	25	25
Uttar Pradesh	88	91	61	69	67
West Bengal	75	72	57	72	67
India	65	66	40	49	46

Source: NSSO (2000)

about 10 % were involved in pump set rentals. Since about 63 % of these rentals occur with dugwells and tubewells, the majority of them must happen as part of water selling activity. This is particularly so in the case of the Indo-Gangetic and hardrock dominated states respectively that are irrigated by tubewells and dugwells. Since the pump set rentals in the 'others' category (i.e., surface water-based lifts or filter points) are substantial only in states such as Assam, Kerala, Madhya Pradesh, Orissa, and West Bengal, the likelihood of them occurring as pure rentals, i.e., independently of water selling, is more in these states than elsewhere. Going by the power source of rented pump sets at the national level, about 25 % of them relying on electric power clearly involve water selling because of the inherent immobility of electric pumps. Even in the case of rented pump sets with diesel power, a substantial proportion can also involve water selling to the extent their mobility is physically limited or economically costly. The overall implication is that pump set rentals occurring as a part of water selling are more than pure renting in the case of all states except Assam, Kerala, Madhya Pradesh, Orissa, and West Bengal.

Table 13.2 shows the percentage of irrigated area UFMC using hired irrigation services in the canal and non-canal areas across the major states in 1997–1998. The reliance on hired services is relatively higher in non-canal regions (49 %) as compared to canal regions (40 %). Across states, the use of hired services is more pronounced in Bihar (69 %), West Bengal (67 %), and Uttar Pradesh (67 %). Notably, in the case of states that were shown to be dominant in water marketing in micro level studies such as Gujarat (38 %), Andhra Pradesh (34 %), Tamil Nadu

(24 %), and Karnataka (17 %), the extent of hired irrigation services is relatively lower. It can be noted that the states with a higher use of hired services are in the Indo-Gangetic region with higher water tables and alluvial aquifers whereas those states with lower use of hired irrigation services are in the hard-rock region with poor aquifers. At the national level, 46 % of the irrigated area UFMC rely on hired irrigation services. This fact taken with the total area UFMC of 150 million hectare (mha) and the share of irrigated area UFMC of 92 mha (NSSO 2000: 15 & 35) would mean that the extent of irrigated area UFMC using hired irrigation services during the survey year of 1997–1998 was 42 mha. This may be an overestimate of the magnitude of water marketing since all these areas may not be using hired services on a regular basis and the hired services may not involve water selling in all cases due to pure pump set rentals. As noted earlier, the latter is actually the case in states such as West Bengal and Orissa as well as in areas relying on surface-based water lifts. This means that we have to use the disaggregated information available in 1976–1977 data for estimating the true magnitude of water marketing at the national level.

With increasing water scarcity and groundwater productivity on the one hand and expanding LIS on the other, not only the scale of water trade must have increased significantly but also its regional composition and energy base must have changed substantially since the mid-1970s. Even if we assume that pump set renting/water selling has tripled since 1976–1977, then, 6.4 million pump sets, representing about 30 % of the total pump sets of 21 million, can be expected to be involved in current rentals and water trade at the national level. If we assume that the same percentage of the total rentals involving water sales (i.e., those rentals occurring in the case of wells and tubewells) observed in 1976–1977, i.e., 63 %, continues today, then, the total pump set rentals involving water selling will only be about 4.2 million. If we make a simple assumption that each pump in the country irrigates, on an average, 2 hectares (ha) of buyers' land per year, the additional irrigation due to WMs would be about 8.4 mha, representing about 14 % of the total area under groundwater irrigation. If we consider hired irrigation services as a whole, i.e., involving both water selling and pure rentals, the irrigation benefits would have been extended to about 13 million ha, representing 22 % of the area under groundwater irrigation. If we assume further an additional output of Rs. 15,000/ha/year, the total value of output due to water selling could be Rs. 126 billion/year and the total value of output due to hired irrigation services would be about Rs. 195 billion/year. These estimated values can be taken as the ballpark figures for the likely value of current water trade/hired irrigation services in India.

13.3 Water Market Environment

Indian WMs are operating within an environment defined by the interplay of technical, resource-related, and institutional factors. To begin with, groundwater dominates with the 60 % share of the total irrigated area of 102 mha in the country at present. The growth in groundwater irrigation was phenomenal during 1951–2007, as it increased by 6.3 times as compared to the increase of surface irrigation of only 3.4 times (Planning Commission 2008). Since irrigation and cropping intensities are higher for groundwater irrigation, its productivity impact is far greater than that for other irrigation sources. Across states, groundwater irrigation dominates in Gujarat (75 %), Maharashtra (63 %), Madhya Pradesh (52 %), Punjab (57 %), Rajasthan (61 %), Tamil Nadu (45 %), and Uttar Pradesh (66 %) (NSSO 2000). While the government does play a catalytic role in groundwater development through the rural electrification program and concessional credit support (Marothia 2003), it is the millions of private farmers, who actually develop and control the groundwater economy in India.

Groundwater irrigation is made possible by about 15 million electric and 6 million diesel pump sets fitted with some 10 million shallow/deep tubewells located mainly in the Indo-Gangetic and deltaic regions, 15 million dugwells spread essentially in the hardrock region covering western and peninsular India, and 0.5 million surface water-based lift systems mostly in the deltaic regions. While there is no current data on the ownership, capacity, and energy use pattern of LIS, an analysis of the data from a national level sample survey conducted during 1976–1977 (see NSSO 1984) suggests that over 90 % of the total LIS are individually owned and about one-fifth of them are jointly owned by two or more farmers because of either sub-division following land inheritance or investment sharing due to cost/risk factors. Diesel pumps are relatively more significant in Uttar Pradesh (69 %), Orissa (86 %), West Bengal (94 %), and Madhya Pradesh (56 %) as they are more suitable in these states with a higher water table and alluvial aquifers. Diesel pumps with a greater mobility are relied on not only as an exclusive means but also as a stand-by mechanism for use during power shortages in many parts of India. The capacity of about 71 % of the pump sets exceeds 5 horse power (hp).

The farm power tariff and supply policies have an important influence on WMs (Mukerjee 2007). The unit rate system where power charges are based on metered consumption is disappearing fast in India. Since the late 1980s, most states have switched from unit rate to a flat rate charged annually on the basis of the information on hp capacity of pump sets that is available with the electricity authorities. Since flat rates are being charged as a lump sum it makes marginal pumping cost almost zero and causes average pumping cost to decline with water sales, which enhances the prospect for water sales. Similarly, the policy of free farm electricity being pursued since the early 1990s in states such as Tamil Nadu and Punjab also encourages water selling as the energy cost of pumping is zero. Unfortunately, most of these positive equity effects get attenuated not only by deteriorating power quality (e.g., load shedding, availability at odd hours, etc.) and limited hours of power availability but also by aquifer depletion.

In practice, the legal and regulatory regimes governing groundwater in India are not much different from those in other countries. While groundwater is popularly considered as a 'common pool' or 'open-access' resource, existing Indian laws seem to link groundwater rights with land rights. For instance, based on the 'dominant heritage' principle implied in the Transfer of Property Act IV of 1882, the Easement

Act of 1882 allowed private usufructuary rights in groundwater by viewing it as an easement inseparably connected to the dominant heritage, i.e., land. When this legal linkage between land and groundwater is operationalized, it amounts to a correlative rights system where an individual's right in groundwater is postulated to be in proportion to his land ownership (Saleth 2004). This proportional form of groundwater rights is either implied or explicitly stated in many subsequent laws and policies. In fact, the 1976 National Commission on Agriculture (NCA) actually recommended such a system for India [see Government of India (GOI) 1976: 23].

Although the National Water Policy of 2002 recognizes the need to limit individual and collective water withdrawals, it fails, however, to specify the required institutional mechanisms. The Model Groundwater (Control and Regulation) Bill of 1992, which postulates a kind of groundwater permit system, also fails to set withdrawal limits (GOI 1992a). However, since the Model Bill requires the mandatory installation of water meters, it does leave room for an eventual introduction of a water quota-based permit system. Unfortunately, this Bill was not adopted by any states, though it did induce some legal/administrative initiatives in states such as Karnataka, Maharashtra, and Tamil Nadu. Prior to the 1992 Bill, Gujarat had a law based on an amendment to the Bombay Irrigation (Gujarat Amendment) Act of 1976 (1979) that limits tubewell depth to a maximum of 45 m (m) but only in the Mehsana region. The ineffectiveness of this law can be gauged by the fact that the actual depth observed for Mehsana region averaged 60–100 m in the early 1990s (Shah 1993). The situation could be much worse today.

There are no explicit policy statements on WMs in India, although there are few actions aimed at discouraging them. For instance, in the initial years, the state electricity boards thought about discouraging WMs since they involved an illegal use of electricity (see Shah 1993: 47). Since the NCA has postulated a correlative system of groundwater rights, it visualized only rental markets for LIS but not WMs. This is so because "if a farmer constructs a private tubewell which yields more water than the share of his holding, then it should be possible for the farmers (i.e., those without their own wells/pump sets) having contiguous holding to avail of their share of water on payment of share cost" (GOI 1976: 23). Based on this reasoning, both the NSSO and the National Council of Applied Economic Research treated WMs as well/pump set rentals or the hiring of irrigation services in their surveys conducted in the mid-1970s and 1990s (see Saleth 1996: 186–187; NSSO 1984, 2000). The Model Groundwater Bill seems to limit WMs when it states that "small and marginal farmers will not have to obtain a permit if the well is proposed to be sunk for *exclusively personal purposes excluding commercial use*" and the commercial use of water could be a basis for refusing permit to *any user* [emphasis added] (GOI 1992a: 3–4).

Despite these legal and policy actions, the control over groundwater at the field level continues to be governed by a *de facto* system of rights[6] where larger farmers

---

[6]On the surface of it, the *de facto* system appears to resemble the appropriative rights system used in the western US. But, unlike the latter, the former is not subject to the 'beneficial use' doctrine nor does it guarantee any legal recognition (see Saleth 1996: 196).

with higher pumping capacity and deeper tubewells have a greater control over the resource than others (Saleth 2004). In the face of weak legal provisions, other regulatory approaches based on well-spacing and depth restrictions as well as power supply manipulations have been tried but with little success. The well-spacing and depth norms are enforced only in areas where annual groundwater withdrawal exceeds 85 % of the annual recharge (Kurien and Sinha 2007).

While the well-spacing norm being enforced prohibits new wells within a radius of 200 m in most parts of India, the norm can be as high as 680 m in areas with deep tubewells and serious depletion (Shah 1993: 11). Similarly, there are also depth restrictions. For instance, in Gujarat, the depth of deep tubewells cannot exceed 100 m. Since these spacing and depth restrictions take effect only when a farmer applies for concessional loan/well permit/electric connection, they restrict mostly the resource poor farmers. While a restricted power supply policy provides some regulatory respite, it is of little value in the face of larger capacity pumps, multiple wells, and the diesel pump option. Thus, current policies reinforce rather than regulate the *de facto* system within which current WMs are operating. As a result, there are no proper institutional mechanisms to regulate groundwater overexploitation in India.

13.4 Economic Characteristics of Water Transactions

Among the economic aspects of Indian WMs, the most important ones deserving our attention are: the method and mode of water payment, the linkages between the WM environment and water rates, and the effects of water scarcity and power tariffs on water rates.

13.4.1 Methods and Modes of Water Rates Payment

As noted already, there has been a sharp decline in WM literature in recent years and even the few recent studies that are available do not provide comparable information on water rates that farmers charge in WMs. But, such information is available from many studies conducted during the 1990s that are useful to provide some idea about the method and modes of water rates prevalent in WMs in different parts of India.[7] Although water rates on an hourly basis are common in most WMs, payments are also based on area irrigated and the number of irrigations. While the hourly rates varied from Rs. 3 to 45 for electric pumps and Rs. 7 to 21 for diesel pumps, the area-based rates varied from Rs. 225 to 3,705/ha. The rates per irrigation varied from Rs. 54 to 62/ha in Bihar and Uttar Pradesh and Rs. 20 to 27/ha in Haryana,

[7] The exchange rate applicable during this period was US$ = Rs. 32.

Punjab, and West Bengal (Saleth 1991: 351). In some cases, water rates also vary by volume. For instance, in Banaskantha district, Gujarat, water has been charged either at a 'full-flow' rate of Rs. 120/h (i.e., Rs. 3.43/hp/h) or at a 'half-flow' rate of Rs. 60/h (Rs. 1.71/hp/h) (Shah and Ballabh 1993: 4).[8] In Vaigai basin, Tamil Nadu, the water rates vary not only by water discharge but also by water quality (e.g., Rs. 10/h for good quality water but only Rs. 5/h for saline water) (Janakarajan 1994: 53 & 56). Water services are also charged on a seasonal basis. For instance, in the Gingee watershed adjoining the Union territory of Pondicherry, water charges per season ranged from Rs. 1,000 to 2,000 in the case of groundnut and blackgram to Rs. 7,000 to 10,000 in the case of rice and sugarcane (Varughese 2012: 68).

Although payments for water are mostly in cash either immediately or at the end of the crop season, payment-in-kind involving a 'water rent' in the form of a given share of the crop output also occurs. This 'water rent' varied from one-third of output in Andhra Pradesh and Tamil Nadu to 50–66 % in Gujarat (Asopa and Dholakia 1983; Shah 1993; Janakarajan 1994). Though rare, water charges are also paid—either in part or in full—in terms of labor (Shah 1993; Janakarajan 1994). Another interesting mode of payment for water is observed in Madurai district, Tamil Nadu where water sales occur in the context of large scale intra-basin water transfers effected through lengthy underground pipelines developed by private initiatives supported, in part, by bank loans. Here, the buyers have to deposit a refundable amount of Rs. 24,700/ha which the sellers use either to service the loans or to buy additional rainfed land in the vicinity. In this arrangement, water is supplied in *lieu* of interest payments (Saleth 1998).[9]

These different pricing methods are used by farmers in different situations to meet specific objectives. For instance, the hourly payment is used for diesel pumps as diesel consumption is linked to hours of operation. In contrast, though hourly rates are more common for electric pumps, especially under unit rate-based power charges, area rates are also used for specific crops or time with a view to minimize monitoring costs, particularly under flat rates for power.[10] While hourly or area-based rates are used in the case of regular water sales, per irrigation rates are charged for sporadic sales for supplemental or conjunctive irrigation. Obviously, the area-based rates, especially with crop sharing arrangement, are more appealing in areas with severe water scarcity and greater production risk. The terms of payments depend on source of power and frequency of irrigation. Immediate payment is

---

[8] Apparently, these rates are used in the context of a pipeline system where 'full-flow' involves the sale of the whole flow to a single buyer whereas 'half-flow' involves sale to two buyers each with an outlet open at the same time.

[9] Since the monthly interest rate charged by private money lenders in the area varies from 2 to 10 %, water payment amounts to Rs. 5,928–29,640/ha/year. The higher amount is not a serious problem as buyers mostly grow a high-value crop such as banana and grapes.

[10] For instance, in Nadia district, West Bengal, the area-based rates are used for paddy but hourly rates for other crops since it is easy to monitor water supply for paddy. Likewise, high monitoring cost and low demand for night irrigation also make area-based rates more appealing in the case of water sales for night time irrigation (Kolavalli and Atheeq 1990: 37–38).

needed for diesel pumps/sporadic water sales whereas monthly or seasonal credit is normally allowed for electric pumps/regular water sales.

13.4.2 Market Environment and Water Rates

Water rates are influenced by water depth, crop pattern, energy source, and power charges that exist in the areas of WMs. Several interesting patterns can be identified irrespective of whether one uses the simple hourly rates or those normalized for pump set capacity. The hourly rates in the three Gujarat cases are far higher (ranging from Rs. 15 to 45) as compared to other cases (ranging from Rs. 3 to 20). The higher water rates in Gujarat reflect the effects of not just the prevalent unit rate-based power charges (i.e., pumping cost) but also the water depth (i.e., water scarcity), pumping capacity (i.e., volume of water), and cropping pattern (i.e., water productivity). Even among the non-Gujarat cases, the level and spread of water rates observed in the hardrock regions (Andhra Pradesh and Tamil Nadu) are higher (Rs. 3 to 20) as compared to those (Rs. 4 to 5) in the Indo-Gangetic regions (Punjab, Uttar Pradesh, and West Bengal).[11] The higher level and spread of hourly rates in the hardrock cases reflect essentially the joint effects of higher water scarcity and wider hydrological variations.

There are a variety of water-based tenancy contracts (see Shah 1993: 51–52). For example, there are two distinct types of such contracts in Kheda district, Gujarat: (a) a two-party contract where the water seller provides irrigation, shares 50 % of the cash expenses (excluding labor costs), and claims 50 % of the output and (b) a three-party contract where a water seller, a land owner, and a laborer share equally the cash expenses as well as crop output. Similarly, in Karimnagar district, Andhra Pradesh, water sales occur within (a) labor contracts, (b) crop sharing contracts, and (c) crop and input sharing contracts. These contracts represent not only an institutional evolution of crop sharing within the context of WMs but also link WMs with other rural input/output markets.

Apart from these water-based tenancy contracts, the pricing methods are also accompanied by certain informal conventions and contractual obligations with considerable implications for water use efficiency and risk-sharing (see Kolavalli and Atheeq 1990: 38–40; Palmer-Jones 1994: 27). The area-based method involving crop shares, which provides lesser incentive for water conservation than the hourly method, allows risk-sharing between buyers and sellers, and involves some informal contractual obligation for sellers to provide irrigation for the whole season. In the case of both the area and per irrigation rates, there are mutually agreed upon conventions (e.g., the level or intensity of irrigation constituting 'full irrigation') to avoid conflicts and water overuse.

[11] As noted earlier, most of the prices were from studies conducted during the 1990s and for the detailed reference for these studies, see Saleth (1998).

13.4.3 Influence of Water Scarcity and Power Charge

While higher rates are often attributed to power tariff and monopoly behavior, water scarcity and productivity-related factors also have a strong influence. For instance, the water rate in north Kheda region has been low because the WM here occurs in canal region with better groundwater supply. In contrast, the rates in Mehsana and Sabarkantha—two regions known for their rapidly receding water table—have been very high. Similarly, the differential pattern of water rates in the hardrock and Indo-Gangetic regions also shows the influence of relative water scarcity. The hourly rates for diesel pumps are more than those for electric pumps due to high diesel and repair costs (Rs. 5.85 to 7.95/h). Regionally, the hourly rates in the Indo-Gangetic region are lower (Rs. 8 to 12) relative to those in the hardrock region (Rs. 7 to 21). The rates in the Indo-Gangetic region are 1.3 to 2.0 times higher than the operating cost while the rates are 2.5 to 3.5 times higher than the operating cost in the hardrock region. Finally, the regional pattern of area-based rates is very similar to that of hourly rates. They are generally higher for water-intensive and long-duration crops like sugarcane and banana but lower for others. Similarly, commercial crops (e.g., cotton and tobacco) have higher rates than food crops (e.g., bajra, ragi, and sorghum). Thus, the area-based rates seem to reflect the effects of relative water consumption and water productivity.

The influence of water scarcity and power charges on water rates is well established (Mukerjee 2007). But, there are also other factors which are equally important in WMs. Water scarcity influences not only the level but also the method of fixing water rates. Unfortunately, there are serious difficulties involved in isolating the exact amount of water payment that is due to scarcity. To understand this, let us decompose the water payment into the following four components: (a) pumping costs, (b) labor costs in pump operation and monitoring, (c) a rental payment for the LIS,[12] and (d) the value of water. As noted earlier, since the private opportunity cost becomes undefined as most water sales occur without any sacrifice of self-irrigation, the opportunity cost principle is of little help in isolating the scarcity value for water. The same is also the case with the productivity approach. It is true that water rates observed across states vary, more or less, directly with both crop productivity and water scarcity. But, this cross-sectional comparison yielding a relative result does not support the idea that water rates, in a given context, fully capture either the productivity or scarcity value of water in an absolute sense. For instance, if the annual area-based water payments are compared with the water productivity estimates used by the Vaidyanathan Committee (GOI 1992b), for most crops except paddy and wheat with payment rates of Rs. 2,223–3,705/ha and banana (Rs. 5,878–9,174/ha) and for most states except Gujarat, the water payment accounts

[12]This component is not to be treated as sunk costs as LIS are often installed with borrowed funds and repair and maintenance require frequent out-of pocket expenses.

for only a small fraction of the relevant water productivity levels.[13] However, if one has information on the first three components noted above, the value component can be obtained as the residual part of water payment.

Although no empirical study provides such decomposed information on water payment, the residual approach can, nevertheless, be used in cases where the rates differ between water deliveries from stationary pumps (i.e., those permanently fitted on sellers' wells) and mobile diesel pumps that extract water from buyers' wells or from other common water sources. Such a differential pricing has been commonly observed with diesel pumps ever since the late 1960s. For instance, in Uttar Pradesh, the hourly rate for stationary pumps was Rs. 3.03 compared to Rs. 2.70 for mobile pumps (Patel and Patel 1969). Similarly, in West Bengal, the hourly rate for water pumped from seller's well was Rs. 14 and that for water pumped from buyer's well or from other common sources was Rs. 12 (Shah 1993: 45). Although the difference between the two rates is only a smaller fraction of water payment, it represents an estimate of the value of water.[14] Although it is difficult to isolate the value in other contexts, it is reasonable to expect that farmers in most contexts will be attributing a certain fraction of the water payment to the water value. Yet the attribution is very small and water rates fail to fully capture the scarcity value of water.

When variable power charges are used, the rates for water are higher than if fixed charges are used for power. A corollary to this is that as the flat rate system leads to more water sales especially to small farmers, which has equity benefits. Free power supply policy also has similar equity benefits. But, in terms of efficiency, the variable rate system is superior to the former. For instance, the Gujarat government switched from unit rate system to flat rate system in mid-1987. Within 4 months, water charges in most parts of the state declined by 27–58 % (Shah 1993: 111). The flat electricity rates charged by different states varied from Rs. 48 to 260/hp/year but the average water price varied only from Rs. 0.40 to 1.10/hp/h (Shah 1993: 112). Although power charges—both their level and method—certainly affect water rates (and even pricing method), the rate changes may have little effect on water sale and purchase decisions of farmers as long as water charges form only a fraction of the marginal value product of water and buyers pay the full pumping costs for the water (Saleth 1996: 164).

---

[13]It is calculated as the difference between the average per hectare productivity levels of irrigated and unirrigated lands. The estimated water productivity levels are: Rs. 3,639 for Gujarat, Rs. 3,370 for Punjab, Rs. 1,555 for Uttar Pradesh, Rs. 4,407 for Andhra Pradesh, Rs. 4,364 for Tamil Nadu, and Rs. 2,457 for West Bengal (see GOI 1992b).

[14]But, when water withdrawals are restricted, for example, by a water quota system, the estimated value for water will become a large and major component of water payment since the opportunity cost of water will be higher because the sellers can sell only if they save some of their quota either through efficient use or non-use. This suggests the need to introduce some form of water quota or water rights system if one wants the water rates to reflect the scarcity value or the opportunity cost of water.

13.5 Institutional Dimensions and Behavioral Patterns

WMs in India display wide variations in terms of organizational features and behavioral patterns. While WMs in north Gujarat have taken almost an agribusiness approach with cash-based transactions complete with cash receipts and purchase records, those in Andhra Pradesh, Tamil Nadu, and even in parts of Gujarat show shades of feudal character involving 'water rent' and unpaid labor services. In between these two extremes fall the WMs of the relatively water abundant Indo-Gangetic and deltaic regions displaying a rather muted form of commercial character. Since the institutional aspects of WMs have important behavioral implications, some of these aspects need to be considered.

13.5.1 Market Size and Structure

Although the geographic locus of WMs is limited by the physical characteristics of water, their size is enlarged, in some areas, by modern water conveyance technologies such as underground pipelines and hoses. But, such efforts in market expansion are confined to a very few regions and can lead to an unbalanced market structure since more buyers are added to the market than sellers. Thus, most Indian WMs are both small and unbalanced as indicated by the average number of buyers per seller. This number varies from 2 to 80 across regions (Shah 1993: 51; Varughese 2012: 68). Typically, the number is higher for Gujarat because of deep tubewells (up to 400 m in depth), heavy duty pumps (30–75 hp), and vast pipeline networks, but lower for other states. Another index of WM size is the extent of water trade captured by the share of water output sold. While this share is about 40 % in Uttar Pradesh (Shankar 1992) and up to 64 % in West Bengal (Kolavalli and Atheeq 1990), it is estimated to be 80–90 % in Gujarat (Shah 1993: 50–51).

In general, the sellers in WMs have more market power than buyers (Rawal 2000; Moitra and Das 2004; Banerji et al. 2010). The weak bargaining position of buyers emerges not only from the unbalanced market with an intense demand side competition but also from the fact that the opportunity cost of an exchange failure is more for the buyer than the seller. The sellers are usually larger farmers and the buyers are smaller farmers.[15] Since the behavioral patterns in WMs with small farm sellers will be distinctly different from that with large farm sellers, it is crucial to know the composition of WMs. Detailed information on the composition of a WM in Uttar Pradesh in terms of a buyers-sellers matrix is given by Shankar (1992: 33). Table 13.3 depicts a typical water sale-purchase matrix in Uttar Pradesh. According to this information, the demand side is dominated by smaller farms with less than 2 ha that account for 81 % of the total area irrigated with purchased water.

[15]There are also exceptions to this pattern as small farmers are also seen on the supply side with a market share ranging from 30 to 45 % (Shankar 1992: 33; Shah 1993: 51).

Table 13.3 Water purchase and sale matrix by farm size, Allahabad district, Uttar Pradesh, 1987

	Size of buyers (ha.)					
Size of sellers (ha.)	<1	1–2	2–3	3–4	4–10	Total
	90.01	36.70	11.77	--	--	138.48 ha.
<1	65.00	26.50	8.50	0.00	0.00	100.00 %
	11.63	10.86	8.47	0.00	0.00	10.19 %
	134.84	69.52	35.36	19.07	13.62	272.41 ha.
1–2	49.50	25.52	12.98	7.00	5.00	100.00 %
	17.42	20.57	25.44	26.30	38.77	20.05 %
	108.08	62.71	35.93	16.26	9.29	232.27 ha.
2–3	46.53	27.00	15.47	7.00	4.00	100.00 %
	13.96	18.55	25.85	22.42	26.44	17.10 %
	75.87	39.74	18.79	5.78	4.34	144.52 ha.
3–4	52.50	27.50	13.00	4.00	3.00	100.00 %
	9.80	11.76	13.52	7.97	12.35	10.64 %
	365.28	129.34	37.12	31.40	7.88	571.02 ha.
4–10	63.97	22.65	6.50	5.50	1.38	100.00 %
	47.19	38.27	26.71	43.30	22.43	42.03 %
	774.08	338.01	138.97	72.51	35.13	1358.70 ha.
Total	56.97	24.88	10.23	5.34	2.59	100.00 %
	100.00	100.00	100.00	100.00	100.00	100.00 %

Source: Shankar (1992: 33)
Note: Figures in each cell are respectively the area irrigated, and row and column-wise percentages. The rows show the percentage of buyers in each size group while the columns show the percentage of sellers in each size group

The supply side is dominated by farms larger than 2 ha that account for 70 % of the area irrigated with purchased water. Curiously, 75 % of the area irrigated with water from small farm sellers belongs to only small farm buyers suggesting a kind of segmentation within WM where small farm sellers deal mostly with small farm buyers.[16] The buyer-seller relation within the same group will be on a more equitable footing and hence, less likely to be exploitative, as compared to when buyers and sellers are unequal and belong to different groups.

13.5.2 Symptoms of Non-competitive Behavior

The WMs in Gujarat and Tamil Nadu appear to have monopolistic or oligopolistic tendencies (e.g., Shah 1993; Janakarajan 1994), but those in Andhra Pradesh, Punjab, Uttar Pradesh, Orissa, and West Bengal appear to be relatively more

[16]This is because small sellers normally with lower capacity wells/pumps but with higher self-irrigation needs (due to intensive land use and irrigation) have less water to spare which can meet only the smaller water demand of small farm buyers.

competitive (e.g., Kolavalli and Atheeq 1990; Shankar 1992). Palmer-Jones (1994) argues that since most WMs are non-contestable, some forms of spatial monopoly become inevitable. Recent studies have also attested to the presence of monopoly and bargaining behavior in WMs observed in the northern and eastern parts of India (Rawal 2000; Moitra and Das 2004; Banerji et al. 2010). The monopoly condition emerges from the fact that large farmers, as sellers, controls most of the water supply in a given area and the bargaining behavior emerges from the fact that buyers with better bargaining strength either crowd out small farm buyers or get better services than others. The non-competitive character of WMs is based on two factors, i.e., water rates being higher than the pumping cost and the presence of price and non-price discrimination.

The fact that water rates are 2–4 times higher than pumping cost cannot always be an indication of monopoly behavior since this measure ignores fixed costs and scarcity value of water, and also presumes marginal cost pricing behavior. While thin and unbalanced WMs often have monopoly potential, the sellers may not actually exercise it in view of social constraints as well as economic linkages emerging from the inter-linked nature of WMs with other rural markets such as those for land, labor, and credit. Moreover, the monopoly potential may get neutralized by state tubewells which not only bring down water rates as in Allahabad district (Shankar 1992: 150) but even reduced water sales as in Deoria district, Uttar Pradesh (Shah 1993: 79). In the case of WMs in West Bengal, village panchayats also play a role in minimizing the exploitative water rates and other discriminatory practices (Rawal 2000). These facts suggest that the operation of WMs is conditioned not just by physical factors related to water supply and cropping pattern but equally also by social and institutional factors such as personal and kinship relationships and local customs and conventions (see Dubash 2002; Naz 2010).

While WMs with single rate structure are dominant, there are also cases of various forms of both price and non-price discrimination. Price discrimination is largely absent in mature WMs. For instance, Shah (1993) observes in Gujarat that even big water companies charge the same rate irrespective of whether the buyer is their member or not. In contrast, the WMs in Vaigai basin, Tamil Nadu exhibit a multiplicity of rates that vary with farmers and locations in the same village (Janakarajan 1994). Price discrimination is also observed even in the case of in-kind payments. For instance, in Banaskantha district, Gujarat, sellers discriminate against certain buyers by requiring differential crop share (25–50 %) depending upon the conveyance distance and cropping pattern (Shah and Ballabh 1993: 12). Unlike price discrimination, non-price discrimination in the form of the quality and timeliness of irrigation service, though difficult to observe, is more widespread. For instance, in the WMs of Tamil Nadu, larger and more regular customers not only get a hidden price concession but also often receive priority service (Narayanamoorthy 1994) while those with unsatisfactory remittance history are either ignored altogether or given low priority (Janakarajan 1994). In most cases, both the price and non-price discrimination not only seem to have some economic basis but also appear to reflect the relative bargaining capacity of the buyers rather than the monopoly behavior of the sellers.

13.5.3 Seasonal Patterns in Water Trade

Although water sales occur throughout the year, the bulk of them are confined to a few months due to seasonalities in water availability, self-irrigation requirements, and water demands. For instance, case studies in Andhra Pradesh and Uttar Pradesh show that 59–66 % of the total area irrigated with purchased water is confined to the rabi season followed by the kharif (26–32 %) and the summer (8–9 %) seasons (Shah and Raju 1988; Shankar 1992). The low water sales during kharif season is caused primarily by more farmers providing their own irrigation water whereas the lesser sales during summer is due partly to lower water availability and partly to lower water demand caused by less-intensive cultivation.[17] This seasonal pattern suggests that whenever own irrigation requirement is high, water sales will be lower and *vice versa*. In addition, better rainfall conditions and surface water availability can also influence the seasonal pattern of water sales.

13.6 Water Marketing: Efficiency, Equity, and Sustainability

Since WMs enhance the value of groundwater, they generate substantial efficiency and productivity gains (Fujita and Hossain 1995). This is illustrated by comparing the area-based charges for groundwater with those for canal water. For instance, during the 1990s, the canal rates vary from Rs. 6 to 1,000/ha depending upon crop, season, and irrigation projects while the groundwater rates vary from Rs. 225 to 9,174/ha depending upon crop, season, and region (Saleth 1996: 35).[18] These higher charges for groundwater not only induce private irrigation investment and fuller use of existing LIS capacity but also provide strong incentive for on-farm water use efficiency. Under the current institutional conditions, however, the efficiency and productivity effects of WMs are relatively stronger among buyers than sellers. This has been observed in Uttar Pradesh (Shankar 1992) and Gujarat (Kolvalli and Chicoine 1989). Similar pattern of better performance of buyers is also observed in Rajasthan (Sharma and Sharma 2006) and Karnataka (Nayak 2007; Manjunatha et al. 2011).[19] However, the efficiency effects of WMs depend much on factors like

[17]In Allahabad district, Uttar Pradesh, most of the summer sales are to meet the water requirements of brick kilns (Shankar 1992: 59–61). Similarly, in Tamil Nadu, water sales around Chennai occur year round to meet urban domestic water needs (Packialakshmi et al. 2011).

[18]The difference is still higher for specific crops especially at the state level. In Gujarat, for instance, while canal water charges for paddy and sugarcane are respectively Rs. 110 and Rs. 830/ha, the comparable groundwater charges are Rs. 2,964–3,705 and Rs. 5,979–9,174/ha. Likewise, in Tamil Nadu, canal water charges for these two crops are in the range of Rs. 49–62/ha and the groundwater charges are in the range of Rs. 225–3,411/ha.

[19]The buyers perform better because they, unlike the sellers, have mostly small farms known for higher cropping as well as land and input use intensities. Likewise, the better water use efficiency

farm size and cropping pattern. Although the relative share of gains can vary by crop, season, region, pricing method, and even, types of buyers, the buyers, as a group, gain several times more than sellers. Since WMs, like all markets, favor rich over the poor, there is potential for income disparity (Singh 2007). Yet, the fact that water buyers are mostly smaller farmers suggests the income distribution potential of WMs. More importantly, by providing reliable water supply, WMs can also reduce uncertainty and provide security (Alder 2008), which is particularly important for small and marginal farmers.

The equity potential of WMs gets further reinforced by the fact that in addition to the direct output impact on small farms, there are also second-round employment and income benefits even for the landless. Such multiplier effects are particularly significant in areas facing water scarcity. However, these intra-generational equity benefits have to be contrasted both with the intra and inter-generational inequity effects of WMs. To understand the latter, let us distinguish between two aspects of intra-generational equity, i.e., equity at the stage of *de facto* water control and equity at the stage of actual water use. As WMs benefit small farmers, they contribute to intra-generational equity at the stage of actual water use. But, as they reinforce and legitimize the *de facto* control of and the appropriation of the rent over water by larger farmers, they contribute to intra-generational inequity among farmers at the stage of water control (Saleth 1994: 165–66). Unfortunately, the regulatory policies like well-spacing and licensing norms actually accentuate this intra-generational inequity as they protect the existing *de facto* control of rich farmers by restricting the entry especially of small farmers.

The inter-generational inequity effects of WMs are intimately linked with their effects on aquifer sustainability. In so far as WMs leads to depletion, they contribute to unsustainability and environmental effects such deterioration in water quality due to seawater intrusion in coastal regions and increasing salinity from deeper layers of the aquifers in some inland regions (Mohanty and Gupta 2002; and Varughese 2012). But, efficient WMs can also contribute to sustainability. For instance, in areas with serious depletion, efficient WMs with economic prices (which cover pumping costs plus the scarcity value of water) can improve resource sustainability in two ways. First, since WMs provide incentive for water conservation, water withdrawals will be lower when compared to the alternative scenario where every farmer has his own wells. Second, an efficient WM could also reduce the over-crowding of wells causing effects similar to well-spacing regulations. It is these two effects that are behind the argument that WMs contributes to sustainability by contributing to a deceleration in groundwater depletion (Foster and Sekhri 2008).

Unfortunately, the economic features and geographical patterns of WMs suggest that their depletive effects can dominate over their efficiency and equity gains. Although it is unrealistic to attribute depletion problems squarely to WMs as the

---

of buyers comes from the fact that they face restriction on their access to groundwater but the sellers do not face such restrictions.

former can occur even without the latter, water selling activities do exacerbate the problem because they occur mostly in areas characterized by scarce water regimes where the rent on water, i.e. the difference between water payment and the actual costs, is higher than in areas with better water endowments. Due to this rent-seeking behavior and the existing institutional vacuum (i.e., the absence of legally and organizationally enforced withdrawal limits as set by some type of water quotas or entitlements), most water sales occur by simply pumping additional water rather than by saving water through efficient use. As a result, not only are the conservation imperatives extremely limited but also water withdrawals have become excessive causing aquifer depletion.

13.7 Concluding Remarks

WMs in India are uneven, localized, and fragmented. They have evolved into a mature institution with substantial efficiency and welfare effects. Although some aspects such as the levels of water rates must have changed, most of the economic and institutional features of WMs that are discussed here are still very much valid. But, the issue as to whether their efficiency and equity gains can compensate for their negative ecological and equity effects can be settled only in the context of each specific region given its resource endowment conditions. Since the prices in WMs do not reflect either the scarcity of water or its productivity, their role in promoting efficiency in water allocation is severely limited. So long as the depletive and inequity effects of WMs outweigh their positive efficiency and equity benefits, the private benefits from water trade may be inadequate to compensate for the social costs in most contexts.

The root cause for the sub-optimality of Indian WMs lies not so much in their economic and organizational aspects but in the legal and institutional vacuum (i.e., the absence of mechanisms to quantitatively fix, enforce, and monitor individual and collective withdrawals) within which they operate at present (Saleth 2004; Gandhi and Namboodiri 2009). A legally instituted and locally managed water quota system defined within an ecologically consistent overall withdrawal limit can provide powerful incentive for water use efficiency and conservation. This could eliminate the negative effects of WMs while magnifying their positive benefits. Because the alleged benefits observed in current WMs form only a fraction of the efficiency, equity, and sustainability gains possible from the 'real' WMs emerging within a well-managed water quota system, the currently observed WMs are only the distant second-best option. Naturally, therefore, the sustainability of WMs as an economic institution in India hinges critically on the speed and effectiveness with which water rights-centered institutional changes are implemented to set and enforce individual and collective water quotas.

References

Alder JH (2008) Warming up to water markets. Regulation 31(4):14–17

Asopa VN, Dholakia BH (1983) The performance appraisal of Gujarat water resources development corporation, vol 1, Summary and recommendations. Centre for Management in Agriculture, Indian Institute of Management, Ahmedabad

Banerji A, Khanna G, Meenakshi JV (2010) Social contracts, markets and efficiency: groundwater irrigation in North India, Working paper no: 183, Centre for Development Economics, Delhi School of Economics, Delhi

Dubash NK (2002) Tubewell capitalism: groundwater development and agrarian change in Gujarat. Oxford University Press, New Delhi

Foster A, Sekhri S (2008) Can expansion of markets for groundwater decelerate the depletion of groundwater resource in rural India?. Available at: http://people.virginia.edu/~ss5mj/water_mkt.pdf

Fujita K, Hossain F (1995) Role of the groundwater market in agricultural development and income distribution. Dev Econ 33(4):442–463

Gandhi VP, Namboodiri NV (2009) Groundwater irrigation in India: gains, costs and risks, Working paper no: 2009-03-08, Centre for Management in Agriculture, Indian Institute of Management, Ahmedabad

Government of India (GOI) (1976) Report of the National Commission on Agriculture, Part V: Resource Development, Ministry of Agriculture, New Delhi

GOI (1992a) Model bill to regulate and control the development of groundwater. Ministry of Water Resources, New Delhi

GOI (1992b) Report of the committee on pricing irrigation water. Planning Commission, New Delhi

Janakarajan S (1994) Trading in groundwater: a source of power and accumulation. In: Moench M (ed) Selling water: conceptual and policy debate over groundwater in India. VIKSAT/Pacific Institute/Natural Heritage Institute, Ahmedabad

Kolavalli S, Atheeq LK (1990) Groundwater utilization in two villages in West Bengal. Indian Institute of Management, Ahmedabad

Kolvalli S, Chicoine DL (1989) Groundwater markets in Gujarat, India. Int J Water Res Dev 5(1):38–44

Kurien J, Sinha AK (2007) Ground water governance issues in irrigation development: a perspective. National Bank for Agricultural and Rural Development, Mumbai. Available at: http://www.nabard.org/fileupload/DataBank/TechnicalDigest/ContentEnglish/GROUND.pdf

Maloney C, Raju KV (1994) Managing irrigation together: practice and policy in India. Sage Publications, New Delhi

Manjunatha AV, Speelman S, Chandrakanth MG, Van Huylenbroeck G (2011) Impact of groundwater markets in India on water use efficiency: a data envelopment analysis approach. J Environ Manage 92(11):2924–2929

Marothia DK (2003) Institutions for common pool resources. In: Pal S, Mruthyunjaya, Joshi PK, Saxena R (eds) Institutional change in Indian agriculture. National Centre for Agricultural Economics and Policy Research, New Delhi

Mohanty N, Gupta S (2002) Breaking the Gridlock in water reform through water markets: international experience and implementation issues for India. Julian L. Simon Centre for Policy Research working paper series, Liberty Institute, New Delhi

Moitra B, Das PK (2004) Groundwater markets in West Bengal, India: emergence, evolution and market structure. The South Asia Network of Economic Research Institutes, Dhaka. Available at: http://coference.jueconomics.in/faculty/papers/Rev_Final_GW_Market_in_WB.pdf

Mukerjee A (2007) The energy-irrigation nexus and its impact on groundwater markets in eastern Indo-Gangetic basin: evidence from West Bengal, India. Energy Policy 35(12):6413–6430

Mukherji A (2008) Spatio-temporal analysis of markets for groundwater irrigation services in India: 1976–1977 to 1997–1998. Hydrogeol J 16(6):1077–1087

Narayanamoorthy A (1994) Who sells more and who sells less in deep bore-well water business in Pudukkottai District, Tamil Nadu: some empirical analysis. J Indian Water Res Soc 14(1–4):51–56

National Sample Survey Organization (NSSO) (1984) Sarvekshana 8(2). Department of Statistics, Ministry of Planning, Government of India

NSSO (1997) Sarvekshana 20(3). Department of Statistics, Ministry of Planning, Government of India

NSSO (2000) Sarvekshana 24(1). Department of Statistics, Ministry of Planning, Government of India

Nayak MR (2007) Groundwater markets in Karnataka: key issues in sustainability. Ph.D. thesis, University of Agricultural Sciences, Dharwad

Naz F (2010) The lens of caste system shaping groundwater market exchange: a case study of rural Gujarat, India. Ph.D. thesis, Centre for Development Research (ZEF), University of Bonn, Germany. Available at: http://www.docstoc.com/docs/87494755/The-Lens-of-Caste-System-Shaping-Groundwater-Market-Exchange--A

Packialakshmi S, Ambujam NK, Nelliyat P (2011) Groundwater market and its implications on water resources and agriculture in the southern peri-urban interface, Chennai, India. Environ Dev Sustain 13(2):423–438

Palanisami K (1994) Evolution of agricultural and urban water markets in Tamil Nadu. In: Rosegrant MW, Schleyer RG (eds) Tradable water rights experiences in reforming water allocation policy. USAID, Arlington

Palmer-Jones RW (1994) Groundwater markets in South Asia: a truncated discussion of theory and evidence. In: Moench M (ed) Selling water: Conceptual and Policy debates over Groundwater Markets in India, VIKSAT and others, Ahmedabad, Gujarat, India

Patel SM, Patel KV (1969) Economics of tubewell irrigation. Centre for Management in Agriculture, Indian Institute of Management, Ahmedabad

Planning Commission (2008) Eleventh five year plan (2007–2012). Oxford University Press, New Delhi

Rawal V (2000) Non-market interventions in water-sharing: case studies from West Bengal. Paper presented in the international conference on democratic decentralization, State Planning Board, Thiruvananthapuram, 23–27 May

Saleth RM (1991) Factors affecting farmers' decision to buy groundwater: empirical evidence from the Indo-Gangetic Region. Indian J Agric Econ 46(3):349–354

Saleth RM (1994) Groundwater markets in India: a legal and institutional perspective. Indian Econ Rev 29(2):157–176

Saleth RM (1996) Water institutions in India: economics, law, and policy. Commonwealth Publishers, New Delhi

Saleth RM (1998) Water markets in India: economic and institutional aspects. In: Easter KW, Rosegrant MW, Dinar A (eds) Markets for water: potential and performance. Kluwer, Dordrecht

Saleth RM (2004) Strategic analysis of water institutions in India: application of a new research paradigm. International Water Management Institute, Research Report 79

Saleth RM, Thangaraj M (1993) Distribution pattern of lift irrigation in India: an analysis by hydrogeological regions. Econ Pol Wkly 28(39):A102–A110

Shah T (1985) Transforming groundwater markets into powerful instruments of small farmers development: lessons from the Punjab, Uttar Pradesh, and Gujarat, ODI Irrigation Network Paper 11d, London

Shah T (1993) Groundwater market and irrigation development: political economy and practical policy. Oxford University Press, Bombay

Shah T, Ballabh V (1993) The social science of water stress: an exploratory study of water management institutions in Banaskantha District, Gujarat. Paper presented in the seminar: Water management: India's groundwater challenge, Ahmedabad, 14–16 December

Shah T, Raju KV (1988) Ground water markets and small farmer development. Econ Pol Wkly 23(13):A23–A28

Shankar K (1992) Dynamics of groundwater irrigation. Segment Books, New Delhi

Sharma P, Sharma RC (2006) Factors determining farmers' decision for buying irrigation water: study of groundwater markets in Rajasthan. Agric Econ Res Rev 19(1):39–56

Singh D (2007) Who gains and who loses in the game of groundwater markets in water scarce regions. Agric Econ Res Rev 20(4):345–360

Varughese S (2012) A study of rural groundwater markets in the Gingee micro watershed. Doctoral thesis, Pondicherry University, Pondicherry, India

Chapter 14
Assessment of the Development of Groundwater Market in Rural China

Jinxia Wang, Lijuang Zhang, Qiuqiong Huang, Jikun Huang, and Scott Rozelle

Abstract Using field survey data collected by the authors, this chapter first describes groundwater markets in northern China that have been developing rapidly in the past two decades. Groundwater markets in the area are informal, localized and mostly unregulated. There is little price discrimination, and institutional characteristics tend to be similar in both high- and low-income villages. The privatization of tubewells is one of the most important driving forces encouraging the development of groundwater markets. Increasing water and land scarcity are also major determinants. The chapter also explores the impacts of the emergence of the groundwater markets on agricultural production – including crop water use and crop yields – and farmer income in northern China. Results indicate farmers that buy water from groundwater markets use less water than those that have their own tubewells. However yields of water buyers are not negatively affected. This is probably because water buyers exert more efforts to improve water use efficiency. Results also show that other things held constant, the crop incomes of water buyers are not statistically different from those of well owners. The chapter also finds that groundwater markets in northern China are not monopolistic, supporting the notion that they offer poor rural households affordable access to irrigation water.

J. Wang (✉) • L. Zhang • J. Huang
Center for Chinese Agricultural Policy, Chinese Academy of Sciences, Jia, No.11, Datun Rd, Anwai, 100101 Beijing, China
e-mail: jxwang.ccap@igsnrr.ac.cn; zhanglj.ccap@igsnrr.ac.cn; jkhuang.ccap@igsnrr.ac.cn

Q. Huang
Department of Agricultural Economics and Agribusiness, University of Arkansas, 221 Agriculture Building, Fayetteville, AR 72701, USA
e-mail: qqhuang@uark.edu

S. Rozelle
Freeman Spogli Institutional Studies, Stanford University, Encina Hall, 616 Serra St, Stanford, CA 94305-6055, USA
e-mail: rozelle@stanford.edu

K.W. Easter and Q. Huang (eds.), *Water Markets for the 21st Century: What Have We Learned?*, Global Issues in Water Policy 11, DOI 10.1007/978-94-017-9081-9_14,

Keywords Groundwater market • China • Water use • Crop yields • Crop income

14.1 Introduction

Groundwater resources are playing an increasingly important role in the economy of northern China. In 2011, on average, 35.5 % of the total water supply (industrial, residential and agricultural sectors) came from groundwater (Ministry of Water Resources of China 2011). Agriculture relies even more heavily on groundwater. As public investment in canal systems waned and deliveries became more unreliable, farmers in northern China began to rely more on small irrigation systems fed by groundwater. The rapid expansion of groundwater irrigation has stimulated the growth of agriculture in northern China (Huang et al. 2006). In the North China Water Resource Survey (NCWRS) survey (described below) sample villages, in 2004, with the exception of rice, at least 70 % of the area sown to grains and other staple crops were irrigated by groundwater (e.g., 72 % of wheat and 70 % of maize, Wang et al. 2007). Groundwater also irrigates most cash crops (e.g., 70 % of cotton, 62 % of oil seed crops and 67 % of the vegetables).

In most rural areas in northern China, central and regional governments have little control over groundwater use. China's National Water Law (China 2002), which was revised in 2002, stipulates that all property rights over groundwater resources belong to the national government, including the right to use, sell and/or charge for water. In practice, however, villages that lie above the aquifers have the de facto rights to groundwater resources. Unlike the US, water rights are not associated with land ownership or historic use. Often they are associated with the ownership of wells. Despite the plethora of laws and policy measures created by government officials, there has been a lack of enforcement (Wang et al. 2007). One of the reasons is the difficulty in regulating millions of small, water using farmers. Another reason is historic neglect. The administration unit that is in charge of groundwater management at the ministerial level is still relatively small. There are far fewer officials working in this division than in other divisions, such as flood control, surface water management or water transfer projects. Moreover, unlike the case of surface water management (Lohmar et al. 2003), there has been no effort to bring the management of aquifers that span jurisdictional boundaries (e.g., counties or provinces) under the control of a single authority that can regulate and coordinate among users in different parts of the aquifer. As a result, few regulations stipulated by upper level government have been implemented at the village level. For example, despite the nearly universal regulation that requires the use of a permit for drilling a well, less than 10 % of the sample well owners in the 2004 NCWRS survey obtained one before drilling. Only 5 % of sample villages had any consideration for well spacing.

With the lack of control from upper level governments, groundwater use is organized and managed at the village level. Before the rural reforms in the 1980s, wells in almost all rural villages were collectively owned and financed primarily by collective retained earnings and additional funding from township governments. Village leaders were largely responsible for arranging for the water resource bureau-run well drilling companies to sink tubewells. Pumps, before the reform all came from either the water resource bureau pump supply company or the state-run local agricultural inputs corporation. As the curator of collective assets, village leaders made decisions on all aspects of water management: when and where to sink the wells, how many wells to sink, and, importantly, how much water to extract during each season. Village leaders often hired a well operator to pump water and deliver to households under their instruction. In most villages individual farmers at most contributed their labor for tubewell construction and maintenance.

Changes brought on by the economic reforms forced local village governments to be fiscally more independent. Many villages, particularly those without lucrative nonagricultural enterprises, eventually faced serious fiscal shortfalls and were unable to invest in agriculture (Lohmar et al. 2003). In addition, due to the fall of the groundwater level and lack of maintenance of pumps, engines and other equipment, a number of collective tubewells became inoperable. As the collective's ability to invest in maintaining existing wells or replacing pumps or sinking new wells declined, farmers began to take its place and the ownership of wells began to shift from collective ownership to private ownership (Wang et al. 2005). This transition took place in the macro environment in which policy makers started to gradually relax the constraints on private activities. In particular, the economic reform has shifted income and control rights of land from the collective to the individual household. The survey conducted by Wang (2000) in Hebei Province showed that in the early 1980s collective ownership accounted for 93 % of all tubewells but diminished throughout the late 1980s and 1990s. During this period the share of private tubewells increased from 7 to 64 %. This is consistent with findings from the NCWRS survey (described below, Wang et al. 2007). In 1995, 58 % of wells in groundwater-using villages were still under collective ownership. By 2004, the share of privately owned wells rose to 70 %, shifting a large part of groundwater management into the hands of private individuals. The shift of tubewell ownership is the result of the establishment of new tubewells rather than ownership transfers of collective tubewells although the absolute number of collective tubewells has declined. The number of private wells sunk by farmers (either an individual farmer or a group of farmers jointly) has increased significantly.

As tubewells have come under the control of private individuals, access to groundwater for those farmers who do not own wells has become a new issue. Groundwater markets have not always existed. In the 1970s and 1980s, when most wells were owned and operated by the collective, in almost all villages simple rules governed water allocations. Under these rules households in the village received an equal share of the total water allocation that was based on the land size. It should be noted that unlike other countries, such as India, land is relatively equally allocated among households in rural China both in terms of land size and soil quality (Benjamin and Brandt 2002). The egalitarian nature of the land distribution provided some rationale for the simple rule of equal water allocation.

Concurrent with the trend of increasing privatization of wells is the development of groundwater markets. Following a pattern similar to that observed in South Asia (Shah 1993, 2009), groundwater markets have begun to emerge in which owners of tubewells sell groundwater irrigation services, mostly to fellow villagers within the village and in some cases to farmers from outside the village. When village leaders (the collective) provided water to villagers, it was done under non-market conditions because any irrigation fees collected went into the village's collective fund, not as compensations to village leaders. In fact, in some villages, the collective provided water free or at a subsidized rate.

The changes in well management have the potential for affecting the rural economy and the nation's water resources. The increased access to groundwater enabled by groundwater markets clearly has the potential to boost agricultural productivity. However, as tubewells have begun to be operated by private individuals and sunk to deeper levels, concern has also arisen that farmers do not have an equal access to groundwater (Meinzen-Dick 1996). Farmers that are buyers in groundwater markets may be forced to use less water because they may have to pay more for water than well owners or farmers serviced by the collective wells. As a consequence, yields and crop income of those farmers may be negatively affected. In addition, policy makers and scholars also debate the question of whether the prevalence of groundwater markets accelerates the decline of water levels in aquifers. Despite the importance of these issues, only a few papers have examined groundwater markets in rural China (e.g., Wang et al. 2005; Zhang et al. 2008, 2010; Huang et al. 2013). This chapter summarizes findings in the previous studies that address these important issues. The focus is on agricultural use of groundwater in northern China. This chapter begins by documenting the development of groundwater markets in northern China and describing the characteristics of groundwater markets. The next section identifies the factors that have led to this development. The third section analyzes the impact of groundwater markets on agricultural productivity, rural incomes and groundwater resources. The final section draws conclusions and discusses policy implications.[1]

14.2 Groundwater Markets in Northern China

Analysis in this chapter is based on two field surveys the authors have conducted. The first survey, the China Water Institutions and Management survey (CWIM) tracks 48 randomly selected villages in Hebei and Henan provinces (Fig. 14.1).

[1]Most materials in this chapter are adapted from the following two articles: Zhang, L., Wang, J., Huang, J., Rozelle, S., 2008, Groundwater Markets in China: A Glimpse into Progress. World Development 36 (4): 706–726.

Zhang, L., Wang, J., Huang, J., Huang, Q., Rozelle, S., 2010, Access to groundwater and agricultural production in China, 97:1609–1616.

CWIM sample provinces:

1. Hebei; 2. Henan;

NCWRS sample provinces:

1. Hebei; 2. Henan; 3. Inner Mongolia;

4. Liaoning; 5. Shanxi; 6. Shaanxi.

Fig. 14.1 Study areas in northern China

The CWIM sample area covers two of the nine major river basins in China. Hebei province covers most of the Hai River Basin and surrounds Beijing. Henan province is located in the middle reaches of the Yellow River Basin. A stratified random sampling strategy was used. The strata are geographic locations, which were correlated with the extent of water scarcity. In Hebei province, one county was randomly selected from each of the three regions: the coastal belt, the most water scarce area of China; the inland belt, an area with relatively abundant water resources since it is next to the mountains in the western part of Hebei province; and the region between the coast and mountains. In Henan counties were randomly selected from each stratum that includes irrigation districts with varying distances from the Yellow River. Locations further away from the river are typically associated with increasing water scarcity. After the sample counties were selected, we then randomly selected 48 villages from these counties. In the CWIM survey enumerators interviewed three sets of respondents: village leaders, randomly-selected households (between 1 and 4 households per village) and randomly-selected well managers. Separate survey questionnaires were designed and used for each set of respondents. The household level data collected in the CWIM survey enable us to analyze the impacts of groundwater market on the crop income of households and plot level water use.

The second survey, the North China Water Resource Survey (NCWRS) covers six randomly chosen provinces: Inner Mongolia, Hebei, Henan, Liaoning, Shaanxi and Shanxi provinces (Fig. 14.1). Similar to the CWIM survey, a stratified random sampling strategy is used. Counties in each province were divided into four water scarcity categories: very scarce, somewhat scarce, normal and mountain/desert. Two townships within each county and four villages within each township were then randomly selected. In total, the survey team visited 50 counties, 100 townships and 401 villages. In the NCWRS survey we only interviewed village leaders due to limitations in time and budget. A more comprehensive version of the CWIM survey village leader questionnaire was used. Data were collected on most variables for 2 years, 2004 and 1995.

14.3 Development and Characteristics of Groundwater Markets

Although almost nonexistent before 1980, by 1995, groundwater markets were present in 9 % of the NCWRS villages that used groundwater and had private wells, defined as wells belonged to farmers (not the collective). Groundwater markets spread at a much faster rate over the next 10 years. By 2004, tubewell operators in 44 % of the sample villages were selling water. At the same time when groundwater markets were expanding spatially, in villages that had groundwater markets, markets become more active. In 1995 water was sold from only 5 % of tubewells; by 2004, however, this number increased to 18 %. In 2004, the average number of tubewells per village was about 75 and water was sold from between 13 and 14 wells in each village. In addition, using the CWIM survey data that contain detailed well-level information, we found that groundwater market activities were dominating the tubewell pumping activities of those farmers-cum-tubewell owners that were selling water. About 80 % of water pumped from private wells was sold in the groundwater market in 1995 and 77 % in 2004. The slight drop from 80 % in 1995 to 77 % in 2004 may be due to the increase in the number of wells, which increased the total available supply of groundwater in the market and at the same time reduced the demand for water because more farmers were pumping from their own wells. Between 1995 and 2004, both the number of wells selling water and the total number of wells increased. In the 68 sample villages that were in the NCWRS, the number of wells selling water increased from 75 in 1995 to 342 in 2004; at the same time, the total number of wells also increased from 1,472 to 1,967.

Although groundwater markets in northern China started later than those in South Asia, they do share some common features. First, almost all groundwater markets in China are *informal*. According to Shah (1993), a water market is informal when transactions between water-selling and water-buying households are done without legal sanction. In other words, farmers buy and sell water without a contract and their oral commitments cannot be adjudicated in a court of law. According to the data, there were zero written contracts covering water sales among participants in the groundwater markets in northern China. Payment is enforced by social norm because sellers and buyers often reside in the same village and often know each other personally. In addition, sellers can refuse to sell water to a buyer in the future if the buyer has not paid for some or all water received in the past. The informal nature is consistent with the general environment in China where the rule of law is still weak. It significantly reduces the transaction cost (such as legal fees to draw up a formal contract and the cost of enforcing the contract) which participants in the markets would have to incur otherwise. This may be one of the reasons why groundwater markets were able to grow at a fast rate in north China.

Second, groundwater markets in northern China are almost always localized. According to Shah (1993), the localized nature of water markets is almost universal due to the constraints on the infrastructure required to transport water. In the survey data in China, water transactions also are mostly limited to households in the same village. In fact, only 6 % of water-selling tubewell owners (and a smaller share of the volume of water that they pump) sell water to farmers in neighboring villages.

Third, groundwater markets in northern China are largely unregulated. In Shah (1993) the word unregulated means the government exercises no direct influence on the functioning of the market. According to NCWRS survey data, fewer than 25 % of villages have any formal regulations in writings about any aspect of groundwater markets (e.g., a price ceiling or the amount pumped). The regulations appear to be largely unenforced. During the field work and interviews with tubewell owners, enumerators almost never encountered a case in which the tubewell owner was constrained by a government regulation; village leaders and tubewell operators were typically unaware that there was any attempt by upper level officials to influence the functioning of water buying and selling.

Fourth, groundwater markets in northern China are largely impersonal. Based on our interviews with village leaders, we found that within villages, only 7 % of water selling tubewell owners charge different prices. In addition, in our survey of the tubewell owners, not one reported that they charged different prices for different types of buyers. Shah (1993) also finds water-selling households in some villages of India do not distinguish among various buyers in terms of price at which they sell water and the quality of service provided. Price discrimination, however, has been observed in other parts of India (Pant 2004) and in other countries such as Pakistan (Jacoby et al. 2004).

Groundwater markets in China do differ in some aspects from those in other countries. In northern China, water sold in groundwater markets is almost always paid for on a cash basis. In India water buyers often provide labor or a share of crop harvest in exchange for water (Shah 2000). The difference in the payment method may be rooted in the difference in land tenure arrangements. In China the ownership of cultivated land belongs to village collectives. Since the household responsibility system was implemented in rural China in the early 1980s which allowed rural households to manage agricultural production on their own initiatives and keep the profits after tax, cultivated land has been allocated relatively evenly to each household within a village. So every farmer in the village has some land he/she can use for agricultural production, although they have no land ownership. However, in South Asia, land allocation is unequal and land ownership varies. There are land lords as well as landless tenants. Tenants often pay the rent to land owners either with labor or a share of their harvest (i.e., through a sharecropping contract). So it is not surprising that water-buying households pay for water in similar ways. Another important difference between groundwater markets in China and South Asia is the way in which electricity is priced. This, too, may have a major impact on the way groundwater markets function. For instance, in many Indian states, electricity is priced on a flat rate basis. In China, however, electricity is priced on a per kwh basis and at market rates (no subsidy). Electricity meters are installed at almost all wells that use electric pumps. Since the pumping cost (and consequently the price of water) is closely related to the depth to water, it reflects mostly the scarcity value of water. Furthermore, in India rural electrification is poor and, hence, many farmers depend on diesel pumps. This may create a configuration of groundwater markets somewhat different from those where there are electric pumps.

14.4 Determinants of Groundwater Markets

Zhang et al. (2008) use econometric analysis to identify the factors that explain why some villages have groundwater markets and others do not. They run two regressions, each using one indicator of groundwater development as the dependent variable: the share of tubewells selling water and the share of water sold. In both regressions, in addition to variables that measure water and land scarcity, a set of control variables is included (Table 14.1). Three policy variables are used. First, fiscal subsidies for tubewells equals one if there was a program of fiscal investment in the village that targeted tubewell construction and zero otherwise. This government program, run by the local Bureau of Water Resources, is primarily targeted at individuals. Second, a similar variable, bank loans for tubewells, is included to control for whether or not there was a loan program through China's banks that gave preferential access to low interest rate loans for investing in tubewells. Unlike the fiscal subsidy program most bank loan programs target local villages and village leaders; the loans are supposed to be used on collective wells. A third variable, well-drilling regulations, controls for the presence of local regulations that would, ceteris paribus, slow down the construction of tubewells. In addition to the policy variables, several other variables are also included. A dummy variable is used, which equals one if the village had adopted technology such as surface (called white dragons in rural China) or underground pipe networks. It is thought that if the cost of delivering water from the tubewell to the field is lower, water markets will emerge more readily. Village income per capita is included to measure the village's socio-economic conditions. In the regression with the share of tubewells selling water as the dependent variable, the share of private tubewells is included as a control variable. In the regression with the share of water sold as the dependent variable, the control variable is a dummy variable that equals one if the tubewell is owned by an individual farmer and zero if the tubewell is owned by a group of farmers (share-holding wells).

The analysis reveals four factors that have fostered water market development (Table 14.1). First, the change of tubewell ownership from collective to non-collective induces the development of groundwater markets. All other things held constant (e.g., village's socio-economic condition, use of water conveyance technology and the policy environment), when the share of the non-collective tubewells in a village increases, the share of tubewells selling water increases. Although causality cannot be inferred, this result shows the correlation between privatization and the rise of groundwater markets. One explanation is that in some villages, private tubewells have risen in response to less service available from collective tubewells either because those wells ran dry or were not maintained. Therefore, in these villages, it would be necessary for farmers to gain access to water from sales from private tubewells.

Second, the development of groundwater markets is highly related to water resource scarcity. There is a clear indication of increasing water scarcity over time. Here water scarcity is measured by the depth to water in wells. In the NCWRS sample villages that have private wells, depth to water in wells fell from 28 m in

Table 14.1 Tobit regression of the determinants of development of markets in China

	Dependent variable: share of tubewells selling water				Share of water sold
	(1)	(2)	(3)	(4)	
Tubewell ownership					
Share of private tubewells	0.183	0.286	0.180	0.286	
	(3.86)***	(7.70)***	(3.83)***	(7.40)***	
Dummy of individual tubewell					0.389
					(4.33)***
Water and land scarcity					
Log of groundwater table	0.003	0.006	0.003	0.006	
	(3.82)***	(5.06)***	(3.81)***	(4.96)***	
Log of groundwater table in 1995					0.008
					(2.01)**
Log of per capita cultivated land	−0.900	−1.036	−0.909	−1.036	−4.745
	(2.39)**	(3.21)***	(2.40)**	(3.10)***	(3.50)***
Policy interventions					
Dummy of fiscal subsidies for tubewell investment	0.051		0.041		−0.121
	(0.46)		(0.38)		−1.58
Dummy of bank loans for tubewell investment	0.065		0.066		0.484
	(0.59)		(0.60)		(3.02)***
Dummy of well-drilling permission regulation	0.116		0.117		0.045
	(3.09)***		(3.08)***		−0.46
Other control variables					
Dummy of adopting water delivery pipes	−0.025	0.008			−0.093
	(0.64)	(0.23)			−0.94
Per capita net income of farmers	−0.000	−0.000	−0.000	−0.000	0.000
	(0.18)	(0.88)	(0.24)	(0.85)	(1.94)*
Constant	−4.257	−3.853	−4.204	−3.918	−2.943
	(3.68)***	(4.74)***	(3.66)***	(4.76)***	(3.34)***
Observations	136	136	136	136	50
Chi-square	35.19	96.41	35.30	94.29	46.37

Data source: Data in the model "share of tubewells selling water" come from authors' survey in 68 randomly selected villages in four provinces (Hebei, Henan, Shanxi and Shaanxi) in 2 years (1995 and 2004) of NCWRS. Data in the model "share of water sold" come from authors' survey in 50 randomly selected tubewells in two provinces (Hebei and Henan) of CWIM. We do not use data from all of the sample villages of the two surveys since the information in the table is conditioned on villages that use groundwater to irrigate and that have private tubewells

[a]Coefficients represent marginal effects; absolute value of z statistics in parentheses

[b]* significant at 10 %; ** significant at 5 %; *** significant at 1 %

1995 to 38 m in 2004. The data show a positive and strong correlation between the depth to water in wells and the amount of groundwater market activity measured as either the share of tubewell selling water or the share of water sold. Regression analysis also reveals the same relationship: in areas in which the depth to water in wells is greater, farmers' demand for water from groundwater markets is higher (relative to obtaining water from one's own well). One explanation is that when the depth to water in wells is greater, the cost of sinking a tubewell is higher, which could keep some farmers from investing in their own tubewells even though they have a high demand for irrigation services. Alternatively (although mainly in a relative sense), it could be that the greater the depth to water in wells, the larger is the size of the optimal tubewell/pump set. In villages with larger tubewells/pump sets, other factors (including size of land holding) held constant, there is less of a need for all farmers to have their own tubewells. In either case, there is some empirical evidence that groundwater markets develop more quickly in villages with scarce water resources.

Third, the data also lend some support to a positive relationship between land pressure and the extent of groundwater markets. Land pressure has increased. Between 1995 and 2004 the average size of land holding per capita for the sample villages fell from 0.12 to 0.10 ha. Regression analysis show that with the decrease of per capita land resources, the share of tubewells selling water has increased and the average tubewell operator sells a greater share of water pumped from his/her tubewell. This result still holds when land pressure is measured by cultivated land per household (instead of per capita). So when the average land holding in a village shrinks, there is more of a tendency for its tubewell owners to sell water. This is probably because with a smaller farm size, households demand less water and are thus less likely to sink their own tubewells. This, however, does not necessarily imply that only small households are buying water. In China, there was not much difference in the size of the land that was allocated to farmers *within the same village* (Benjamin and Brandt 2002). Therefore, the positive relationship between land pressure and groundwater markets activities is largely driven by inter-village differences. This means that it is in villages that have mostly small households that have more sales, as opposed to villages with mostly large households. This distinguishes the market in China from those in other countries, particularly those in South Asia.

Finally, if the tubewell is owned by an individual (a single household), a higher share of water is sold, compared with shareholding tubewells. Since the demand by the individual farm household for water from its own well is more likely to be less than that of all the members of the shareholding tubewell, a positive relationship would be expected, due to the excess capacity available for sale.

Most of these findings are consistent with international experience. For example, Shah (1993) descriptively shows that the availability of water resources, the scale of irrigation technology and the extent of land fragmentation are correlated with the rise of groundwater markets. Strosser and Meinzen-Dick (1994) argue that the depth of groundwater table and the population density of a community are the important factors affecting groundwater markets.

14.5 Impacts of Groundwater Markets

This section examines the impacts of groundwater markets on groundwater resources, agricultural productivity and crop income. The household level data from the CWIM data are used. Plot or household is used as the unit of analysis. Wheat is the major crop grown on most plots during the winter season (planted during the previous October and harvested in June) in Hebei and Henan provinces. In our sample, about 94 % of the sample plots (or 97.6 % in terms of sown area) only grew wheat in the winter season. Only a small percentage of the sown area was allocated to other crops including beans, legume and cash crops such as oil crops and vegetables. After wheat is harvested, either maize or cotton (competing summer crops) is grown in the summer season (planted in June and harvested in October). In both Hebei and Henan provinces, the rotation of first wheat and then maize or cotton is the most common cropping pattern. In Henan province, rice is another major crop grown in the summer season. Most cash crops are also grown in the summer season. Wheat production relies more heavily on irrigation than other crops in the region. This is because the growing season for summer season crops (June to October) coincide with the rainy season in the region while that of wheat does not. For example, in years with abundant rainfall, corn could potentially be 100 % rainfed. There is little or no overlap between the irrigation of wheat and that of summer season crops since those crops are usually planted after wheat has been harvested. Since wheat is the major crop that relies on irrigation in the region, we only use the data on the plots that grew wheat in 2004. By doing so, we hold the type of crop constant and also amass the largest number of observations.

Consistent with the findings from the NCWRS data, the CWIM data show that farmers in the North China Plain have three ways to access groundwater. About 47 % of households were still using groundwater from collective tubewells in 2004. The remaining 53 % of households are pumping water from private wells. Among them, about 30 % of households irrigate their crops from their own tubewells. The remaining 23 % buy water through markets.

14.5.1 Impact on Water Use

The CWIM data show that compared with other ways to access groundwater, farmers who gain access through groundwater markets use less water to produce wheat. In 2004, farmers who buy groundwater to irrigate wheat use 9 % less water than farmers who use water from their own tubewells (3,241 versus 3,571 m^3 per hectare). The level of water use by water buyers is also 11 % lower than that by farmers relying on water from collective tubewells (3,660 m^3).

The results remain when we restrict the comparison to be only within villages. In about 40 % of villages, farmers can access groundwater in more than one way. In some villages, one group of farmers irrigates wheat from their own tubewells and another group irrigates their wheat with purchased groundwater. In some cases, a single household has two plots that are in separate locations and the household has sunk a well next to one plot but needs to buy water to irrigate wheat grown on the other plot. When comparing the two groups of farmers or two types of plots, those farmers getting irrigation from their own tubewells use 12 % more than farmers buying water from markets. In addition, in other sample villages some farmers gain access to irrigation from collective tubewells while others purchase their irrigation from groundwater markets. Those farmers in the villages that use water from collective tubewells use 35 % more than farmers that buy water from groundwater markets. When regression analysis is used that controls for the characteristics of villages, households and plots, the difference in water use between farmers that use water from their own wells and farmers that rely on collective wells disappears (Table 14.2). In the regression analysis, the key variables are the two dummy variables that measure the various ways of accessing groundwater: the first equals one if farmers irrigate their plots by buying water; the second equals one if farmers irrigate their plots by pumping water from collective tubewells. The base group is those farmers that use water from their own tubewells.[2]

Importantly, regression results still show that water use falls for farmers that buy water from groundwater markets compared with those that have their own tubewells (Table 14.2). So why is it that farmers that buy water use less water? One reason may be that farmers that purchase water pay more for their water. If so, they would have an incentive to reduce water use. Compared with farmers that pump from their own tubewells or depend on water delivered from collective tubewells, farmers that buy water have higher outlays for their water. The cost of water buyers pay to irrigate wheat is 0.39 yuan per cubic meter, which is more than two times the cost of pumping water well owners incur.[3] When the comparison is restricted to be within villages, the results are the same: water buyers pay more than other farmers that do not depend on groundwater markets for irrigation. Because of this, it is reasonable to expect that farmers that purchase their water on groundwater markets will use water differently than those farmers that have their own tubewells.

The empirical results discussed here are also relevant to the investigation of the impact of groundwater markets and more generally the privatization of wells on

[2]Other control variables are also included. The first group of variables measures the village's production environment such as the share of irrigated area serviced by groundwater and the degree of water scarcity in the village. The second group of variables measures household characteristics such as age and education of the household head. Finally, we also control for plot characteristics such as plot size, soil type and the distance of the plot from home. Access to groundwater, however, suffers from potential endogeneity, because there may be some unobserved factors that affect both water use and the way farmers access groundwater (e.g., water yields of the aquifers). The Instrumental Variable (IV) estimation is used in order to control for the potential endogeneity of access to groundwater. The instrumental variables are two policy interventions variables that measure the way in which policy markers have attempted to intervene in China's groundwater markets. In our field work and during interactions with officials in the local Bureaus of Water Resources, officials told us that they believed that these government programs were done on a basis that is not related to the water use in the village; village leaders and farmers almost never were aware that they could influence these programs. Personal relationships (between officials with control over subsidy/loan programs and village leaders) often was one of the most cited basis for granting a subsidy or a loan to a village leader or farmers (Luo and Kelly 2004). Therefore, the instrumental variables, fiscal subsidies and bank loans for tubewells, are most likely to be exogenous. There is no reason to believe that they have any independent effect on water use except through their influence on the way in which farmers gain access to groundwater.

[3]Yuan is the unit of currency used in China. One dollar was about eight yuan in 2004 and about seven yuan in 2008.

Table 14.2 Impact of groundwater markets on crop water use, crop yield and farmer income

	Log of water use per hectare for wheat	Log of wheat yield per hectare	Crop income per capita	Total income per capita
Buying water from private tubewell (1 = yes; 0 = no)	−0.340 (1.65)*		84.249 (0.05)	−718.512 (0.34)
Using water from collective tubewell (1 = yes; 0 = no)	−0.424 (0.97)		2,305.948 (1.51)	861.595 (0.44)
Production inputs				
Log of water use per hectare		0.022 (0.44)		
Log of labor use per hectare		−0.066 (1.37)		
Log of fertilizer use per hectare		0.134 (2.49)**		
Log of value of other inputs per hectare		0.105 (2.40)**		
Production environment				
Share of village irrigated area serviced by groundwater	−0.315 (1.18)	0.148 (1.22)	437.095 (0.74)	169.110 (0.23)
Village water scarcity indicator variable	0.155 (1.82)*	0.014 (0.30)	−102.536 (0.34)	−215.973 (0.56)
Household characteristics				
Age of household head	0.051 (0.83)	−0.002 (0.11)	22.576 (0.31)	54.391 (0.60)
Age of household head, squared	−0.001 (0.95)	0.000 (0.25)	−0.053 (0.07)	−0.384 (0.37)
Education of household head	−0.014 (0.67)	0.003 (0.31)	−59.787 (1.19)	42.633 (0.67)
Area of plot	−1.088 (1.91)*	−0.371 (1.66)*		
Number of plots per household	−0.003 (0.17)			
Population of household	0.063 (1.74)*			
Arable area per capita of household			9,412.560 (3.69)***	6,123.917 (1.89)*
Plot characteristics				
Loam soil	−0.004 (0.03)	0.040 (0.70)		
Clay soil	0.069 (0.61)	0.115 (2.13)**		
Distance to home	−0.163 (1.26)	−0.097 (1.91)*		

(continued)

Table 14.2 (continued)

	Log of water use per hectare for wheat	Log of wheat yield per hectare	Crop income per capita	Total income per capita
Water saving technology				
Share of surface or underground channel	−0.275			
	(2.26)**			
Flood irrigation (1 = yes; 2 = no)	−0.108			
	(0.98)			
Production shocks				
Yield reduction due to production shocks		−0.015		
		(10.44)***		
County dummy	–	–		
Constant	7.932	6.860	−2,017.856	−644.721
	(6.12)***	(9.20)***	(1.19)	(0.30)
Observations	120	140	200	200
R^2	0.37	0.61	0.10	0.09

Absolute value of z statistics in parentheses; * significant at 10 %; ** significant at 5 %; *** significant at 1 %

China's groundwater resources. Partly because the shift to private well management during 1990s coincided with the rapid decline of water levels in aquifers, some scholars have blamed private well management for the accelerated decline in groundwater levels in northern China (Zhang and Zhao 2003). When wells are managed by the collective, the authority associated with village leaders entails the presence of some governance structure in the groundwater sector, which is often missing in most groundwater economies including India (which is now the largest groundwater economy worldwide, Shah 2009). Village leaders, as the custodian of the village's asset including water resources, may have incentive to conserve groundwater for future use. In contrast, when wells are controlled by farmers, the incentive of a well owner to conserve water may be limited. Given the typical large number of wells in groundwater-using villages, the incentive diminishes rapidly as the number of competitors increases. Even if the well owner wants to regulate water use, he is just one person in a village of water users, and does not have the same authority as village leaders and thus would be less effective in influencing his fellow villager's water use. As a result, it is entirely plausible that unregulated pumping by well owners could result in the tragedy of the commons.

The empirical results discussed above, however, show that the difference in water use between farmers that depend on collective wells and farmers that depend on private wells (either as buyers or sellers) are not statistically significant. In other words, there is little difference between collective well management and private well management in terms of their effects on groundwater. When trying to explain this result, Huang et al. (2013) shows that the hydrology of the aquifers plays a key role. If water in an aquifer is accessible not only to the village above the aquifer but also to neighboring villages, the water level in one village may be affected by the pumping of neighboring villages (or nearby cities) and vice versa. If this is the case, then the aquifers underlying the different villages are connected. In villages using connected aquifers, instead of being assured that water not used in this period is available in future periods, they now need to worry about what their neighboring villages will do because water left in the aquifer this period may be pumped away by them and thus no longer available in future periods. In such cases even village leaders do not have incentive to conserve water. Huang et al. (2013) test this hypothesis by including a dummy variable that equals one if a plot is in a connected village in the regression with plot level water use as the dependent variable. The regression results show that only in villages that are hydrologically isolated from other villages do we observe a higher level of water use by farmers that depend on private wells for irrigation. Farmers that pump from private wells use 70 % more water than those pump from collective wells and the difference is statistically significant at 1 %. Given that a large share of the villages (more than 60 %) are connected, it is not surprising to find no difference between collective well management and private well management.

14.5.2 Impact on Agricultural Productivity and Crop Income

The previous section shows that farmers that purchase their water on groundwater markets use less water than farmers that have their own tubewells. As a result, crop yields and income of water buyers may also be negatively affected. Data show that if farmers irrigate wheat with water purchased from groundwater markets, the average yield is 4,843 kg/ha, which is slightly lower (by 1 %) than that of well owners. A simple t test shows that the difference is not statistically significant. Compared with farmers that depend on collective tubewells, the average wheat yield of water buyers is lower by 8 %. The result is not significant (at the 5 % level) either. The results are still the same when comparison is restricted to farmers within the same village: wheat yields of water buyers are lower but the difference is not statistically significant.

The results of regression analysis (Table 14.2, second column) are also consistent with the descriptive analysis: although water use per hectare falls for farmers that buy water from groundwater markets, yields do not fall significantly.[4] Thus, even though those who buy water from groundwater markets use less water, wheat yields are not negatively affected. While we are not able to prove why empirically, observations during our field work suggest that this is because those that buy water

[4]Wheat yield is regressed on water use per hectare and other production inputs including the amount of labor per hectare measured in man days, fertilizer measured as expenditure per hectare and expenditures on other inputs such as harvesting services. The regression also includes the same set of variables as in the regression on water use above to control for village, household and plot characteristics. We also added a variable that represents production shocks, measured as the farmer-estimated percentage reduction in yields due to floods, droughts or other negative events. The impact of groundwater market on crop yield is measured through its impact on water use.

may be working harder at not wasting water. During our discussions with farmers, we are repeatedly told that because they pay more for their water, farmers that buy water from private tubewell owners pay strict attention to when the water is being applied.

The descriptive analysis indicates that groundwater markets may have a negative effect on the income of farmers that buy water. Per capita crop income for water buyers is 902 yuan, which is only 61 % of that of tubewell owners (1,482 yuan) and 77 % of that of farmers getting irrigation from collective tubewells (1,168 yuan). However, when regression analysis is used, the estimated coefficient on the groundwater market variable is not statistically significant in either the crop income or the total income equations.[5] This means that when other factors are held constant, compared with tubewell owners and farmers that buy water from collectively managed wells, the income of those that buy water from groundwater markets is not lower.

14.5.3 Do Groundwater Markets Help the Poor?

As groundwater markets become increasingly important, it is necessary to understand whether groundwater markets are helping or hurting the poor, and how they affect rural China's income gap. Elsewhere in the world, research has shown that groundwater markets can be equity enhancing. For example, Meinzen-Dick (1996) shows that groundwater markets in Pakistan has improved the equity of groundwater use by making water available to small landowners, tenants and younger households, the group of farmers that are least likely to own tubewells. This may also be the case in China. The results discussed above show that both rich and poor farmers participate in the groundwater market. The data indicate that groundwater markets benefit farmers that are small, less educated, and older. The per capita land area of water-buying households is 0.13 ha, slightly smaller than that of water-selling households (0.15 ha) but the difference is not statistically significant. The average years of schooling of the head of water-buying households is 5.5 while that of water-selling households is 6.3 and the difference is statistically significant. The average age of household head of water-buying households is higher by 2.4 years (50 versus 47.6) and the difference is statistically significant.

Whether groundwater markets benefit the poor is likely related to the structure of the markets (that is, whether they are monopolistic or competitive). The poor should be able to benefit more when markets are competitive than when faced with a single

[5]In the regression on crop income or total income, control variables are similar to those in the regression on water use with two differences. Variables that measure plot characteristics are excluded since this is a household level regression. Total land size and household size are replaced by their ratio: arable land per capita. The same instrumental variable strategy is used to address the endogeneity problem.

seller. To measure the degree of competition, following the work of Lerner (1970), Shah (1993) argues that the ratio of water price to total variable cost can be used as a fairly good indicator of the level of monopoly power. Following this approach, we calculate the ratio of water price to total variable cost to examine the structure of groundwater markets in China. Our data yields a "competitive" ratio of 2.2, ranging from 1.2 to 3.3. More than 70 % of tubewells have ratios lower than 2.5. Hence, if the low ratio of water price to total variable cost does, in fact, measure competition, there is evidence that groundwater markets in China are relatively competitive. In contrast, in some areas in India such as eastern Uttar Pradesh, this ratio is as high as 3.6 (Kumar 2009).

Another way to assess the degree of competitiveness of China's groundwater markets is by comparing within-village price variations to between-village variations. Since groundwater markets are localized and most transactions are among farmers in a single village, if markets were competitive, we would expect prices to vary mostly between villages, not within villages. Indeed, this is what we observe in the data. For example, in one village the price of water from one tubewell is more than 3.4 times that from one tubewell in another village. However, within any of our villages, the highest price difference is only 50 %. In 75 % of the sample villages, water price differences among tubewells selling water within villages are much smaller than that. This is consistent with the findings of other researchers. After controlling for the influence of other factors, Kajisa and Sakurai (2003) found that the variation of water prices in their sample in Madhya Pradesh of India mainly comes from regional differences, leading them to the conclusion that groundwater markets are not monopolistic. In our sample villages that have both collective tubewells and private tubewells selling water, we found that water price of collective tubewell is only slight lower than in the private groundwater markets. On average, the difference in the price of water between collective tubewells and private tubewells is less than 15 %. Also, we found no statistically significant difference in the price of water between private tubewells owned by individual farmers and those owned by a group of farmers (shareholding tubewells).

Our data provided two additional pieces of evidence that support the non-monopolistic nature of groundwater markets. First, we looked at profits from selling water. With our data, we were able to estimate both the fixed and variable costs of pumping and selling water. Accordingly, our results demonstrate that (even when we do not consider the value of family labor that is used to pump and sell water), profits are generally small.

Second, we also looked at the number of well operators selling water and at water delivery conditions. Shah (2000) suggests that when wells are sunk in a fairly dense manner, and when there are lined conveyance structures in a village, there is less of a probability that a single seller will have monopoly power and that the price of water will be relatively more competitive. Using this approach with our survey data, we find that in almost all villages, there are many tubewell operators selling water, not just one. Furthermore, the adoption rate of efficiency-enhancing conveyance technologies (surface plastic pipes or hoses) by farmers in groundwater irrigation regions of northern China is high, at over 70 %, partly because these technologies are not expensive. The adoption of surface pipes greatly increases the ability of farmers to choose the tubewells from which they want to buy water. Based on these analyses, it seems that groundwater markets in northern China almost certainly are not monopolistic, supporting the notion that they offer poor rural households affordable access to irrigation water.

14.6 Concluding Remarks

Our results provide strong evidence that groundwater markets in northern China have developed in terms of both their breadth (the share of villages in which there are groundwater market activity) and depth (the share of water which the average water-selling tubewell owner sells to others on a market basis). Groundwater markets in northern China are informal, localized and mostly unregulated. There is little price discrimination, and institutional characteristics tend to be similar in both high- and low-income villages.

While much of the results are suggestive that groundwater markets are largely self- organizing and unregulated, there does appear to be a role for the state. The findings show that when the government facilitates individuals' and shareholding groups' access to capital, and when they are not subject to local regulations, there is a greater level of groundwater market activity.

In terms of the effect of groundwater markets on access to water for low-income households, our research shows that poor households have been involved in both the supply and demand side of the markets, which is somewhat different from what has been observed in other parts of the world where groundwater markets have emerged. This may be because well-functioning, competitive markets that will expand access to resources for the poor require a relatively unregulated market environment, as well as agents that have access to a minimum amount of land and capital resources. In the case of China, almost all households have land and the government has instituted programs offering loans and grants to those wanting to sink a well. In addition, the incomes of most farmers were already high enough to allow some farmers to gain access to enough capital for investment (and to sufficient liquidity) that they were able to afford to buy water when it was provided in a competitive market environment. When groundwater markets emerge in such an environment, buyers and sellers can both benefit, and overall access to water can raise production levels and the welfare of all participants. In places where land and capital resources are less equitably distributed, this may not occur.

Further evidence that groundwater markets expand irrigation access to the poor comes from our results showing that households that buy water from groundwater markets are poorer than households that sell water on the market. Our research shows that farmers who purchase their water on the market pay more on a per cubic meter basis than farmers who either have their own tubewell or have access to water from collectively owned wells. They also use less water. However, crop yields do not fall, nor is there any measurable negative effect on income. Since Huang et al. (2006) have shown that irrigation has a positive impact on agricultural production and rural incomes, and we have shown above that the households accessing water through groundwater markets are able to maintain agricultural production equal to that of households that access water through other means, it follows that groundwater markets have a positive impact on the incomes of those who participate. Moreover, since low-income households are the primary purchasers, groundwater markets can be said to decrease regional income inequality through their disproportionate positive impact on low-income households.

Our research also has important policy implications. They indicate that farmers respond to incentives: when farmers have to pay more for water, they take measures allowing themselves to save water while maintaining crop yields. Groundwater markets thus represent a simple way to increase water efficiency without materially hurting either production or incomes. As water in China becomes more scarce, and water efficiency needs to be increased, allowing the emergence of groundwater markets may be an efficient way to provide irrigation services.

Assuming that farmers who rely on groundwater markets are unable to access water elsewhere, groundwater markets should lead to an increase in groundwater use and an expansion of irrigated area. While this accrues financial benefits to the individual farmers, it raises concerns as to the long-run sustainability of such a scheme. Despite the relatively efficient water use of farmers who purchase water on groundwater markets, their increased water usage may still be contributing to a fall in the groundwater table. If this is the case, should groundwater markets be abolished? We say no. Instead, water pricing policies should be promoted to control the drawdown of the water table. This would encourage greater water efficiency across the entire irrigating population – instead of simply among those who have no other access to irrigation water – while continuing to afford poorer farmers the access to groundwater that would otherwise be unavailable to them. Thus, the pro-poor benefits that come from increased access to irrigation would be maintained, while the potential negative impact on the water table would be at least partially offset by increased water savings among all users. Of course, policy makers will also benefit from studies that examine groundwater institutions and rules of water allocation which can assist in explaining the impacts of groundwater markets (Aarnoudse et al. 2012; Bluemling et al. 2010).

References

Aarnoudse E, Bluemling B, Wester P, Qu W (2012) The role of collective groundwater institutions in the implementation of direct groundwater regulation measures in Minqin County, China. Hydrogeol J 20(7):1213–1221

Benjamin D, Brandt L (2002) Property rights, labor markets and efficiency in an economy in transition: the case of rural China. Can J Econ 35(4):689–716

Bluemling B, Pahl-Wostl C, Yang H, Mosler HJ (2010) Implications of stakeholder constellations for the implementation of irrigation rules at jointly used wells – cases from the North China Plain, China. Soc Nat Resour 23(6):557–572

China (2002) Water law of the People's Republic of China. Beijing, China

Huang Q, Rozelle S, Lohmar B, Huang J, Wang J (2006) Irrigation, agricultural performance and poverty reduction in China. Food Policy 31(1):30–52

Huang Q, Wang J, Polasky S, Rozelle S, Liu Y (2013) The effects of well management and the nature of the aquifer on groundwater resources. Am J Agric Econ 95(1):94–116

Jacoby H, Murgai R, Rehman S (2004) Monopoly power and distribution in fragmented markets: the case of groundwater. Rev Econ Stud 71:783–808

Kajisa K, Sakurai T (2003) Determinants of groundwater price under bilateral bargaining with multiple modes of contracts: a case of Madhya Pradesh, India. Jpn J Rural Econ 5:1–11

Kumar DM (2009) Water management in India: what works, what doesn't. Gyan Publishing House, New Delhi

Lerner AP (1970) Principles of welfare economics. Augustus M. Kelley, New York

Lohmar B, Wang J, Rozelle S, Huang J, Dawe D (2003) China's agricultural water policy reforms: increasing investment, resolving conflicts and revising incentives, vol 782, Agriculture information bulletin, economic research service. USDA, Washington, DC

Luo X, Kelly D (2004) Economic growth under unlimited supply of cheap "migrant labour": origins, characteristics and prospects of China's economic growth model in the last decade. Paper presented in the international conference "globalization, smallholder farmers, and market integration". Nanjing

Meinzen-Dick R (1996) Groundwater markets in Pakistan: participation and productivity, vol 105, Research reports. International Food Policy Research Institute, Washington, DC

Ministry of Water Resources of China (2011) Water Resources Bulletin 2011. Beijing, China. http://www.mwr.gov.cn/zwzc/hygb/szygb/qgszygb/201212/t20121217_335297.html. Accessed May 19, 2014

Pant N (2004) Trends in groundwater irrigation in eastern and western UP. Econ Pol Wkly 39(31):3463–3468

Shah T (1993) Groundwater markets and irrigation development: political economy and practical policy. Oxford University Press, Bombay

Shah T (2000) Groundwater markets and agricultural development: a South Asian overview. In: GWP, Pakistan water partnership, proceedings of regional groundwater management seminar, October 9–11, 2000, Islamabad, pp 255–278

Shah T (2009) Taming the anarchy: groundwater governance in South Asia. Resources for the Future, Washington, DC

Strosser P, Meinzen-Dick R (1994) Groundwater markets in Pakistan: an analysis of selected issues. In: Moench M (ed) Selling water: conceptual and policy debates over groundwater markets in India. VIKSAT-Pacific Institute-Natural Heritage Institute, Ahmedabad

Wang J (2000) Innovation of property right, technical efficiency and groundwater irrigation management. Ph.D. Thesis, Chinese academy of agricultural sciences, Beijing, China

Wang J, Huang J, Rozelle S (2005) Evolution of tubewell ownership and production in the North China plain. Aust J Agric Resour Econ 49(2):177–195

Wang J, Rozelle S, Blanke A, Huang Q, Huang J (2007) The development, challenges and management of groundwater in rural China. In: Giordano M, Villholth K (eds) The agricultural groundwater revolution: opportunities and threats to development. CAB International, Cambridge

Zhang X, Zhao Y (2003) Existing problems and counter measures on exploitation and utilization of groundwater in Hengshui city. Groundwater 25(2):87–89

Zhang L, Wang J, Huang J, Rozelle S (2008) Groundwater markets in China: a glimpse into progress. World Dev 36(4):706–726

Zhang L, Wang J, Huang J, Huang Q, Rozelle S (2010) Access to groundwater and agricultural production in China. Agric Water Manage 97(10):1609–1616

Chapter 15
Design and Implementation of Markets for Groundwater Pumping Rights

Nicholas Brozović and Richael Young

Abstract Groundwater is an important resource for agricultural and urban water users. In a number of regions around the world, there is rapid change in water management institutions as a result of the impacts of groundwater use on neighboring wells, streams, and groundwater-dependent ecosystems. Increasingly, regulations are based on quantification, monitoring, and enforcement of irrigation rights. Under these conditions, allowing water users to trade pumping rights is a cost-effective mechanism to reduce the costs of regulations on water users. Indeed, despite high transaction costs, nascent markets for tradable groundwater pumping rights have emerged. This chapter describes the history, current institutional context, and economic framework of markets for groundwater pumping rights. In particular, we compare key differences in design and management between groundwater pumping rights markets and surface water markets. We provide a case study that compares groundwater trading to alternate water allocation systems in the Republican River Basin in the United States, an area with active interstate water conflict.

Keywords Tradable permits • Environmental management • Monitoring • Enforcement • Transboundary conflicts • Spatial externalities

15.1 Introduction

Groundwater is an important resource for agricultural and urban water users, and represents about a quarter of freshwater withdrawals worldwide. Typically, groundwater use is unmonitored and unregulated. However, there may be negative consequences of groundwater use on neighboring wells, on adjacent stream flow,

N. Brozović (✉) • R. Young
Department of Agricultural and Consumer Economics, University of Illinois at Urbana-Champaign, 326 Mumford Hall, 1301 W, Gregory, Urbana, IL 61801, USA
e-mail: nbroz@illinois.edu; young40@illinois.edu

K.W. Easter and Q. Huang (eds.), *Water Markets for the 21st Century: What Have We Learned?*, Global Issues in Water Policy 11, DOI 10.1007/978-94-017-9081-9_15,

on endangered aquatic species and groundwater-dependent ecosystems, and on the future availability of water supplies for growing populations. Concerns about groundwater pumping externalities are manifested in ongoing litigation over water resources and rapidly changing water management institutions. For example, in the United States, restrictions on agricultural groundwater use to protect stream flow have been implemented or considered in several western states, including Colorado, Nebraska, and Texas. Increasingly, regulations are based on the quantification of allowable groundwater pumping rights and irrigated land area. These regulations are often combined with mandatory metering of all wells and strong enforcement actions against water users found to be in violation of their permissible pumping.

Under conditions where groundwater pumping is constrained, monitored, and enforced, allowing water users to trade pumping rights is a cost-effective mechanism to reduce the costs of regulations on water users. Indeed, despite high transaction costs, nascent markets for tradable groundwater pumping rights are emerging. In this chapter, we consider the design and implementation of systems for reallocating groundwater pumping rights. First, we discuss the motivation and development of recent groundwater management institutions that support market-based reallocation systems. Second, we consider particular issues that must be addressed in the design of groundwater trading systems, and how these compare with issues pertinent to surface water markets. Next, we present a simple theoretical model to help us better understand the basis of groundwater permit trading, with a particular focus on dealing with externalities related to surface water-groundwater interaction. Following this, we present an overview of current groundwater trading worldwide. Finally, we provide a case study of groundwater trading in the Republican River Basin of Nebraska, an area with active interstate water conflict, in order to demonstrate the potential benefits of a market-based reallocation system.

15.2 Motivation for Groundwater Trading

There has been rapid innovation in groundwater management institutions in the last decade. Two main concerns that underlie recent changes in groundwater policies are long-term aquifer depletion and surface water-groundwater interaction.

First, as a result of sustained groundwater pumping, saturated thicknesses and well yields have been reduced in many regions. In some cases, reductions in aquifer viability have been severe enough that there has been a gradual transition from irrigated back to dryland agriculture (e.g. parts of western Kansas). In other areas, irrigation is still possible but it is clear that the current intensity (in terms of acreage and application rates) will not be maintained in the future (e.g. parts of north Texas). Concerns over the ability of future generations to continue profitable irrigation have led to some groundwater management districts considering and implementing self-regulation and self-enforcement to reduce aggregate pumping. It should be noted that although concerns about the aggregate stock of water available for future pumping has led to groundwater regulation, well interference between adjacent pumping wells has not been an important driver of policy change in recent years. One reason for this may be that – at least as irrigated agriculture is practiced in much of the developed world – well spacing is large enough that the pumping externality imposed on neighboring well owners is relatively small compared to drawdown induced by a well's own pumping.

Second, surface water-groundwater interaction has also been a major driver for changes in groundwater policy. When water is pumped in proximity to a stream, the groundwater and surface water components of the hydrologic system interact in such a way that pumping will move water out of the stream and into the well (Glover and Balmer 1954; Jenkins 1968; Wallace et al. 1990; Sophocleous 2002). This process is known as stream depletion. Stream depletion occurs when the cone of depression of a pumping well intersects a stream. As a result of stream depletion, groundwater pumping may measurably reduce surface water flows. Surface water flows are the subject of both transboundary legal challenges over river basin allocations and potential environmental impacts to instream habitat. Thus, through stream depletion, groundwater users may face stringent oversight and regulation not over their impacts on groundwater, but on their impacts on stream flow (Kuwayama and Brozović 2013). For example, concerns over stream depletion have led to the introduction of regulations on groundwater use in a number of transboundary river basins in the United States, including the Pecos River (between Texas and New Mexico), Arkansas River (Kansas and Colorado), and the Republican River (Kansas, Nebraska, and Colorado). The adverse effects of stream depletion on instream habitat and endangered species have led to regulatory action in Nebraska, Wyoming, Colorado, Idaho, Texas, and California, among others.

The social and political pressure for groundwater management may come from federal, state, or local levels. However in general, local groundwater organizations and institutions are the ones that are developing, implementing, and enforcing new management mechanisms. This has led to an extremely large variability in the kinds of management tools and mechanisms that are being used to address the problem of excessive aggregate pumping levels.

15.3 Design of Groundwater Trading Schemes

There are several concerns that need to be addressed in the design and implementation of groundwater trading schemes. Some of these concerns are unique to groundwater trading and do not occur in the management of surface water markets. On the other hand, there are also concerns for surface water markets that are less relevant for groundwater markets (e.g. see Young 1986; Saliba 1987; Saliba and Bush 1987; and Chong and Sunding 2006 for detailed descriptions of the operation of and issues associated with surface water markets).

15.3.1 Monitoring and Enforcement

In areas where binding, enforced groundwater use regulations do not exist, efficient trading is not possible because the necessary conditions for an efficient market system will not be met. However, well metering and reporting are mandatory in a growing number of groundwater management areas. For example, a large portion of the US states of Kansas and Nebraska requires that all irrigation wells are metered and pumping reported annually, while groundwater management districts in other states such as Texas are increasingly requiring meter installation. Metering is also found elsewhere in the world, including in Australia and New Zealand, as well as in some river basins in China (Easter and Liu 2007; Webber et al. 2008).

Monitoring of groundwater use is only meaningful to resource management to the extent that there is enforcement when violations occur. Where reporting of meter data is voluntary and without sanction, there is little incentive to provide timely or accurate readings. Conversely, in some groundwater management districts, paid district employees do the meter reading, with fines for broken meters and severe penalties for violators. For example, in 2010, the Upper Republican Natural Resources District in Nebraska revoked groundwater pumping rights, estimated to be worth in excess of $3 million, for several groundwater users who had attempted to increase their water use illegally through bypassing their well flow meters. In Australia, meters are similarly read by salaried government employees, with large penalties for violators.

It should be noted that even if metering is not present, it may be possible to establish groundwater transfer systems. For example, if the total irrigated area within a water district is constrained to be less than the total area potentially available for cropland within a district, then the right to irrigate units of land can be reallocated using a market system. Systems of this nature operate in the Platte River Basin in Nebraska. However, the aggregate pumping resulting after reallocation will be subject to uncertainty. Moreover, there is still a need to enforce limits on the irrigated areas for such systems to succeed.

15.3.2 Transaction Costs

The transaction costs associated with water markets may be very large (Young 1986; Saliba 1987; Chong and Sunding 2006). For example, if there is no formal market then search costs will be high. Similarly, the costs associated with brokerage and with monitoring and enforcement may be large. However, the existence of nascent groundwater trading in areas without formal market mechanisms in place implies that transaction costs are less than the gains from trade for some wells.

15.3.3 Consideration of Spatial Externalities

Groundwater pumping leads to a variety of spatial externalities (Brozović et al. 2010; Kuwayama and Brozović 2013). The spatial reallocation of groundwater rights may change the distribution and magnitude of pumping externalities; this is often the explicit purpose of groundwater management. Trading schemes may be designed in order to encourage trades that reduce the magnitude of externalities, or to prevent unwanted impacts on third parties.

For example, current groundwater trading schemes in Nebraska use trading ratios that adjust for the difference in stream depletion between locations of buyers and sellers of groundwater rights (NE DNR and URNRD 2010; NE DNR and MRNRD 2010; NE DNR and TBNRD 2012; Kuwayama and Brozović 2013). Consequently, when moving a unit of water to a location that induces more stream depletion than the original location, less than a unit of water may be transferred. The effect of trading ratios is to create idiosyncratic prices in the permit market that are a function of both buyer's and seller's hydrologic characteristics. We will return to the optimal design of such policies in the theoretical analysis below. However, we can note that currently implemented stream depletion adjustment factors are generally not first-best from an economic standpoint as they are unidirectional: groundwater use moving to a higher stream depletion area is discounted to account for the increased impact, but no additional incentive is provided to move water use from a higher-impact to a lower-impact area and no adjustments are made for such trades.

In areas where the major concern is to reduce long-term aquifer depletion or to limit aggregate water use in critical stream or aquifer or recharge areas, zonal trading may also be implemented. For example, trading in some Australian groundwater markets (e.g. the Lower Lachlan and Murrumbidgee in the Murray-Darling Basin) is subject to zonal restrictions where pumping rights may be transferred out of critical areas (defined as those with the highest density of existing wells) but may not be transferred into critical areas. Similarly, in the Middle and Upper Republican Natural Resources Districts in Nebraska, trading is restricted to defined sub-areas so that the distance between the original point of groundwater pumping and the point to which water pumping is transferred is limited. For example, in the Upper Republican Natural Resources District, the original and new points of abstraction must fall within a 6-mile by 6-mile area.

The groundwater pumping externality is spatial and a function of local hydrologic properties (Brozović et al. 2010). However, in general, potential well interference between wells has not been an impediment to trading. This may be because well spacing regulations are sufficient to prevent large spatial externalities between adjacent wells. An exception is Kansas, which has appropriative water rights for groundwater, as opposed to the more common correlative groundwater rights. Any trade of groundwater in Kansas must not result in detrimental impacts on a senior right. This appropriative rights system, which assigns seniority to water rights, is seen as an impediment to trading. As a result, one area of Kansas (the Sheridan-6 Local Enhanced Management Area) has equalized the seniority of its water rights to allow reallocation to occur.

15.3.4 Conveyance Issues

In general, groundwater permit trading does not involve the transfer of actual water, but only the right to pump water, so that no additional conveyance system is needed. This is in contrast to many surface water quantity markets, where conveyance constraints may be a major issue. In areas with existing binding constraints on groundwater use, those agricultural producers that are interested in purchasing additional pumping rights generally already have excess pumping capacity and should be able to take advantage of purchased permits without any additional capital investment. An exception where conveyance is an impediment to trading is in Texas, where under current groundwater law, trading is allowed but the buyer is expected to pump the water at the location of purchase, on the seller's land. Portions of land overlying the Edwards Aquifer are an exception to this rule, where trading is allowed to change the location of pumping as it is assumed that the area encompassing all potential transfers is small enough that impacts on third parties will not be altered significantly by transfers.

15.3.5 Consumptive Water Use

Understanding whether reallocation of groundwater will increase or decrease consumptive water use is an important consideration for evaluating policies (Thompson et al. 2009). However, existing groundwater permit trading schemes use applied water and not consumptive use as the unit of trade. This is in contrast to surface water markets, where it is not unusual for only the consumptive portion of water used to be transferable. The main reason for the difference is likely pragmatic: well metering measures applied water, and quantifying consumptive water use at a field level is challenging. Moreover, in many cases the same irrigation technology is being used on both buyers' and sellers' fields (typically center pivot systems in the western United States). As a result, differences in consumptive use between buyers and sellers may not be large. In surface water markets where water is moved outside of basins, or between agricultural and urban water users, the need to estimate consumptive use is thus greater.

15.3.6 Other Considerations

There are a number of further issues that need to be considered in the design of groundwater markets. In some cases, conservation offsets have been applied to groundwater transfers to reduce aggregate water use. The conservation offset functions as an additional transaction cost to transferring water rights that creates a divergence between buyers' and sellers' prices for traded water that will thin the potential market. For example, Kansas currently requires a 10 % conservation offset for most groundwater transfers, and some groundwater management districts have imposed additional offsets. A 10 % conservation offset implies that 1 acre-foot of water sold would result in 0.9 acre-feet of water transferred to the buyer. Depending on how the transfer is structured, the costs of a conservation offset may be borne by the buyer, by the seller, or may be split between them.

If pumping constraints are not binding on everyone, there may be a potential problem with trading as rights holders could sell the slack portion of their quota without altering their own water use behavior. As a result, aggregate water use would increase as a result of trading (Palazzo and Brozović 2014). This phenomenon is identical to that of 'hot air' observed in carbon markets ('hot air' refers to allocated emissions permits that are in excess of current emissions and therefore may be sold into a market at any positive price). Thus, additional constraints on transfers that limit them to the metered historical use may be necessary. For example, the Upper Republican Natural Resources District has imposed such limits to prevent the transfer of slack quotas.

A related issue is carryover of pumping permits between years. As crop water demand varies enormously based on climate, it is desirable to provide groundwater users with some flexibility of how permits are used across time. Groundwater management areas in both the United States and Australia allow intertemporal banking, or carryover, of unused allocations, though the amount that may be carried over is restricted in some cases.

15.4 Conceptual Basis for Groundwater Trading

Agriculture is the largest user of groundwater worldwide, and most current regulations that allow groundwater trading are focused on agricultural water use. More rigorous overviews of the economics of agricultural water use in general, and of the economic value of groundwater, are provided elsewhere (e.g. Carlson et al. 1993; Qureshi et al. 2012). Here, our focus is on an analysis of policies for trading of agricultural groundwater pumping rights.

Water is an input to agricultural production. The value of the marginal product (VMP) of water is declining with applied water, and in the absence of regulation, a producer will apply water until the VMP (net of the marginal variable cost of pumping) equals zero. However, if a binding constraint on applied water is introduced, the VMP of water will be positive. A prerequisite for water trading, whether based on surface water or groundwater, is that at current allocations, at least some producers must be constrained so that they have positive VMPs for applied water (e.g. Saliba and Bush 1987; Chong and Sunding 2006). If producers are heterogeneous, then an equal allocation of water for all producers will produce heterogeneity of VMPs for those producers that are constrained. If all producers have identical production functions, gains from trading can still be generated so long as the VMPs at the initial allocations vary (this condition is equivalent to variability in the initial allocations).

Thus binding constraints on available groundwater to pump are necessary but not sufficient for trading to occur; heterogeneity of VMPs is a necessary condition for trading to generate gains for producers. If there is no heterogeneity of VMPs of applied water, then even if producers face severe water constraints, allowing trading will not generate benefits.

15.5 Theoretical Analysis

We present a simple model of agricultural groundwater use for the purpose of developing key intuition about the optimal design of groundwater trading policies. In particular, we present a general model that abstracts from aquifer dynamics and strategic behavior (e.g. Saak and Peterson 2007; Athanassoglou et al. 2012). Our focus on aggregate pumping levels and stream depletion as the relevant constraints matches the focus of current regulations, which generally do not explicitly model either well interference or longer-term aquifer dynamics. Incorporating more realistic spatial dynamics provides qualitatively similar results with more complex optimality conditions (e.g. Brozović et al. 2010; Kuwayama and Brozović 2013).

Consider J wells pumping water from an aquifer during N separate increments of time. For each well j, pumping a quantity u_j^n at time $n = 0, \ldots, N$ produces net benefits given by $B_j(u_j^0, u_j^1, \ldots, u_j^N)$ where $\partial B_j/\partial u_j^n \geq 0$ and $\partial^2 B_j/\partial(u_j^n)^2 < 0$ for all n. For simplicity, we assume that individual well benefit functions are independent, so that pumping at one well will not affect pumping at other wells (this means that we are ignoring well interference between wells and also aquifer dynamics).

We assume that the basic goal of groundwater management is to choose a set of pumping paths to address issues of either aquifer depletion or stream depletion. The general management problem is then given by

$$\max \sum_{j=1}^{J} B_j\left(u_j^0, u_j^1, \ldots, u_j^N\right) \;\; \text{s.t.}\, \Phi\left(u_1^0, u_2^0, \ldots, u_{J-1}^t, u_J^t\right) \leq \overline{\Phi}(t) \;\; \forall t$$

where $\Phi(u_1^0, u_2^0, \ldots, u_{J-1}^t, u_J^t)$ is a transfer function relating the full pumping path at all wells to the relevant pumping constraint $\overline{\Phi}(t)$ at time $t \leq N$. Note that discounting of future benefits may be incorporated into the benefits from pumping, $B_j(u_j^0, u_j^1, \ldots, u_j^N)$, without problem. Next, we consider cases where the primary goal of groundwater regulation is (i) to reduce aquifer depletion and (ii) to reduce stream depletion.

15.5.1 *Case 1: Aquifer Depletion*

If the primary concern driving regulation is aquifer depletion, then the goal of regulation is to reduce pumping in each year below a given amount. In this case, the transfer function Φ simplifies to the sum of pumping at all wells in each year. With the additional assumption that the benefit function is separable by year, $B_j\left(u_j^0, u_j^1, \ldots, u_j^N\right) = \sum_{t=0}^{N} \tilde{B}_j\left(u_j^t\right)$ and the management problem becomes

$$\max \sum_{j=1}^{J} \tilde{B}_j\left(u_j^t\right) \quad \text{s.t.} \sum_{j=1}^{J} u_j^t \leq \overline{\Phi}(t) \forall t$$

For this simple model, the first-order conditions for the problem can be used to show that $d\tilde{B}_i\left(u_i^t\right)/du_i^t = d\tilde{B}_j\left(u_j^t\right)/du_j^t \forall i, j, t$. Thus, an optimal allocation of water across all constrained groundwater users is one that equates their marginal benefits at each point in time. A frictionless tradable permit scheme will by definition achieve this allocation, with a permit price equal to the marginal benefit of water across all locations (e.g. Montgomery 1972; Sunding et al. 2002; Jaeger 2004). The more binding the constraint on total water used, the larger will be the equilibrium permit price.

15.5.2 Case 2: Stream Depletion

Stream depletion caused by groundwater pumping may be related to transboundary legal obligations or to instream impacts on habitat. In the former case of transboundary surface water obligations, the intent of regulation is generally to reduce pumping in order to limit cumulative stream depletion over a fixed interval such as a year or multiple years. In the latter case of instream habitat, regulations are intended to maintain minimum streamflow requirements throughout the year. Here, we will consider the design of tradable permit systems to address cumulative stream depletion problems (Kuwayama and Brozović 2013). The extension to maintain streamflow throughout the year is left to future work (but see Han 2011).

A key feature of surface water-groundwater interaction is that stream depletion is a spatial and dynamic process and that because groundwater is a diffusional system, it is also subject to lagged effects (Glover and Balmer 1954; Sophocleous 2002). Thus, the impact of ongoing pumping on streamflow needs to consider the pumping history rather than just pumping in the current period. A general equation for stream impact from groundwater pumping at any time T after the start of pumping, for a well at a distance d from a stream, is then

$$\sum_{j=1}^{J} \sum_{s=0}^{T} \sum_{n=0}^{s} u_j^n \phi\left(d_j, T - n\right) = \Phi(T)$$

By assumption, the stream depletion externality is linear in pumping (this is known as the principle of superposition in hydrology; Domenico 1972; Freeze and Cherry 1979). Because of this, the transfer function ϕ may be interpreted as the marginal externality of pumping at time n occurring at time $T > n$. Thus, the equation represents the sum of lagged impacts occurring *at time* T from all pumping that occurred at or before time T.

Hydrologic stream response functions can be used to model the exact relationship between pumping and stream flow, accounting for the fact that significant time lags exist between pumping decisions and the consequent stream depletion, and that the magnitudes of these time lags depend primarily on the distance, d, between wells and nearby streams. Both analytical and numerically-derived methods are currently in use in implemented regulations to determine stream response to groundwater pumping. In addition to their use for designing groundwater regulations, analytical methods are also widely used by practitioners for general assessments of stream depletion.

Several US states have determined areas where groundwater is hydrologically connected to adjacent streams. In some cases, entire watersheds are given a designation of connectivity, but in others, a combined spatial and temporal definition is used. For example, the Nebraska Department of Natural Resources (DNR) has implemented the "10/50 rule" (Nebraska DNR 2007). The rule defines separate zones, and therefore potential regulations, for wells based on whether or not groundwater pumping over a period of 50 years will include at least a 10 % contribution from an adjacent stream. In some cases, Nebraska has also applied a 28/40 rule. This rule defines zones based on wells expected to pump at least 28 % of their water from an adjacent stream over a 40-year time period (Nebraska DNR 2004). As stream depletion increases with both time and proximity to a stream, all else equal, the 10/50 rule is more stringent than the 28/40 rule and will cover a large area adjacent to streams where stream depletion is a concern.

Where detailed numerical groundwater models (e.g. MODFLOW) are available, these have been used to determine the impact of pumping on stream depletion. In Nebraska, numerical methods have been used in the Republican River Basin and the Big Blue River Basin (MODFLOW-based), and in the Platte River Basin (COHYST-based). Elsewhere, analytical and graphical methods based on the Glover-Balmer equation have been used (Glover and Balmer 1954; Jenkins 1968; Nebraska DNR 2007).

In Kansas, analytical methods have been used to determine whether additional groundwater is available for appropriation. For the Lower Republican River Basin and Belleville Formation in Kansas, the Jenkins method (a graphical approach based on the Glover-Balmer equations; Jenkins 1968) has been used to estimate the cumulative volume of stream depletion that occurs in one year after the day pumping begins for an application to appropriate groundwater to see whether the new appropriation is acceptable (Barfield 2013).

In addition to their current use for designing groundwater regulations, analytical methods are widely used by practitioners for general assessments of stream depletion. For example, the Glover and Balmer method has been employed in

Colorado to evaluate the current and projected stream depletion impacts of water pumped and discharged during coalbed methane production (Papadopulos and Associates and Colorado Geological Survey 2007).

Hydrologists have derived stream response functions for use in different hydrologic settings; the analytical solution by Glover and Balmer (Glover and Balmer 1954) is one of the simplest analytical solutions but because it has been widely applied in a policy context, we will discuss it here. Stream depletion caused by well j after t years of pumping at a constant rate, measured in acre-feet per year, is given by the Glover-Balmer equation as

$$\phi\left(d_j, t\right) = u_j \operatorname{erfc}\left(\sqrt{\frac{d_j^2 S}{4\tau t}}\right)$$

where d is the distance between well and stream, S is the aquifer storage coefficient, τ is the aquifer transmissivity (units are square feet per year), t is the time in years since the start of pumping, and erfc is the complementary error function. This equation can be modified to account for seasonal pumping; other versions are available for more complex surface water groundwater interactions such as partially penetrating wells or streambed clogging (e.g. Hunt 1999, 2012).

Given the management problem above and the Glover-Balmer equation, the first-order conditions for the problem can be used to show that $\partial B_j\left(u_j^0, u_j^1, \ldots, u_j^N\right) / \partial u_j^i - \lambda \sum_{s=i}^{T} \phi\left(d_j, T - s\right) = 0 \forall i, j$, where λ is the Lagrange multiplier. It follows that for an interior optimum:

$$\frac{\partial B_j\left(u_j^0, u_j^1, \ldots, u_j^N\right) / \partial u_j^i}{\sum_{s=i}^{T} \phi\left(d_j, T - s\right)} = \frac{\partial B_l\left(u_l^0, u_l^1, \ldots, u_l^N\right) / \partial u_l^k}{\sum_{s=k}^{T} \phi\left(d_l, T - s\right)} \forall i, j, k, l$$

Thus, the ratio of the marginal benefit from pumping to the marginal externality caused by pumping should be equal across all well locations. The Lagrange multiplier λ may then be interpreted as the effective (present value) permit price. If the marginal damage of the externality is equivalent for all firms, this outcome can be induced with marketable permits that are traded on a one-to-one basis, where marginal abatement costs of all firms will equal marginal damage multiplied by λ (Palazzo and Brozović 2014). Conversely, if the marginal benefit function is the same at each pumping location, so that $\partial B_j(u_j^0, u_j^1, \ldots, u_j^N)/\partial u_j^i = \partial B_l(u_l^0, u_l^1, \ldots, u_l^N)/\partial u_l^k = B'(u^0, u^1, \ldots, u^N)$, then wells closer to the stream will always be more constrained than wells further from the stream i.e. $u_j^i < u_l^i$ for all i and $T > i$ if $d_j < d_l$ (Kuwayama and Brozović 2013). To show the latter result, consider two wells j and l with $d_j < d_l$. Then it must be that $\sum_{s=i}^{T} \phi\left(d_j, T - s\right) > \sum_{s=i}^{T} \phi\left(d_l, T - s\right)$ for all i and $T > i$ (e.g. consider the Glover-Balmer equation). The result follows immediately from the optimality conditions, as it implies that $B'(u_j^0, u_j^1, \ldots, u_j^N) > B'(u_l^0, u_l^1, \ldots, u_l^N)$, and because $B'' < 0$, it must be that $u_j^i < u_l^i$.

15.6 State of Groundwater Trading Schemes

Below, we briefly consider the state of groundwater trading schemes worldwide, as of early 2013. Note that the rate of institutional innovation is quite high, and so it is expected that this list will change relatively quickly. The focus is on groundwater permit trading between agricultural water users where each well provides water for, and is operated by, a single owner and as a result of trading, the location of pumping changes. This setting is primarily a developed world one. There are also groundwater markets in the developing world, for example in India, Pakistan and China, where one well serves multiple small producers, and these producers may participate in market-like mechanisms to obtain the water they need from the well owner (e.g. Shah 1993; Saleth 1994; Meinzen-Dick 1994; Zhang et al. 2008). However, such groundwater markets will not be considered here (see Chaps. 13 and 14 for a discussion of such markets).

15.6.1 Nebraska

Oversight of groundwater in Nebraska is undertaken by Natural Resources Districts (NRDs), which are local government agencies. Groundwater regulation in the Republican River Basin (Fig. 15.1) has been driven by interstate litigation between Kansas, Nebraska, and Colorado over the surface water allocations to each state from the Republican River (McKusick 2002). All wells in the Nebraska portion of the Republican River Basin are metered, with mandatory annual reporting and moratoria on new wells. The Upper Republican NRD completed metering in 1982, and the remaining NRDs completed metering in 2005. There are pumping quotas in place with carryforward provisions. Current updates of the integrated management plans for three of the NRDs in the Republican River Basin, the Upper (UR) and Middle (MR) Republican and Tri-Basin (TB) (Republican River portion) NRDs, allow for some trading of groundwater pumping rights (Nebraska (NE) DNR and URNRD 2010; NE DNR and MRNRD 2010; NE DNR and TBNRD 2012). The NRDs that allow trading each have slightly different rules that constrain trading. For example, in the Upper Republican NRD, trades must stay within an area equal in size to a township (36 mile^2). In the Middle Republican NRD, trading is limited to groundwater users within certain distances from streams. In years in which the Middle Republican NRD is concerned about meeting its stream depletion targets under the Republican River Compact, trading is suspended. In all cases, there is an adjustment for differences in stream depletion if pumping rights are transferred to a location where stream depletion is greater than the original location. However, if pumping

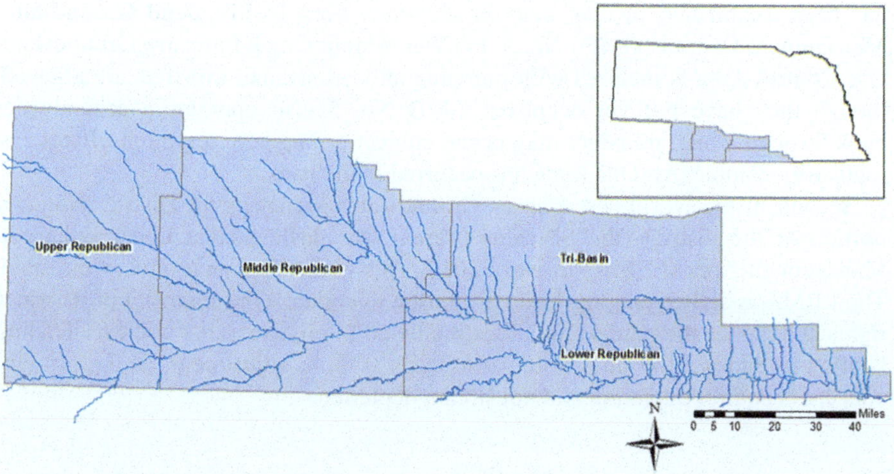

Fig. 15.1 Location map of the Natural Resources Districts within the Nebraskan portion of the Republican River Basin

rights are transferred to a location with lower stream depletion than the original location, no adjustment to the rights takes place. As a result, adjustment for stream depletion as currently implemented in the Republican River Basin does not correspond to the optimality conditions for trading derived above, and is not first-best. The unidirectional nature of the adjustment is driven in part by the NRDs' desires to maintain aggregate pumping levels in their districts as well as to reduce stream depletion. Under the settlement with Kansas, stream depletion is calculated over a 50-year horizon. There are currently no formal markets for groundwater in the basin, but even so several hundred trades occurred over the period from 2005 to 2012.

In the Platte River Basin in Nebraska, groundwater regulation is driven by stream depletion impacting endangered species habit. The Twin Platte, Central Platte, and Tri-Basin (Platte River portion) NRDs allow transfers of groundwater pumping rights. These NRDs are not currently metered, but have certification of irrigated acres and allow the transfer of this acreage. Stream depletion is calculated over a 50-year horizon and transfers are usually adjusted if acreage is transferred to a location with higher stream depletion than the original location. There are also additional spatial limits on trading, such as constraints that trades cannot move water upstream (Twin Platte NRD) or outside of specified zones (Tri-Basin NRD).

15.6.2 Kansas

Kansas is unusual in having appropriative, rather than correlative, rights for groundwater. This creates a potential hurdle for groundwater trading as any transfer cannot impact a senior rights holder (Barfield 2013). However, groundwater trading has been established in two areas of the state. First in Big Bend Groundwater Management District (GMD) No. 5, the Wet Walnut Creek Intensive Groundwater Use Control Area is metered with pumping allocations, and transfers are allowed, though they have not yet occurred. GMD No. 5 also operates a groundwater bank through which transfers may occur, subject to large conservation offsets and regulatory complexity. One trade has occurred in the bank.

Second in the Northwest Kansas Groundwater Management District No. 4, a portion of the district (the Sheridan-6 area) was designated a Local Enhanced Management Area (LEMA) in early 2013. This is the first such area in the state. The LEMA is self-regulating, and has chosen to equalize the seniority of its water rights and reduce the total water allocation by 20 % relative to historic use. Trading is allowed and will be on a volumetric basis without adjustment, as the primary concern is aquifer depletion and not stream depletion.

15.6.3 Australia

Groundwater law in Australia is governed by the National Water Initiative, an agreement that all states have signed (GHD et al. 2011; Skurray et al. 2012). In principle, the Initiative requires a move towards tradable systems of groundwater rights. In particular, in management areas that are viewed as overallocated (currently all major shallow aquifers), the only way to obtain a new right to pump groundwater is through trade with an existing rights holder.

In Eastern Australia, and particularly in the Murray-Darling Basin, groundwater markets are well developed. All wells are metered with annual pumping allocations and intertemporal banking allowances that depend on location and water rights seniority. There are severe penalties for overpumping, though producers can purchase additional water rights if they reach their allocation limit. The most junior rights holders have no allocation at all, so that well owners must acquire water rights in order to pump anything. In general, most agricultural producers have access to both surface water and groundwater, and groundwater markets are secondary and much smaller than surface water markets. Groundwater basins with active trading include the Lower and Upper Lachlan, the Murrumbidgee, and the Artesian Basins. Trading rules vary between basins. For example, the Lower Lachlan zone allows trading between any wells, whereas the Murrumbidgee is divided into three zones based on irrigation intensity, with trading only allowed within or out of some zones. To date, there has been limited groundwater trading in Western Australia, often related to urban expansion (Skurray 2012).

15.6.4 Other Regions

Several other western US states, such as Arizona, California, and Texas, have local groundwater trading programs (Western Governors' Association 2012). These programs often involve trading for urban development. For example, the Edwards Aquifer Authority in Texas has implemented well permitting and metering programs and allows transfers of the right to pump up to 1 acre-foot/acre of certified irrigated land (EAA 2012). Both permanent transfer and lease markets exist. In California, there are a number of groundwater banking programs that store (underground) excess surface water in wet periods and draw from the aquifer during dry periods. The recharged groundwater can be sold and conveyed through the existing water supply infrastructure. However, the banking programs in California do not involve the transfer of groundwater pumping permits between individuals (see Chap, 5 for more information on California examples). New Zealand also allows trading of groundwater pumping permits, although trading has been very limited to date.

15.7 Case Study: Effectiveness of Groundwater Trading to Reach Stream Flow Goals

As shown above, the most efficient way to achieve a given reduction in aggregate water use is with a tradable permit scheme. The gains from a tradable permit system depend on the heterogeneity of marginal benefits at the initial allocation. For any given application, it is important to understand whether the gains from the introduction of a market-based system for groundwater regulation are quantitatively important, and the potential gains relative to alternate regulations.

We now consider a comparison of alternate policies to reach stream flow goals in an agricultural watershed with widespread groundwater pumping for irrigation. The case study area is the Upper Republican Natural Resources District (URNRD) in Nebraska (Fig. 15.1). This groundwater management district overlies the High Plains aquifer and must reduce aggregate groundwater pumping to meet stream depletion goals as a result of interstate litigation (McKusick 2002). All 3,200 wells in the district are metered and groundwater pumping rights at each well are quantified and enforced. Pumping is only allowed on certified irrigated acres, and there is a moratorium on new wells or acres, so the maximum irrigated land area is fixed.

We use profit functions calibrated for each well that consider the joint land use, crop choice, and applied water decision (Martin et al. 2007; Palazzo and Brozović 2014). We use several types of spatial data at the well level, including information on acreage irrigated by each well, depth to water and well yield, soil type, crop evapotranspiration requirements, and irrigated and dryland crop yields for corn, wheat, soybeans, and sorghum (the major crop types in the study area). For each set of well-specific parameters we use a nonlinear optimization to determine crop choice, land use, water application, and expected profits (Palazzo and Brozović 2014). The baseline water availability for the analysis is the current regulation for the URNRD, 13 acre-inches per year for each certified irrigated acre. Then, we sequentially reduce available water for each well in order to estimate the marginal benefits of water use in irrigation and the resulting foregone profits.

Adjustment to reductions in water availability is allowed at both the extensive and intensive margins (English 1990; Palazzo and Brozović 2014). Next, the set of profit functions can be used to compare the tradeoffs between aggregate water use reductions, resulting improvements in stream flow, and producers' foregone profits. In particular, we analyze three different kinds of policies, each of which has been implemented in the watershed of interest: pumping quotas, irrigated land retirement, and tradable pumping permits. In all cases, we consider the current water allocation in the URNRD as the baseline for measurement. For each of the alternate policies described below, we reduce aggregate water use by varying amounts and then compare the total foregone profits required to attain that water use.

15.7.1 Pumping Quotas

Aggregate water use may be reduced by reducing the pumping quotas at each well equally (i.e. reducing the allocation of water for each certified irrigated acre in the district). From an economic point of view, quotas are not a cost-effective method to reduce aggregate water use if producers have heterogeneous marginal products of water use. However, quotas are generally viewed as an equitable regulation as they are imposed equally on water rights holders.

The URNRD has used pumping quotas since it completed well metering in 1982. Over time, the quotas have been reduced and application limits are currently at 13 in. per year for certified acres (65 in. over 5 years with carryover allowed).

15.7.2 Irrigated Land Retirement

A land retirement program operates through existing land markets. Farmland with the right to irrigate is purchased and then the irrigation rights are retired. Formerly irrigated land moves to dryland agriculture. Thus, aggregate water use within the district is reduced by an amount equal to the total pumping rights associated with the purchased land. Equivalently, the irrigation right may be purchased and retired by itself, separately from the land. From an economic standpoint, land or water retirement programs will be a relatively expensive solution as they generally operate on the extensive and not the intensive margin. Moreover, a limited range of rights may be available for acquisition at any point. However, the transaction costs of rights retirement may be low as only one landowner at a time is involved. Rights retirement programs may be targeted based on the cheapest land (reducing irrigated acreage fastest), the cheapest water (reducing aggregate pumping fastest), or by stream depletion (shutting down wells with the highest marginal externality first). We consider all three of these land retirement targeting options. For the analysis, we use the Stream Depletion Factor (SDF) to estimate the stream depletion externality. Following URNRD rules, the SDF is the proportion of water pumped from a well that is drawn from an adjacent stream. SDFs used in the analysis were calculated over a 50-year time horizon with seasonal pumping for irrigation and the Glover-Balmer equation parameterized with hydrologic data for the Republican River Basin (Kuwayama and Brozović 2013). As stream depletion is modeled as an additive process, the SDF is also the marginal externality of an additional unit of pumping.

In recent years, the URNRD has had an active land retirement program, spending \$10 million to purchase 3,300 acres of irrigated land in 2011 (in the Rock Creek area) and joining with the Middle and Lower Republican NRDs and the Twin Platte NRD to purchase almost 20,000 acres of land for \$83 million in 2012 (in a project called N-CORPE). In both cases, the NRD intends to construct stream augmentation projects that will link wells to nearby streams directly with a pipeline. This will allow pumping of groundwater directly into the river to provide compliance with interstate compacts in drought years. Note that as pumping itself induces stream depletion, stream augmentation is a temporary measure that has been controversial within the community as it can be viewed as depleting the limited groundwater resource without increasing agricultural production through irrigation.

15.7.3 Tradable Pumping Permits

Tradable permits allow equalization of the values of marginal products, with or without adjustment for stream depletion, and are a cost-effective method to achieve any given water use reduction or stream depletion target (Kuwayama and Brozović 2013; Palazzo and Brozović 2014). In this case, the marginal benefits of pumping are equalized across all traders, where different levels of marginal benefit correspond to different aggregate water uses. Tradable permit schemes are voluntary and will benefit both buyers and sellers. Here, we assume frictionless trading, noting that metering, quantified allocations and enforcement are already in place. We consider both trades that are unadjusted for stream depletion where the unit of transfer is quantity of water (here called 'simple') and first-best trades that are adjusted for stream depletion, where the unit of transfer is quantity of stream depletion (here called 'complex'). For the simple trading scheme, marginal benefits of pumping water are equalized across all wells without any adjustment for differences in the spatial stream depletion externality. This corresponds to trading across the district at a single market price, equal to the marginal benefit. For the complex trading scheme, the marginal benefits at each well are normalized by the expected impact on stream depletion (as described in the theoretical development above). In this case, each well faces an idiosyncratic price for pumped water, corresponding to a single market price for expected stream depletion resulting from pumping. As in the analysis of targeted land retirement, we use estimated Stream Depletion Factors to quantify the stream depletion externality (Kuwayama and Brozović 2013).

The URNRD currently allows transfers of groundwater pumping rights (NE DNR and URNRD 2010), and there is an adjustment for stream depletion that is unidirectional (i.e. total water use is not allowed to increase even if water moves

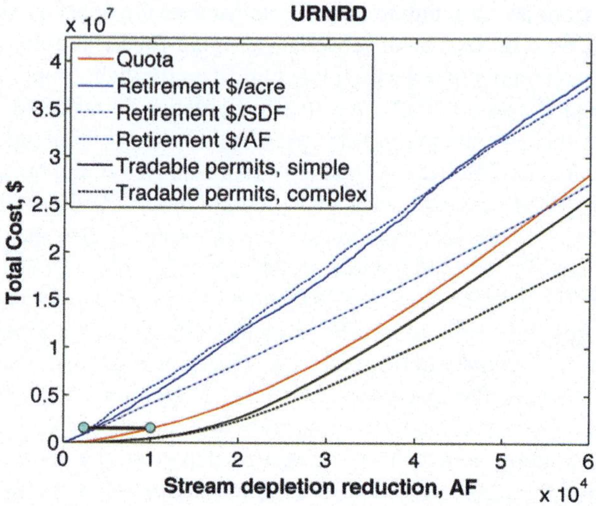

Fig. 15.2 Comparison of the cost-effectiveness of alternate policies for reducing stream depletion impacts in the Upper Republican Natural Resources District, Nebraska. Total costs (*the vertical axis*) are annualized. The *two dots joined by a line* represent the estimated annualized cost of the URNRD's Rock Creek project, with costs and depletion reductions shown for irrigated acreage retirement (*left dot*) and retirement together with stream augmentation pumping (*right dot*). SDF is the Stream Depletion Factor, defined as the proportion of water pumped from a well that is drawn from an adjacent stream over a 50-year time horizon with seasonal pumping (Kuwayama and Brozović 2013)

away from a stream). In the past, simple trading has also been used where the unit of transfer is quantity of water. Though this is an informal market, over the period from 2005 to 2012, over 100 trades occurred.

15.7.4 Results

By definition, when stream depletion reduction is the policy target, a complex tradable permit scheme that adjusts for stream depletion will be the cost-effective method of achieving any instream target (Fig. 15.2). Perhaps surprisingly, the simple tradable permit scheme is also cost-effective for small reductions in stream depletion. The simple permit scheme does not adjust for stream depletion, but in the URNRD there are a number of wells with high stream depletion impacts but very low value of the marginal product of water. These wells will be sellers in any groundwater market, whether there is an adjustment for stream depletion or not.

The uniform quota is also more cost-effective than the land retirement schemes until large reductions in stream depletion are needed, when the land retirement policy targeted on stream depletion performs better. The URNRD's Rock Creek land retirement and stream augmentation project is similar in cost to the predicted costs of land retirement (left hand dot on the horizontal line in Fig. 15.2), and with the stream augmentation in operation but without considering the energy costs of pumping, is similar to a reduced quota in cost-effectiveness.

15.8 Conclusion

Over the last decade, concerns about the impacts of ongoing groundwater pumping have led to rapid innovation in groundwater management institutions. Long-term aquifer depletion and surface water-groundwater interaction underlie most recent changes in groundwater management policy. The increasing adoption of well metering and allocation of groundwater pumping rights have allowed nascent tradable groundwater pumping permits to emerge. As groundwater management is often undertaken at a local level, developing institutions tend to be idiosyncratic, based on local hydrologic conditions and organizational structure. This chapter has described the history, current institutional context, and economic framework of markets for groundwater pumping rights, as well as considering some of the specific challenges present when establishing groundwater markets. It is very likely, however, that tradable permit systems for groundwater management will become much more prevalent in coming years as the long-term impacts of groundwater pumping continue to grow.

Acknowledgments This material is based upon work supported by the National Science Foundation under grant No. EAR-0709735, by USDA grant No. AG 2012-67003-19981, and by USDA-NIFA Hatch project number #ILLU-470-369

References

Athanassoglou S, Sheriff G, Siegfried T, Huh WT (2012) Optimal mechanisms for heterogeneous multi-cell aquifers. Environ Resour Econ 52:265–291. doi:10.1007/s10640-011-9528-0

Barfield DW (2013) Kansas Department of Agriculture, Division of Water Resources, Topeka, Kansas

Brozović N, Sunding DL, Zilberman D (2010) On the spatial nature of the groundwater pumping externality. Resour Energy Econ. doi:10.1016/j.reseneeco.2009.11.010

Carlson GA, Zilberman D, Miranowski JA (eds) (1993) Agricultural and environmental resource economics. Oxford University Press, New York

Chong H, Sunding DL (2006) Water markets and trading. Annu Rev Environ Resour 31(2):239–264

Domenico PA (1972) Concepts and models in groundwater hydrology. McGraw-Hill, New York

Easter KW, Liu Y (2007) Who pays for irrigation: cost recovery and water pricing? Water Policy 9:285–303

Edwards Aquifer Authority (2012) Edwards Aquifer Authority rules. http://www.edwardsaquifer.org/files/download/2b4674cbea20e5b

English M (1990) Deficit irrigation I: analytical framework. J Irrig Drain Eng 116(3):399–412

Freeze A, Cherry J (1979) Groundwater. Prentice-Hall, Englewood Cliffs

GHD, Hamstead Consulting, O'Keefe V (2011) A framework for managing and developing groundwater trading, waterlines report, National Water Commission, Canberra

Glover R, Balmer G (1954) River depletion resulting from pumping a well near a river. Am Geophys Union Trans 35(3):468–470

Han J (2011) Spatial dynamic groundwater regulation to ensure adequate instream flow for Salmon. Unpublished MS thesis, University of Illinois, Department of Agricultural and Consumer Economics

Hunt B (1999) Unsteady stream depletion from ground water pumping. Ground Water 37(1):98–102

Hunt B (2012) Review of stream depletion solutions, behavior, and calculations. J Hydrol Eng. doi:10.1061/(ASCE)HE.1943-5584.0000768

Jaeger WK (2004) Conflicts over water in the Upper Klamath Basin and the potential role for market-based allocations. J Agric Resour Econ 29:167–184

Jenkins CT (1968) Techniques for computing rate and volume of stream depletion by wells. Ground Water 6:37–46

Kuwayama Y, Brozović N (2013) The regulation of a spatially heterogeneous externality: tradable groundwater permits to protect instream flows. J Environ Econ Manage. doi:10.1016/j.jeem.2013.02.004

Martin D, Supalla R, McMullen B, Nedved S (2007) Water optimizer: a decision support tool for producers with limited water. University of Nebraska-Lincoln Departments of Biological Systems Engineering and Agricultural Economics, Lincoln, Nebraska

McKusick V (2002) State of Kansas v. state of Nebraska and state of Colorado: joint motion of the states for entry of proposed consent judgement and approval and adoption of final settlement stipulation, Supreme Court of the United States, 2002

Meinzen-Dick R (1994) Private tubewell development and groundwater markets in Pakistan: a district-level analysis. Pakistan Dev Rev 33(4):857–869

Montgomery D (1972) Markets in licenses and efficient pollution control programs. J Econ Theory 5(3):395–418

Nebraska Department of Natural Resources (2004) COHYST technical committee. The 40-year, 28-percent stream depletion lines for the COHYST area west of Elm Creek. Nebraska, Lincoln

Nebraska Department of Natural Resources (2007) 2008 annual evaluation of availability of hydrologically connected water supplies, Lincoln, Nebraska

Nebraska Department of Natural Resources, Middle Republican Natural Resources District (2010) Integrated management plan, Middle Republican Natural Resources District, Curtis, Nebraska, p 28

Nebraska Department of Natural Resources, Tri-Basin Natural Resources District (2012) Integrated management plan, Tri-Basin Natural Resources District, Holdrege, Nebraska, p 30

Nebraska Department of Natural Resources, Upper Republican Natural Resources District (2010) Integrated management plan, Upper Republican Natural Resources District, Imperial, Nebraska, p 24

Palazzo A, Brozović N (2014) The role of groundwater trading in spatial water management. Agric Water Manage. doi:10.1016/j.agwat.2014.03.004

Qureshi ME, Reeson A, Reinelt P, Brozović N, Whitten S (2012) Factors determining the economic value of groundwater. Hydrogeol J 20(5):821–829. doi:10.1007/s10040-012-0867-x

S. S. Papadopulos and Associates, Inc., Colorado Geological Survey (2007) Coalbed methane stream depletion assessment study – Raton Basin. Colorado Geological Survey, Boulder

Saak A, Peterson J (2007) Groundwater use under incomplete information. J Environ Econ Manage 54(2):214–228

Saleth RM (1994) Groundwater markets in India: a legal and institutional perspective. Indian Econ Rev 29(2):157–176

Saliba BC (1987) Do water markets "work"? Market transfers and trade-offs in the southwestern states. Water Resour Res 23:1113–1122

Saliba BC, Bush DB (1987) Water markets in theory and practice: market transfers, water values, and public policy. Westview, Boulder

Shah T (1993) Groundwater markets and irrigation development: political economy and practical policy. Oxford University Press, Oxford/New York

Skurray JH, Pandit R, Pannell DJ (2012) Institutional impediments to groundwater trading: the case of the Gnangara groundwater system of Western Australia. J Environ Plan Manage. doi:10.1080/09640568.2012.716368

Sophocleous M (2002) Interactions between groundwater and surface water: the state of the science. Hydrogeol J 10:52–67

Sunding DL, Zilberman D, Howitt R, Dinar A, MacDougall N (2002) Measuring the costs of reallocating water from agriculture: a multi-model approach. Nat Resour Model 15(2):201–225

Thompson CL, Supalla RJ, Martin DL, McMullen BP (2009) Evidence supporting cap and trade as a groundwater policy option for reducing irrigation consumptive use. J Am Water Resour Assoc 45:1508–1518

Wallace R, Darama Y, Annable M (1990) Stream depletion by cyclic pumping of wells. Water Resour Res 26(6):1263–1270

Webber M, Barnett J, Finlayson B, Wang M (2008) Pricing China's irrigation water. Global Environ Change 18(4):617–625

Western Governors' Association (2012) Water transfers in the west, The Western Governors' Association, Denver, Colorado, 80202

Young RA (1986) Why are there so few transactions among water users? Am J Agric Resour Econ 68:1143–1151

Zhang L, Wang J, Huang J, Rozelle S (2008) Development of groundwater markets in China: a glimpse into progress to date. World Dev 36(4):706–726

Chapter 16
Western Water Markets: Effectiveness and Efficiency

Christopher Goemans and James Pritchett

Abstract Most rivers throughout the western U.S. are fully appropriated. New municipal, industrial, recreational and environmental water demands will likely be met by reallocating water out of agriculture, the region's largest user of water. The question is: how best to do so? Water markets have long been advocated by many as the answer to this question. This chapter begins with an overview of water allocation law and water markets in the West including a discussion of the various alternative market-based reallocation mechanisms being considered. A summary of recent literature on market activity in the West is followed by a detailed look at transactions in the Colorado River Basin.

Keywords Reallocation • Recreational demands • Agriculture • Water law • Colorado Basin

16.1 Introduction

The Western United States grew its population faster than any other region in the United States between 2000 and 2010, and this rapid growth is forecasted to continue (Mackun et al. 2011).[1] Actual and anticipated population growth increases municipal and industrial (M&I) demands for reliable water supplies. An increasing appetite for reliable water is not limited to municipal consumption. Increasing water

[1]The western United States includes: Alaska, Arizona, California, Colorado, Hawaii, Idaho, Montana, Nevada, New Mexico, Oregon, Utah, Washington, and Wyoming.

C. Goemans (✉) • J. Pritchett
Department of Agricultural and Resource Economics, Colorado State University, 1200 Center Ave Mall, B312 Clark Building, Fort Collins, CO 80523-1172, USA
e-mail: Chris.Goemans@Colostate.edu; james.pritchett@colostate.edu

K.W. Easter and Q. Huang (eds.), *Water Markets for the 21st Century: What Have We Learned?*, Global Issues in Water Policy 11, DOI 10.1007/978-94-017-9081-9_16,

needs in the energy and agriculture sectors as well as for ecosystem services are also anticipated throughout the West (Smith and Pritchett 2010). At the same time that total water demand is increasing, climate models suggest a hotter West with some states facing decreased precipitation and more intense drought cycles (Udall 2013). Many western aquifers are being depleted at unsustainable rates. Water is scarce for all users.

A plausible argument can be made that all existing water supplies in the West are allocated to current users; indeed, only limited "new" water is available for development. Conserving municipal and agricultural water is expensive, and this 'saved' water will only satisfy a small portion of growing demands.[2] Moreover, municipal planners are reluctant to use saved water exclusively for new urban development, concerned that increased efficiency will decrease the effectiveness of mandatory conservation programs during drought.

Significant new supplies from conservation or firmed supplies from large-scale, federal water storage projects are unlikely to develop. As a result, the voluntary sale of water rights from agricultural water right holders to M&I suppliers is inevitable. Simply put, some of the water used to irrigate crops will be redirected to other uses.

While reallocation is inevitable, the mechanisms and institutions guiding water transfers are evolving. Permanent market transactions have been the popular default mode of transfer, but the outright purchase of water rights and the drying of agricultural lands can be expensive and politically contentious. Leases, drought-year options, interruptible-supply agreements and lease-back agreements are all examples of innovative water transfer mechanisms (Smith and Pritchett 2010) that may be more politically viable. An important contribution of economics is describing the efficiency of water transfer institutions, and then advancing the public policy debate by describing the distributional impacts of transfers. With this in mind, this chapter provides an overview of water allocation in the West, discusses different institutional mechanisms for reallocating water and examines recent trends in water markets across states.

16.2 Water Allocation in the West

Loosely speaking, three layers of laws govern the allocation of water in the West.[3] Interstate compacts largely determine the surface water allocation of river systems *across* states, and the underlying allocation mechanisms vary widely according to

---

[2]Saved water implies improvements in conveyance and application efficiency or saved consumptive use. In some states, water law allows for some saved water to be claimed and put to beneficial use, but other states do not allow for conveyance and application savings to be re-used as these are return flows already claimed by others.

[3]A set of laws dictating the use of water on Federal and Indian lands also exist. While they impact the availability of water they are not immediately relevant to the focus of this chapter.

the negotiated compact agreement. Examples include allocations based on historical flow measurements (the Colorado River Compact), recent climatic conditions (Rio Grande Compact), and priority (South Platte River Compact).

Water allocation *within* a state's boundaries is primarily implemented under the Doctrine of Prior Appropriation, which serves as the overarching surface water allocation mechanism for all states in the West (Bell and Taylor 2008).[4] In this case, the individual states administer a system of legally adjudicated water rights, and the details of implementation vary by jurisdiction. Ownership of a water right grants the owner the right to divert water, subject to availability, for a specified beneficial use at a specified location and time, in perpetuity. Water is allocated based on the seniority of each right (i.e., when it was first established) so that the owners of the most senior water rights are guaranteed water before owners of junior rights receive their allocation. The security/reliability of one's supply is therefore determined based on seniority of the water right. Senior water rights are more valuable in water markets and more likely to be the target of acquisition for municipal water providers.

Under the Doctrine of Prior Appropriation, water rights represent property that is tradable and separable from the land/activity in which it is used. Organizations often own a portfolio of water rights and collectively provide for water storage and conveyance. Mutual ditch companies and irrigation districts are two such organizations and these allocate the majority of water in the West (Getches 2009). Mutual ditch companies are incorporated businesses that are responsible for the financial and operational activities of distributing water to its shareholders. An individual shareholder's right to use water comes indirectly via their ownership in the company. Irrigation districts, including conservancy and conservation districts, are typically associated with US federal projects. Notable examples of irrigation districts include southern California's Imperial Irrigation District that provides water and energy to its members under the direction of bylaws and operational procedures established by a board of directors, and Northern Water, an irrigation district in northern Colorado that administers flows from the Colorado Big Thompson project. Unlike ditch companies, irrigation districts are government entities where the district, not the shareholders, own the water rights.

Prior appropriation determines an irrigation district's water supplies for a given year, but the water distribution amongst shareholders is based on the by-laws of the organization. Ditch companies typically tie water deliveries to stock ownership, and these deliveries are not based on a priority apportionment – a wider range of rules exists. Some districts apportion supplies based on ownership of the number shares in the project (e.g., the Colorado Big-Thompson project), others determine allocations annually based on the needs of particular types of use (e.g., Fryingpan-Arkansas Project).

[4]The particular application of Prior Appropriation varies from state to state, with California and New Mexico also utilizing Riparian water law to some extent. For an overview of the different forms of Prior Appropriation see Getches (2009).

16.3 Water Markets and Reallocation in the West

The allocation mechanism for the Doctrine of Prior Appropriation is first-in-time, first-in-right – owners of water rights with the earliest appropriations guaranteed their needs are satisfied before others when water is limited. The water rights filed first are deemed "senior" water rights and receive priority over "junior" rights. Irrigated cropping made the first foray into water development in the West, as a result the vast majority of senior water rights where initially granted to, and are still owned by, agricultural producers. Agriculture diverts 80 % of surface water flows in the West and more than 90 % of its consumptive use (Kenny et al. 2009).

Rapid growth in municipal and industrial water demand will create a gap between existing supplies and forecasted demands. In-stream flow demand for environmental and recreational use is also increasing. As a result, water suppliers seek opportunities to acquire new water.

Re-timing surface water flows via storage is one means of addressing growing water demand in the West, but large-scale water-supply projects are often controversial and expensive, so the recent water policy focus has shifted toward conservation and the reallocation of existing supplies (Donohew 2009). Most policymakers agree that accommodating new demands will involve a voluntary redistribution of water from agriculture and into M&I, environmental and recreational uses (Western Governors Association 2012). Two arguments are made in support of redistribution. The first argument posits that reallocating a relatively small proportion of water currently allocated to irrigated cropping can lead to a significant increase in the amount of water available for other uses (Nichols et al. 2001) given agriculture's dominance in water right ownership.

The second position is rooted in economic efficiency and argues that the prior appropriation apportionment fails to match water to its highest valued use. The value of water fluctuates as new uses emerge, or as water availability and reliability are impacted by a changing climate, market dynamics, evolving institutions, et cetera. In an idealized setting, markets reallocate water to the highest valued use as dictated by the existing conditions. Absent markets that respond to change, the doctrine of prior appropriation allocates water to the most senior water rights first, regardless of the use value. Proponents of water markets argue that observed differences in the marginal value of water in agriculture compared to that in M&I are evidence of an inefficient allocation of the resource. The latter value is often more than an order of magnitude higher than the former (Howe et al. 1986; Nichols et al. 2001).

Economists believe water markets are an effective instrument for fostering reallocation (Bell et al. 2008). Advocates prefer water markets because of their potential to:

- reallocate water and water rights based on individual decision making in a decentralized setting,
- incentivize conservation as water rights owners and water users face the opportunity costs of ownership and use,
- result in the efficient allocation of water rights and water, reallocating water from low-to-high-valued activities (Howitt and Hansen 2005).

An allocation of water resources is said to be economically efficient if there is no redistribution of the resource that will create greater benefits for society. In an idealized setting, markets result in efficient resource distribution by matching the resource to its highest valued use. The market mechanism is relatively straightforward – rights are exchanged when perceived water values diverge between owners and buyers, and this value difference exceeds the transaction costs. Buyers will purchase when the value of using the resource exceeds or is equal to its expense, and sellers choose to sell when the price exceeds the benefits of retaining the resource. All agents act in their best interests, and repeated voluntary exchanges eventually match the resource to the highest valued use. Better redistributions of the resource are not possible from society's perspective assuming perfectly competitive market conditions; e.g., where transactions costs are limited, information is symmetric, no one party has market power and transactions do not result in externalities.

Of course, these conditions do not exist in most settings. The characteristics of water create less than ideal circumstances for trading. In order for buyers and sellers to gain from a transaction, the property rights of water needs be enforced. This can be challenging, as the complex, underground movement of water can make it difficult to track and measure. Transporting water is challenging because of its bulk and weight, and water is hard to store because it evaporates into the air and percolates into the ground. Water infrastructure is not always available to permit timely transport of water, and the costs of navigating state laws impede transactions. In spite of this, water markets do exist in different forms.

16.3.1 What Are Water Markets?

Water markets generically[5] describe a wide range of institutions (Brown 2006), and these may be categorized as markets for water rights and markets for water.[6] These two differ primarily in that, for the latter, the seller retains ownership of the asset. From this point forward in the chapter, "water-right markets" or "permanent transfers" will describe markets/transactions in which the property right (e.g., a water right or ditch company share) is being permanently transferred to a new use/user, while "leasing markets" or "temporary transfers" refer to transfers where water, not the property right, is being bought or sold. Water markets refer generically to all market-based reallocation mechanisms.

[5]The Western Governors Association defines a water transfer as: "A water transfer is a voluntary agreement that results in a temporary or permanent change in the type, time, or place of use of water and/or a water right." (WGA, 2011, p. vii)

[6]Kenney (2005) observes that when developing interstate compacts and when implementing them that "... states have determined that water marketing is inappropriate and, consequently, it does not occur." (Kenney 2005, p. 173).

16.3.1.1 Permanent Transfers: Water Right Markets

Water-right transactions include both direct and indirect transfers of ownership. Direct transfers involve buying the actual right to divert, and indirect transactions occur when a water user buys shares of a ditch company to gain water resources but the ditch company retains the right to divert. This distinction is important because the rules governing direct transfers are established by the states, whereas the rules governing the buying/selling of shares are governed by the ditch company's bylaws. As might be expected, the statewide rules on water transfers are more complex, unwieldy and costly compared to organizational bylaws of a conservation district or ditch company. These differences help explain patterns in water transaction size, timing and intensity as is discussed later in the chapter.

Direct transfers of ownership become more complex when the point of diversion needs to be changed or when the use of water will be different than before. A water right grants its owner the ability to put water to beneficial use in a manner consistent with the original appropriation. That is, if the original use is irrigated cropping, the buyer may continue it in that use and quantity. In cases where the buyer wishes to divert the water for a different use, the buyer must apply to the appropriate administrative body to have the water right changed.[7] As a result, direct transfers may involve two steps: the purchase of the water right and, potentially, the change of use.

Approval for a change in use varies from state to state, but is conditional on demonstrating the proposed change will not injure the right of others to divert. Injury includes both decreasing the quantity of water available for diversion and increasing the likelihood of a senior call.[8] While the water right retains its original seniority date, the no-injury requirement effectively means that the amount of transferable water is limited to that which has historically been consumed and not the amount historically diverted (Kenney 2005).

Several important points are worth noting about this two-step process. First, these two actions do not need to take place in any particular order – the right could be changed first, in anticipation of the sale, or the right could be sold first. Second, the actual change in use could occur at a much later date than when the water rights are sold. It is common for municipal providers to purchase water rights in anticipation of future growth, but lease water back to the original water right owner in the interim (Howitt and Hansen 2005). For example, the city of Thornton made substantial

[7]Generally speaking, the purchase of a water right preserves the seniority date and the quantity of the right. The quantity of the transfer is limited to the historical consumptive use of the water by its owner rather than a diversion amount (Kenney 2005, p. 172). Historical consumptive use may be determined in court proceedings or another jurisdiction. Determination of historical consumptive use preserves return flows to downstream users, but can make the transaction more costly, and may introduce institutional risk into the exchange for owners whose presumed consumptive use may be different than the court determined amount.

[8]When a senior "call" is on the river, junior water right holders are not allowed to divert until senior water right holders needs are met.

purchases of water rights in northern Colorado in the early 1990s. More than 20 years later, all of the water is still used to irrigate cropland in northern Colorado. The impacts of these transfers on water allocation may not be seen for decades after they have occurred.

Transactions involving shares in a ditch company or conservancy district do not necessarily change the water right ownership. Negotiations occur between potential buyers and the individual shareholder. Similar to the sale of a water right, the purchaser of a share is restricted to using water in a manner consistent with the water right. Trades involving similar users (e.g., agriculture to agriculture) do not require state approval; such transactions are only subject to the by-laws of the organization. However, transferring shares from one use to another (e.g., agriculture to M&I) includes two steps: the sale of the share and the change of the water right.

16.3.1.2 Temporary Transfer of Use: Leasing Markets

Sellers may lease water rights to other users, but still retain ownership of the right for future use. Leases can be organized into three types: water banks, single/multi-year leases and interruptible water supply agreements.

16.3.1.2.1 Water Banks

Water banks provide a mechanism for reallocating water on a short-term, spot transaction basis. The water bank is similar to an agricultural commodity exchange: it serves as a facilitator of exchanges by matching buyers and sellers, provides services to enable transactions including determining the type of right that may be banked and adherence to regulations, and is a clearinghouse for those transactions. Potential sellers deposit water into a bank making it available to potential buyers. The clearinghouse tracks offer and bid information and ensures transactions are completed satisfactorily. The water banks may also solicit water deposits and market water to potential buyers. While water banks exist in almost every state in the West, significant differences exist in their implementation. Water banks are categorized based on:

- Organizing agency: water banks are often formed to fulfill a specific need such as maintaining supplies during drought, creating in-stream flows for critical habitats or augmenting flows for later use. As might be expected, the organizing entity of the bank may be a federal agency (e.g., US Bureau of Reclamation), state government, special district or organization of interested parties.
- How price is determined: Water banks may post a fixed price that is intended to clear the market. This approach works best when water rights of similar type and use are available to be traded otherwise the administrator may have to make adjustments or create variable prices. In contrasts, auctions may be used in order to determine a market-clearing price in a more decentralized fashion. Double-blind sealed-bid auctions have been used, as well as open-outcry auctions.

- Contract Types: Contract types may be unique to a particular water bank and its function. As an example, supplier contracts are used to organize specific entitlements in a water bank, storage contracts allow for deposit of water rights in a physical storage facility and contingent claim contracts permit the buyer to use water from the bank when specific circumstances (e.g., drought) are present (Clifford, Landry and Larsen-Hayden 2004).

16.3.1.2.2 Water Leases

The previously described water banks represent a centralized exchange for water rights in which many buyers and sellers participate. In contrast, water leases are private treaty agreements between an individual buyer and seller in which the water right owner agrees to lease a specified amount of water (or number of shares) according to contract stipulations. Leasing arrangements differ from water banks in that agreements typically involve bilateral negotiations and are made in advance of the water being available. Bilateral negotiation allows contracts to be customized and facilitates risk sharing. Typical contract stipulations include:

- Contract Length: The length under which the lease terms and conditions apply.
- Price per Unit: Water volume may be allocated on a share or other unit basis, so a fixed price is typically negotiated on a unit basis. The price listed price might include an inflation multiplier.
- Option to Lease Back: Some lease arrangements are organized around a contingency, and the contingency may allow the lessor the first right to use the water if it is not needed by the lessee.

A special type of water leases are interruptible supply agreements. These contracts are written as multiple year agreements, but water delivery is made on an as-needed basis. A specific example of an interruptible supply agreement are option agreements in which the lessee pays a baseline fee, either lump-sum up-front or annualized over the course of the contract, to gain an option to use a water right, but need not exercise the option each year. If the option is exercised, the lessee pays an additional, pre-negotiated fee to exercise the option. If the option is not exercised, the lessor has the first right to use the water. Interruptible supply contracts generally contain provisions for payment when the option is (is not) exercised, a date of prior notification for years in which the option is exercised and a maximum number of years in which options can be exercised. The fees paid by the lessee to secure the option provide the lessor a secure revenue stream. In return, the lessee receives a guarantee that they will be able to acquire additional supplies when needed at a pre-negotiated price. Few IWSA have been negotiated to date and insufficient data exists to comment on the relative magnitude of the different payments (i.e. upfront, annual and exercise fees). The relative split between the baseline and option fees would depend on the risk preferences of the parties involved.

Similar to water rights markets, significant differences exist in the legal environment surrounding leasing markets across geographic locations. These differences are largely attributable to differences in prior appropriation across states and the rules/regulations adopted by ditch companies and conservancy districts.

16.3.1.2.3 Water Exchanges

On occasion, water diverters seek to change the location of a diversion or its timing using a mechanism called an exchange. An exchange allows an upstream diverter to take water a downstream diverter would otherwise receive, if the water is replaced at the time, place, quantity, and suitable quality the downstream diverter enjoyed before the exchange. The four critical requirements for a water exchange are:

- the source of substitute water supply must be upstream of the senior diversion calling the water,
- the substitute water supply must be equivalent in amount and of suitable quality for the downstream senior,
- substitute water must be from legally available flows,
- the water rights of others cannot be injured when implementing the exchange (Hobbs 2004)

16.3.1.3 Limitations of Existing Water Markets

Can water markets provide efficient and equitable outcomes amidst the practical realities of water storage, transport and use in the West? Markets for natural resources are generally seen as desirable based on the presumption that allowing participants to trade will result in efficiency gains (Godby 1996). These beliefs are often based on expectations derived from an analysis of a single market where perfectly competitive conditions exist. In reality, water markets are incomplete, subject to high transaction costs, and dominated by a few large buyers. Moreover, the impacts of water transfers to third parties are a concern (WGA 2012).

Transaction costs have long been cited as a significant impediment to water market activity (e.g., Howe et al. 1986; Saliba 1987). Transaction costs include locating a willing buyer/seller, negotiations, navigating institutional requirements (permits, water court proceedings, etc.), and the physical expense of water collection, storage and treatment (McCann and Easter 2004). Significant transaction costs reduce the frequency of water transfers, and make it difficult to match water supplies to changing uses. At risk of not having water when needed, M&I suppliers often purchase water rights significantly in advance of needs and seek the most senior rights available, which can be relatively expensive to purchase. High transaction costs also limit the participants in the marketplace, and market power may reduce the provision and expense of water supplies in the marketplace. In an evolving West, transaction costs result in slowing and/or preventing the development of water markets in a number of areas (Donohew 2009).

Temporary transfers may provide a more efficient means of matching supply and demand for western water, but generally speaking temporary transfers have been subject to the same institutional barriers and high transaction costs as those found with permanent transfers. If transaction costs are the same, buyers prefer, and are sometimes legally required to acquire, the long term security of permanent transfers. Moreover, M&I suppliers may actually purchase significantly greater quantities of water rights than their anticipated needs because they purchase to meet minimum needs during extreme drought events. M&I users may also be required by law to secure reliable water supplies. Yet, well-functioning water leasing markets that facilitate exchange from agricultural to M&I users do not exist in all states.

Substantial transaction costs are incentives for water buyers to "buy and dry" agricultural water rights when changing use. The buy-and-dry practice involves purchasing agricultural rights, preferably in large quantity, from a holder (e.g., a ditch company), applying for a change in use, and if successful, de-watering the irrigated farmlands completely. The land is then fallowed, shifted to a dryland crop or restored to native vegetation. For water buyers, a buy-and-dry regime is a lower-cost approach to securing water than other alternatives including shifting only a portion of purchased water rights from their original use since it involves only the one basic transaction and is easy to verify.

Buyers and sellers of the water rights are adequately compensated in a water transfer otherwise they would not engage in the transaction. Yet, these same water transfers can have negative third-party effects. Third parties affected by water transfers include rural communities in which irrigated agriculture is a large share of the economic base, the ecology and species that depend on irrigated agriculture for habitat, urban interests, ethnic communities and Native Americans, non-agricultural rural communities and federal taxpayers (Natural Resource Council 1992). Concerns surrounding impacts on rural communities and the environment have motivated institutional changes designed to encourage the use of temporary transfers as a means of meeting future water demands. The perception is that state laws are evolving to make permanent and temporary transfers easier (Howitt and Hansen 2005). This is especially true of a variety of alternative institutions designed to provide alternative market avenues to permanent markets (WGA 2012).

The previous sections are set in the context of water transfers as a means of reallocating water among users in the West. Burgeoning population and its associated demands mean that finding effective reallocation mechanisms are important. Market transactions are one opportunity of redistributing water rights. Transactions may be permanent or temporary, however, given similar transactions costs and M&I providers preferences for reliability, reallocation from agricultural to M&I uses has traditionally occurred via water rights markets. Reallocation is occurring, and the analysis of transactions found in the next section give some insight into the effectiveness of current institutions.

16.4 Trends in Market Activity

16.4.1 Previous Literature

Economic analysis of western water markets is constrained by a lack of quality data. Studies with a broad geographic scope have been limited to summarizing market activity (e.g., number of trades and direction of trades), with detailed analysis of market functionality occurring via case studies of particular markets. Comprehensive market transaction data is difficult to come by, in large part, because trades are frequently conducted in informal settings with participants often hesitant to reveal transaction information. It is often the case that even when detailed data on water right transactions is available, information on price is not (Brookshire et al. 2004). Alternatively, when information on price is available detailed information on the water right transacted is not.

The majority of studies that have analyzed market activity across the West have utilized data from the *Water Strategist* which, until 2010, published monthly transaction data.[9] This data set, while the most comprehensive[10] available, suffers from several limitations. First, the *Water Strategist* did not provide information on all transactions that took place, only listing self-reported transactions (either by the organization or parties involved). It is not known to what extent the *Water Strategist* listings are representative of all transactions. Brewer et al. (2008) compared transaction data available from various California agencies to those reported in the *Water Strategist*. The author found that the *Water Strategist* listings were comprehensive during some periods and not in others. It is likely that data sets based on the *Water Strategist* are fairly representative in those markets operated by government agencies or organizations such as Northern Water.[11] Brozovic et al. (2002) suggest that a considerable amount of informal leasing occurs within groups and/or ditch companies (e.g., agriculture use to agriculture use), and this may not appear in the *Water Strategist*, so it may be that agriculture to M&I transactions are overstated in the data set. Second, the *Water Strategist* data provides limited detail beyond estimates of quantity, price, and a general descriptor of the buyer and seller (e.g., irrigator, developer, etc.). Detailed information on, for example, location, priority, transaction costs, and the transferable quantity (in terms of consumptive use) is not available.

---

[9]This includes a number of studies, including this one, whose analysis is based on data compiled from the Water Strategist by researchers at the Bren School of Environmental Science & Management at the University of California, Santa Barbara. This data set is available online at: http://www.bren.ucsb.edu/news/water_transfers.htm.

[10]As noted by Brewer et al. (2008), the Water Strategist has historically advertised itself as "the only source of published information on water transactions in the West."

[11]Evidence of this is the fact that the *vast* majority of transactions reported are associated with leases or sales of C-BT shares.

In spite of limitations, the *Water Strategist* remains the most comprehensive data source for water transactions in Western states. Table 16.1 provides a list of recent studies that have used these transactions to analyze water market trends.

Previous analysis of market activity, including those summarized in Table 16.1, generally find that the majority of trades involve agricultural buyers and municipal sellers. However, significant variation in prices exist across time, location, and by type of transaction. This variation reflections differences in demand and supply conditions across space and time, as well as, the large gap between the marginal productivity of water in agriculture and the willingness to pay of M&I and environmental users.

While the studies summarized in Table 16.1 are in agreement when it comes to their findings on the variability in pricing and the general direction of trades, differences exist in terms of the relative importance of permanent versus temporary transfers in terms of the quantity of water transferred and the number of transactions. For example, Brown (2006) finds that 95 % of water transferred was done so via leases, whereas Brewer et al. (2008) conclude that 57 % of water transferred was via water rights transactions. Discrepancies such as these are largely a result of differences in how individual authors "cleaned" the Water Strategist data, how they chose to bundle individual transactions,[12] and how quantities of water are "annualized" for multi-year leases and permanent transfers.

16.4.2 Analysis of Recent Transactions: A Case Study of the Colorado River Basin

The Colorado River Basin (CRB) is a representative western river basin from which transactions can be examined. The CRB is one of the most critical sources of water in the West, spanning seven US states and two states in Mexico (Fig. 16.1). This river's remarkable reach includes providing water to more than 30 million people, irrigating nearly four million acres of agricultural land, and serving as the limiting resource for at least 15 Native American tribes, seven National Wildlife Refuges, four National Recreation Areas, five National Parks in the US and a Biosphere Reserve in Mexico. The river's energy powers more than 4,200 MW of electrical

[12]Two examples highlight these differences. First, Brown (2006) categorized transactions involving water banks where the original seller and eventual buyer were not reported as belonging to a "water management agency". Water management agencies include federal or state government agencies, conservancy districts, and water districts, associations and companies. Brewer (2008), on the other hand, attempted to infer based on the description of the transaction and past knowledge, the nature of the buyer and seller in these cases. Second, Brown, unlike Brewer, combined all similar Colorado- Big Thompson transactions within a particular month, into a single case reducing the number of transaction overall (by approximately 30 %) and skewing (relative to Brewer) some statistics relating to the size and number of transactions by type.

Table 16.1 Empirical studies of water transactions in western states

Study and overview	Major findings
Howitt and Hansen (2005): Summarize market activity for 14 western states over the period 1999–2002	**Type of contract, frequency and size** 90 % of water transferred via leases Leases dominated trading activity (volume) in all states but CO and NV **Prices** Sale prices were significantly higher in CO, NV, and NM than in other states Municipal and Industrial buyers pay higher prices than agricultural and environmental buyers **Buyer and seller types** Majority of trades involve agricultural sellers and municipal buyers Increase in purchases for environmental uses
Brown (2006): Summarizes market activity including quantity and price by state, year, and type of transfer for 12 western states over the period 1990–2003	**Type of contract, frequency and size** 95 % of water transferred via leases Number of leases exceeded number of water rights transactions overall and in all but four states (AZ, CO, NM, and UT) Extreme variability in market activity existed across states; two-thirds of transactions occurring in CA, CO, and TX Increasing trend in the number of leases No statistically significant trend in number of water rights transactions **Prices** Significant variation in lease prices existed across states with median prices < $12 per acre foot in ID, OR, UT, and WY and > $55 in AZ, CA, and NW Significant variation in water rights prices existed across states with median prices < $120 per acre foot in ID and > $2,500 in CO, NM, and NV **Buyer & seller types** Municipalities were the predominant buyers of water rights with agriculturalist being the predominant sellers Public agencies and irrigators were the predominant lessors; lessees being evenly distributed across all activity types
Brewer et al. (2008): Summarizes market activity including quantity and price by state, year, and type of transfer for 14 western states over the period 1987–2005	**Type of contract, frequency and size** More than 79 % of water leases are for one-year or less One-year leases account for 47 % of all leased water 57 % of water transferred was via water rights transactions Increasing trend in number of water rights transactions, no such trend exists in one-year leases Active markets primarily exist in states with growing urban demands; the vast majority of transactions occurring in CO CA and TX accounted for half of all leases

(continued)

Table 16.1 (continued)

Study and overview	Major findings
	Prices
	Significant variation in lease prices existed across states with median price for one year leases ranging from \$4.43 (ID) to \$44.53 (CA) per acre foot
	Significant variation in water rights prices existed across states with median prices from \$111.37 (OR) to \$2,693.38 (CO) per acre foot
	Municipal buyers pay significantly higher prices than agricultural buyers
	Buyer & seller types
	The majority of transactions (leases and water rights) involve agricultural sellers and municipal buyers
	Ag to municipal trades occur primarily through water rights transactions; this is increasingly the case
	Within agricultural transactions are predominantly short-term leases
	Increasing trend in Ag to Municipal transfers; no such trend in Ag to Ag or Municipal to Municipal transfers
Basta and Colby (2010): Summarize market activity in most activity states (CA, CO, NM, and TX) and for the Northwest and Intermountain regions[a] over the period 1987–2007	**Type of contract, frequency and size**
	Increasing trends in the number of sales and leases in all states/regions except NM
	Average volume per transaction are decreasing
	Volume of water sold via water rights transactions decreased in all states/regions except TX
	Volume of water transacted via leases increased in all states/regions except the
	Intermountain
	Prices
	Average water rights prices per acre foot were highest in CO; the highest lease prices were in the Intermountain region
	Considerable variability in year-to-year sales and lease prices exists in all states/regions
	Sales and lease prices were non-decreasing in all states/regions

[a]Market activity in these states was done due to limited activity. The Northwest included Idaho, Oregon, and Washington. The Intermountain region included Arizona, Nevada, Utah, and New Mexico

capacity for households and industry. However, the river is at risk because increasing water demands and climate change are jeopardizing water security.

Water is a scarce resource in the Colorado River Basin and the rights to its use, for all intents and purposes, are fully allocated. As water demands change, water rights will need to be transferred in order to meet these changes.

For CRB farmers and ranchers, water rights are an asset that generates financial returns from irrigated cropping. In this context, water is a capital asset, and the decision to sell an asset depends importantly on anticipated profits. The anticipated

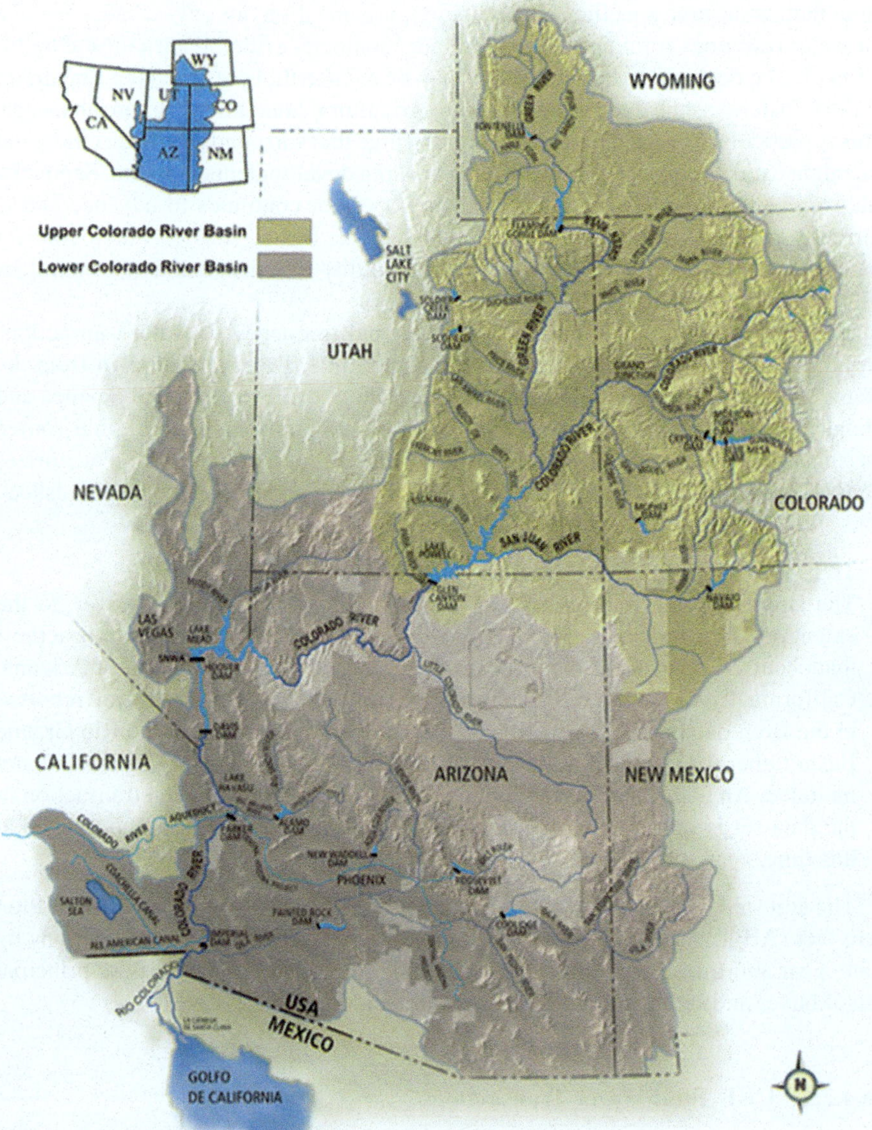

Fig. 16.1 Colorado River Basin. http://www.gcdamp.gov/aboutamp/crb.html

profits from irrigated agriculture often, but not always, form a reservation price for water right holders who may choose to sell. In contrast, the offer price of the water buyer depends on their own projected revenues and costs that include the conveying, storing and treating the resource. Voluntary transactions occur when the offer price exceeds the reservation price.

Expected profits are not the only ingredient in a water transfer decision, and two other factors deserve specific attention. First, urbanization not only creates demand for water resources from agriculture, but also competes with the agricultural sector for land. The decision to transfer water may be an afterthought when the true driver of the transfer was the decision to develop agricultural land into an urban landscape. This is particularly true near the urban-rural fringe that surrounds growing cities, and we might expect transfers to accelerate as housing developments blossom. Secondly, the average age of a farm and ranch owner/operator continues to advance and is currently greater than 62 years in the Colorado River Basin. The decision to transfer water may have as much to do with adding liquidity to retirement assets as it does with agricultural profitability.

The following section focuses on the size and frequency of water transactions in the Colorado River Basin between 1988 and 2008. Data is the drawn from the Water Transfer Database housed at the Bren School of Environmental Science and Management at the University of California, Santa Barbara. The original source for this data base is the *Water Strategist*. The Bren research team categorizes transactions ranging from 1987 to 2009 and reports them in an Excel spreadsheet. This study has reduced the transactions accordingly so that:

- Transactions are limited to the period from 1988 to 2008.
- The Bren database records transactions throughout the West; however, to the extent possible, transactions reported in the current case study reflect transactions that occurred in the CRB. This means that transactions are included for Arizona, California, Colorado, Nevada, Utah and Wyoming. Transactions that are reported in the Bren database for New Mexico all appear to take place in the Rio Grande Basin rather than the Colorado River Basin, so New Mexico transactions are not included. An additional research opportunity is to verify that each transaction in the data set is in the Colorado River Basin, but this has not been completed at this time.

The adjusted dataset includes 3,291 transactions and approximately 23.5 million acre feet (AF) transacted. The following section is an overview of transactions by state, year and transaction type (sale, lease and exchange), as well as the principal use of buyer and seller (agriculture, environment, municipal).

16.4.2.1 CRB Total Water Transactions

All transactions used in this analysis are illustrated in Fig. 16.2, which lists the total acre feet transacted in each year with column bars, and the number of transactions within a year is represented by a line. The number of transactions follows an increasing trend from 1988 to 2008, but the volume of acre feet traded within each year ebbs occasionally and then increases again. These cycles tend to follow periods of economic growth with particular peaks of acre feet traded in 1991, 1994, and 2000. A smaller volume of water has been traded in recent years, in spite of a peak number of transactions, suggesting the average size of transactions is falling.

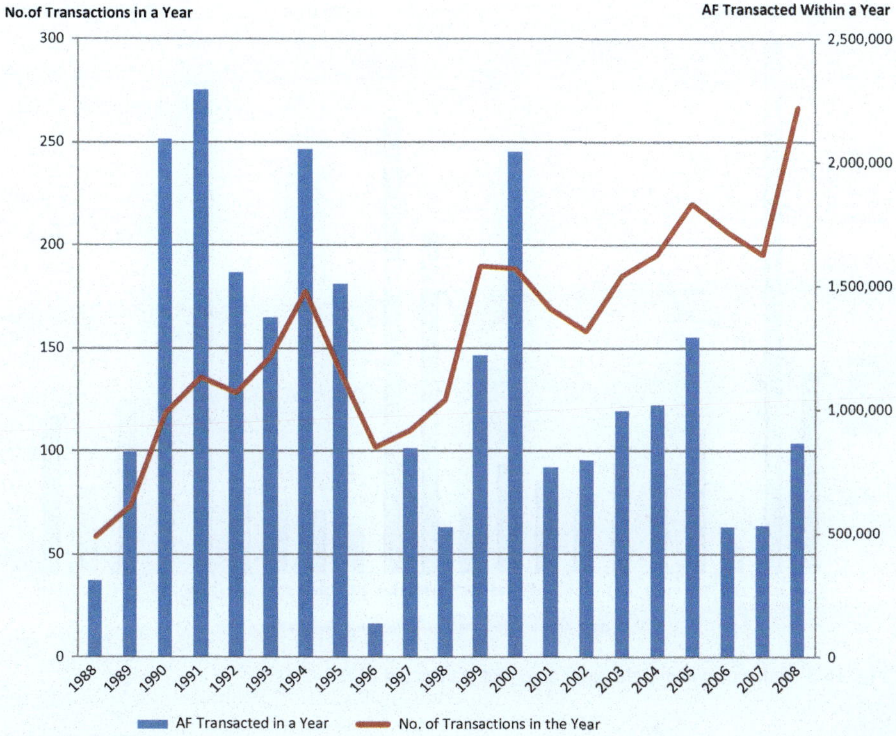

Fig. 16.2 Yearly volume and number of selected transactions (1988–2008)

Individual state transactions follow a similar pattern. In most states, the number of yearly transactions has been increasing since 2003. Intensive transaction activity is observed in 1990–1995 in Utah, in Colorado from 1993 to 1995, California from 1991 to 1992 and Arizona from 1994 to 1995. The differences in timing may reflect regional differences in rates of municipal development. The differences do not seem to correlate to noteworthy periods of poor profitability in irrigated cropping that occurred in the years 1997, 1998, 1999, 2002, and 2006. In this sense, it may be that buyers rather than sellers were driving these voluntary transactions.

16.4.2.2 CRB Sales, Lease and Exchanges

The Bren database categorizes transactions into three types: sales, leases and exchanges. In the time period 1988–2008, sales constitute the largest share of the number of transactions (68 % but only 11 %) of the volume of acre feet was transacted via sales. Leases comprise 25 % of the transactions, but 85 % of the volume of transactions. Leases are of varying lengths, and while a few instances of 100-year leases exist, 78 % of leases are for one year, and nearly 90 % of all leases

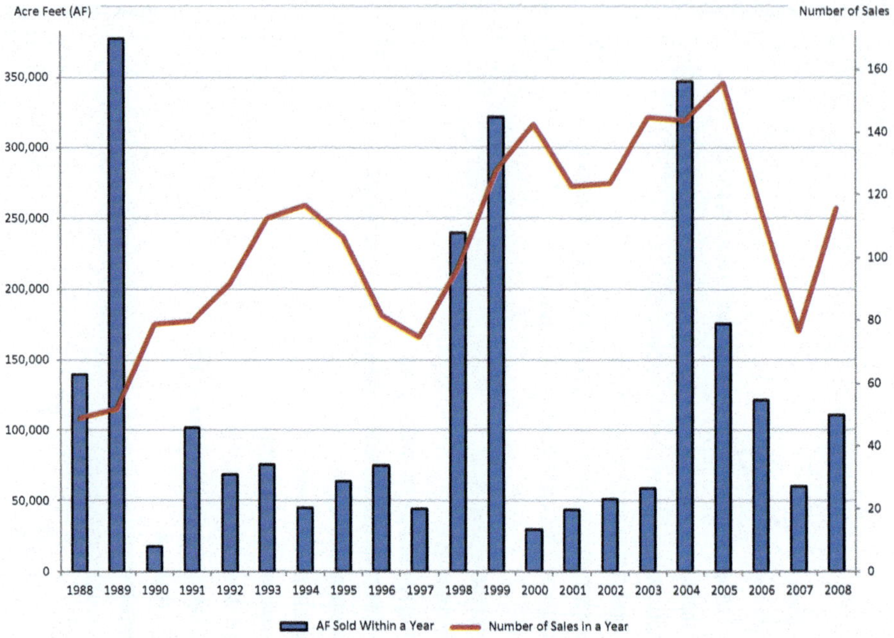

Fig. 16.3 Sales of water rights and acre feet transacted (1988–2008)

are 10 years or less. Exchanges represent only 3 % of the number of transactions and roughly 6 % of the volume of acre feet transacted.

Examination of transactions over time suggest that the number of sales are occurring less frequently than before, and leases and exchanges are relied upon more frequently. Figure 16.3 represents the number of water right sales within a year and the amount of acre feet transacted. Most recently, the number of sales has declined, unlike the number of leases and exchanges that have taken place (Figs. 16.4 and 16.5). Perhaps the decrease in sales reflects a moderating rate of urbanization, reduced expectations of future demands, greater diversification in water supply sources, and the beginning of a liquidity crisis in public finance.

Lease transactions are increasing in number, especially after 2004, but the volume of water transacted has decreased substantially since the mid-1990s except for 2000 (Fig. 16.4). Exchanges, which make up a small proportion of all transactions, increased substantially in 2005–2007 (Fig. 16.5). Future analysis might be able to uncover factors driving the shift toward shorter-term transactions in lieu of outright purchase of water rights. It may be that actual water demand has not met forecasted expectations, either because of decreasing migration to the West, or because municipal users have been able to moderate demand through conservation. It may also be that it is becoming more costly to acquire water via sales because of increasing costs to store, convey and ship water particularly when large-scale investment in infrastructure is needed. Lastly, the risk of failing to

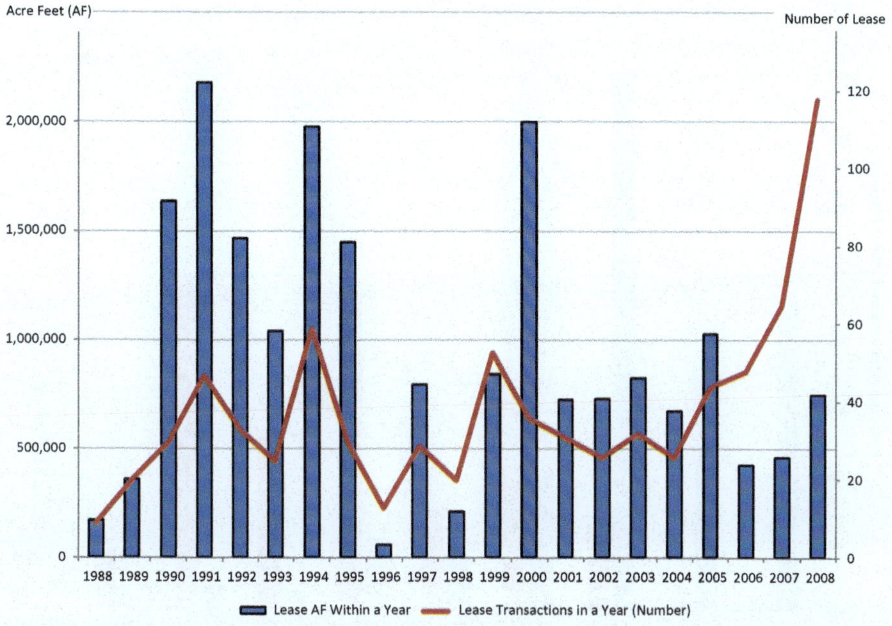

Fig. 16.4 Number of leases and acre feet leased within a year (1988–2008)

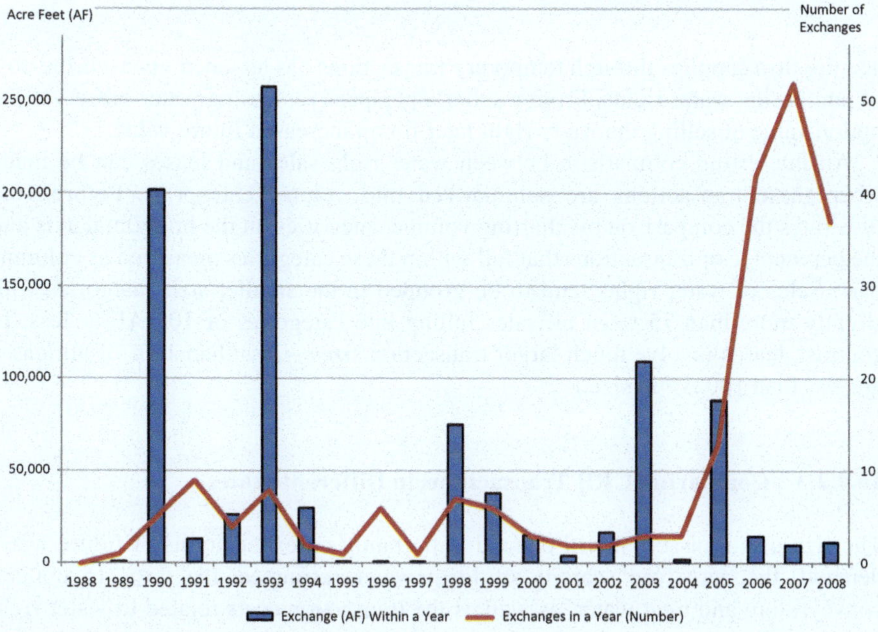

Fig. 16.5 Number of exchanges and acre feet of exchanges with a year (1988–2008)

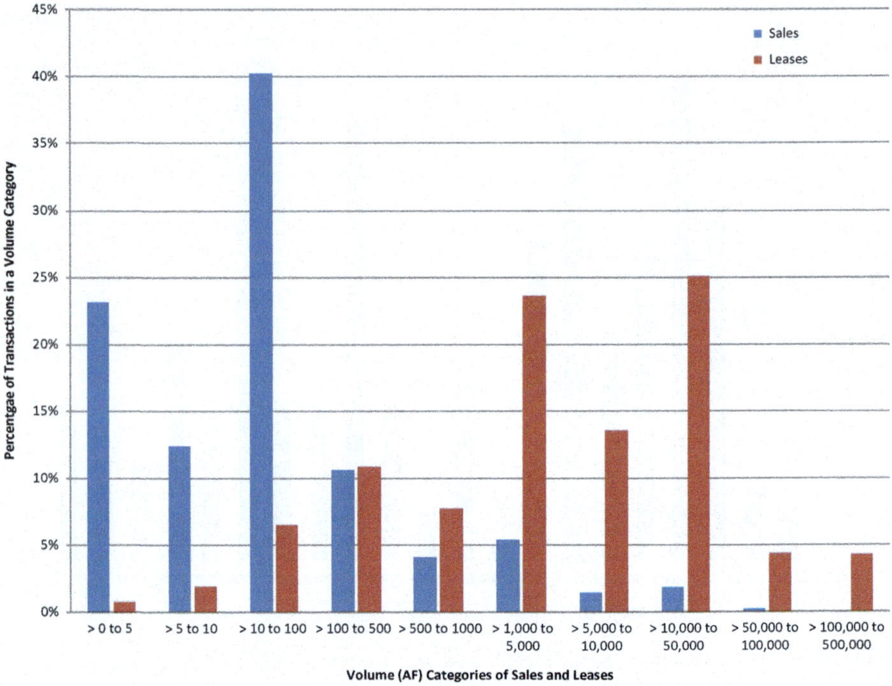

Fig. 16.6 Sales and lease transactions by acre feet category (1988–2008)

acquire firm supplies through temporary transactions has lessened vis-à-vis the cost of purchasing water rights. From a seller's perspective, leasing water preserves the option value of selling the water right later for an increased future value.

An interesting comparison between water right sales and leases can be made when these transactions are summarized into volume categories. Figure 16.6 illustrates the comparison by charting volume categories on the horizontal axis and the percentage of transactions that fall within these categories measured as columns bars. Sales of water rights tend to be grouped in the smaller size categories with slightly more than 75 % of all sales falling into categories of 100 AF or less. In contrast, leases involve much larger transaction sizes – less than 10% of all leases are less than 100 AF in size.

16.4.2.3 Comparing CRB Transactions in Different States

The size of a transaction will depend importantly on expectations of future water demand, the perceived scarcity of local water resources, the costs to collect, convey, store and treat water, as well as the transaction costs related to water right

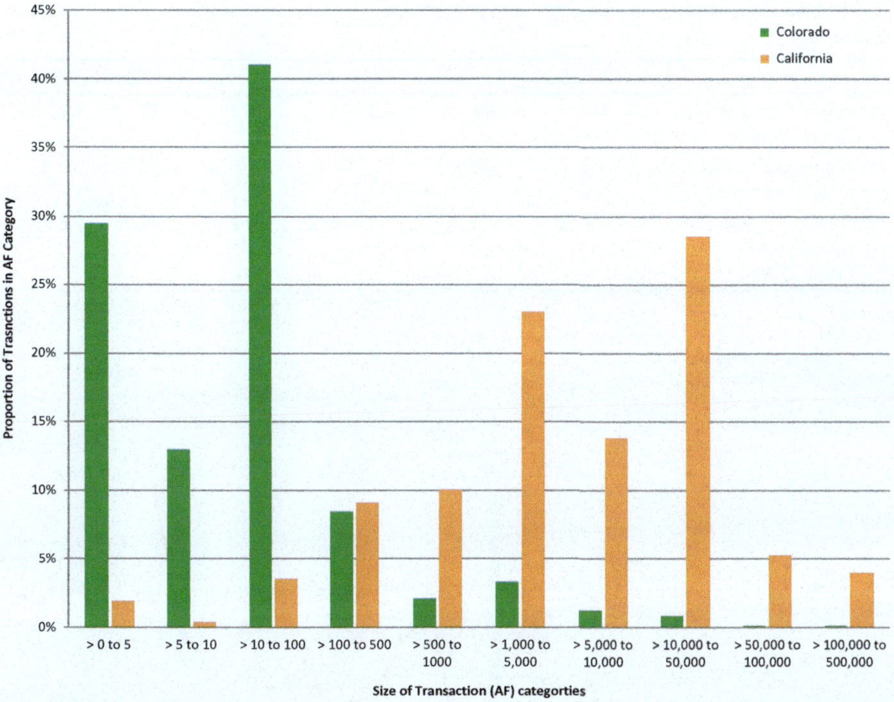

Fig. 16.7 Colorado and California water right transaction by AF size category (1988–2008)

adjudication or change of use. It is no surprise that the size of transactions (e.g., sales, leases and exchanges) might be different between states that have different rates of urbanization, different climates, disparate concentrations of water rights among holders (i.e., ownership is far more fragmented in Colorado when compared to California) and distinct legal institutions. An example of this is found in Fig. 16.7, which illustrates water right transactions in Colorado and California according to volume category.

Transactions in Colorado tend to be much smaller than those experienced in California, perhaps because the vast majority of Colorado transactions occur under Colorado Big-Thompson (CB-T) project. Within the CB-T, water rights may be exchanged with relatively low transaction costs because return flows do not have to be considered, which increases the frequency of transactions and makes smaller transactions more economically palatable. Of the transactions examined in this study, approximately 64 % occurred in Colorado and 20 % occurred in California. However, far more acre feet were transacted in California relative to Colorado and other states primarily because of California's growing municipal demands (Table 16.2).

Table 16.2 Percentage of transactions and percentage of acre feet transacted in each state (1988–2008)

	Arizona	California	Colorado	Nevada	Utah	Wyoming
Number of transactions (1988–2008)	218	644	2,113	177	77	61
Percentage of transactions (1988–2008)	7 %	20 %	64 %	5 %	2 %	2 %
Percentage of acre feet transacted (1988–2008)	36 %	53 %	6 %	1 %	2 %	2 %

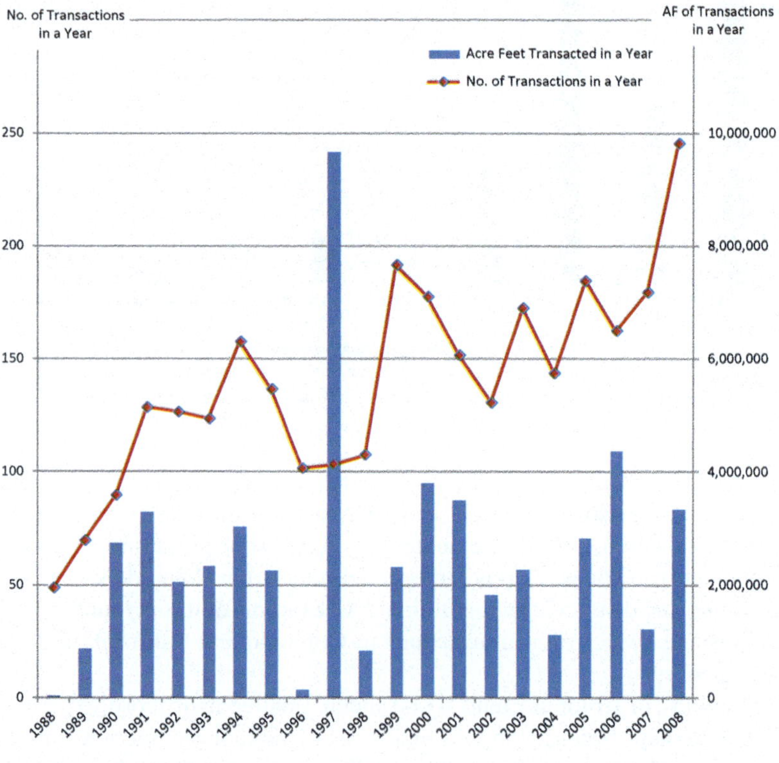

Fig. 16.8 Number and acre feet of transactions in which agriculture is the supplier

16.4.2.4 Water Transactions in Which Agriculture Is the Supplier

Agriculture continues to divert and use the vast majority of water in the West and the Colorado River Basin. With supplies fully appropriated, reallocation among users is one means of meeting increasing demands among agricultural, environmental and municipal interests. Agriculture was the source of at least 82 % of the transactions recorded in the Bren database between 1988 and 2008, and agriculture supplied 68 % of the acre feet of water that was transacted. Figure 16.8 indicates the pattern

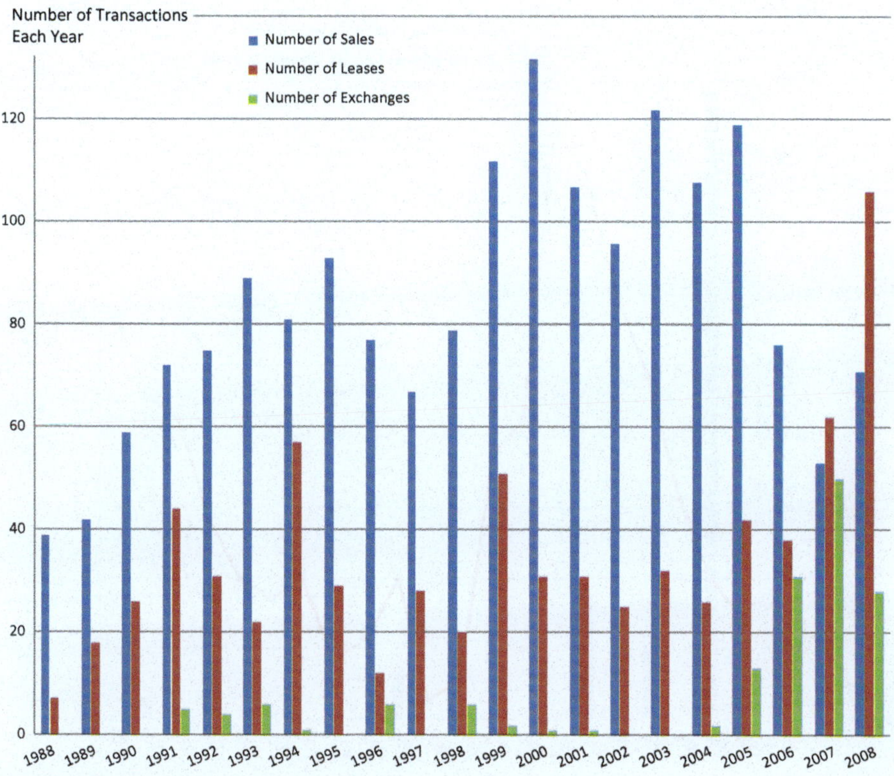

Fig. 16.9 Number and type of transaction when agriculture is supplier

of transactions through time in which agriculture provided water to other parties, including agriculture interests.

While it is true that transactions sourced from agriculture are increasing in recent years (Fig. 16.8), it should be noted that transfers do not necessarily imply a movement away from agricultural use. Of the transactions in which an agricultural water right holder was the supplier, about 44 % of the acre feet that were transacted went to an agricultural entity followed by municipal use (29 %) and environmental use (26 %).

The previous transactions included sales, leases and other exchanges of agriculture water rights. As might be expected, sales are the most frequently reported type of transaction, but over time leases and exchanges have been gaining popularity (Fig. 16.9). The incidence of each agricultural transaction type varies by state. Water sales are the most popular form of transaction in Colorado and leases occur more frequently in California and Wyoming.

Of particular interest are sales of water rights from agriculture to municipal use. These voluntary transactions may be the result of increasing urbanization in the West and are linked to the reduced acreage in irrigated cropping. Data represented in

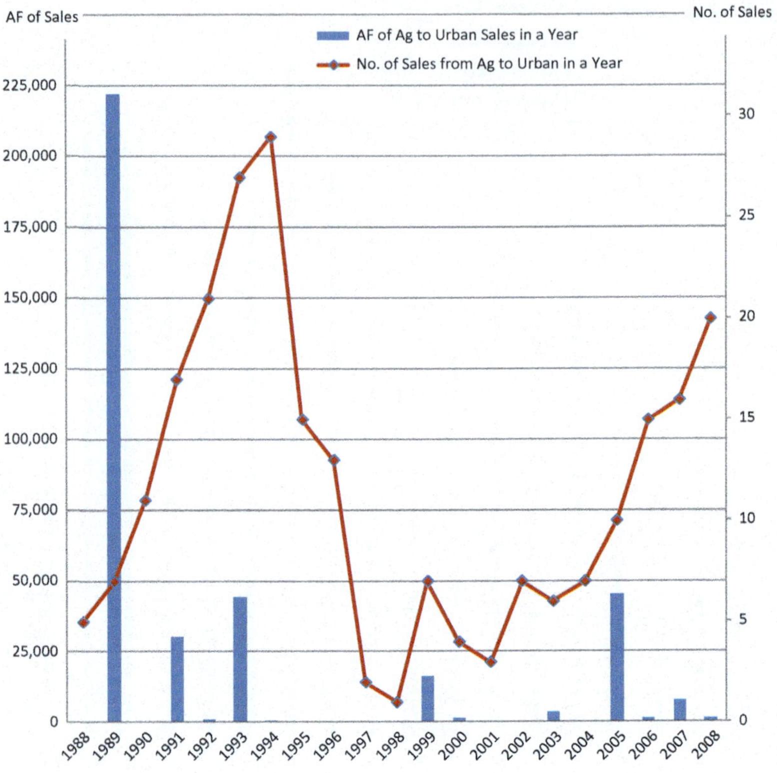

Fig. 16.10 Volume and number of transactions from agriculture to municipal water right holders

Fig. 16.10 represents agriculture to municipal water transfers as measured in terms of acre-feet transferred and the number of sales recorded by the *Water Strategist*. A significant decrease occurs in both measures in the mid to late 1990s, and while not a definitive explanation, the source of decline may be associated with a slowdown in western municipal economies, significantly improved revenues for western agriculture as a result of historically high prices in 1996, and front loaded municipal purchases in anticipation of future growth. This drop is followed by an increasing trend in number of sales from agriculture to urban interests from 1998 to 2008, but the acre feet transacted is declining. Perhaps municipal suppliers are beginning to seek less permanent transactions for meeting increasing demands as the cost to transfer water becomes more expensive. Likewise, planners' expectations for population growth may be adjusting downward, and these water providers are seeking a broader portfolio for securing water supplies that include conservation and reuse.

16.4.2.5 Summary of Water Transactions in the CRB

The Colorado River Basin is a vital resource of water for agricultural, environmental and municipal interests. Increasing demands among users and climate variability are driving a reallocation of use that will persist for some time. This chapter provides some insights to stakeholders and policymakers about the current status and trends of water use in the basin with particular attention to agriculture, the largest water diverter and consumptive user. Summary points include:

- Water transactions are increasing in the CRB, but the average size of transactions is declining. The number and volume of water transacted varies according to state, in part due to differential rates of urbanization and institutional structures.
- Sales are the most frequently used transfer mechanism, but a greater volume of water is transferred using leases. The trend is toward increased leasing and decreasing use of sales as a transfer mechanism.
- Agriculture is the predominant water right holder in the CRB, and agriculture water right holders are most often the supplier in a transaction. Agricultural users are the most frequent receivers of water in transactions, but agricultural to municipal transactions are increasing although the volume of transactions are not.
- Care needs to be taken when linking the ownership of a water right to the use of that right. This chapter reports ownership by agricultural, environmental and municipal use, but it need not be the case that the water is used for that purpose. As an example, 66 % of water units in the Colorado Big-Thompson project are owned by municipal entities, but 60 % of water deliveries are made for agricultural use. In Colorado at least, water designated as a "municipal" use may actually be used for irrigation.

References

Basta E, Colby B (2010) Water market trends: transactions, quantities, and prices. Appraisal J 78(1):50–66

Bell C, Taylor J (2008) Water laws and policies for a sustainable future: a Western state's perspective. June 2008. Western States Water Council. http://www.westgov.org/wswc/laws%20&%20policies%20report%20(final%20with%20cover).pdf

Brookshire DS et al (2004) Market prices for water in the semiarid West of the United States. Water Resour Res 40:9

Brown TC (2006) Trends in water market activity and price in the western United States. Water Resour Res 42:9

Brozovic N, Carey JM, Sunding DL (2002) Trading activity in an informal agricultural water market: an example from California. J Contemp Water Res Educ 121(1):14

Clifford P, Landry C, Larsen-Hayden A (2004) Analysis of water banks in the Western states. State of Washington, Department of Ecology. Publication number 04-11-011. Available at: https://fortress.wa.gov/ecy/publications/publications/0411011.pdf

Donohew Z (2009) Property rights and western United States water markets*. Aust J Agric Resour Econ 53(1):85–103

Getches DH (2009) Water law in a nutshell. Thomson West, St. Paul, 492 pages

Godby RW (1996) An experimental economic examination of market power in emission permit markets. Nov. 1996

Hobbs G (2004) A citizen's guide to Colorado water law. Colorado Foundation for Water Education. Available at: http://www.yourwatercolorado.org/

Howe CW, Schurmeier DR, Shaw, Douglas W (1986) Innovative approaches to water allocation: the potential for water markets. Water Resour Res 22(4):439–445

Howitt R, Hansen K (2005) The evolving western water markets. Choices 20(1):5

Kenney DS (2005) Prior appropriation and water rights reform in the Western United States. In: Lessons for institutional design: 167. International Food Policy Research Institute, Washington, DC

Kenny JF, Barber NL, Hutson SS, Linsey KS, Lovelace JK, Maupin MA (2009) Estimated use of water in the United States in 2005: U.S. Geological Survey Circular 1344, 52 p

Mackun P, Wilson S (2011) Population distribution and change: 2000 to 2010. 2010 US Census Briefs. C2010BR-01. Accessed at http://www.census.gov/prod/cen2010/briefs/c2010br-01.pdf

McCann L, Easter KW (2004) A framework for estimating the transaction costs of alternative mechanisms for water exchange and allocation. Water Resour Res 40:W09S09. doi:10.1029/2003WR002830

National Research Council (1992) Water transfers in the West: efficiency, equity, and the environment. The National Academies Press, Washington, DC

Nichols PD, Murphy MK, Kenney DS (2001) Water and growth in Colorado. Natural Resources Law Center, University of Colorado School of Law, Boulder

Saliba BC (1987) Do water markets "work"? Market transfers and trade-offs in the southwestern states. Water Resour Res 23(7):1113–1122

Smith M, Pritchett J (2010) Agricultural/urban/environmental water sharing: innovative strategies for the Colorado River basin and the West. Colorado Water Institute special report series no. 22. Colorado State University. http://www.cwi.colostate.edu/watersharing

Udall B (2013) Water: impacts, risks, and adaptation. In: Garfin G, Jardine A, Merideth R, Black M, LeRoy S (eds) Assessment of climate change in the Southwest United States: a report prepared for the National Climate Assessment. A report by the Southwest Climate Alliance. Island Press, Washington, DC, pp 197–217

Western Governors Association (WGA) (2012) Water transfers in the West: projects, trends and leading practices in voluntary water trading. http://www.westgov.org/initiatives/water/373-water-papers. Dec 2012

Chapter 17
The New Role for Water Markets in the Twenty-First Century

K. William Easter and Qiuqiong Huang

Abstract We have learned a great deal since the end of the 1980s. Water markets have shown the potential to help countries and regions deal with their growing water shortages. However, to be effective water markets need to have wide acceptance by stakeholder and be designed so that water can be traded at low transaction costs. Third party impacts tend to be the most difficult issues that must be addressed when water markets are being designed. This involves both downstream users and environmental impacts. After over two decades of trial and error Australia has done a good job of adjusting their institutional arrangements to address most third party impacts. Luckily local water markets tend not to have many third party impacts as illustrated by the village level markets in Oman.

Keywords Water shortages • Reallocation • Transaction costs • Environmental concerns • Climate change

17.1 Introduction

This chapter reviews what we have learned over the past quarter century regarding the future possibilities for water markets to help address our growing water scarcity problems and what we can do to make them more effective and acceptable. As we have seen there is real resistance to using water markets to reallocate water

K.W. Easter (✉)
Department of Applied Economics, University of Minnesota, 1994 Buford Avenue, 231 Ruttan Hall, St. Paul, MN 55108-6040, USA
e-mail: kweaster@umn.edu

Q. Huang
Department of Agricultural Economics and Agribusiness, University of Arkansas, 221 Agriculture Building, Fayetteville, AR 72701, USA
e-mail: qqhuang@uark.edu

K.W. Easter and Q. Huang (eds.), *Water Markets for the 21st Century: What Have We Learned?*, Global Issues in Water Policy 11, DOI 10.1007/978-94-017-9081-9_17,

(Chap. 12). Yet several countries have over time adjusted their institutions so that water markets are effective and accepted as a good tool for allocating or reallocating water (see Chaps. 6, 9 and 10). Other areas such as California still need to adjust their institution so that markets can be used more extensively to reallocate water (Chap. 5). This points to one of our key findings that it is not a simple task to develop effective institutions for water markets because as Chap. 2 points out it is a "wicked" problem. Possible third party impacts, including environmental damage and downstream water shortages, need to be addressed before formal water markets are implemented on a large scale especially if water is to be transferred between river basins or watersheds.

This book has the luxury to add the experience of Oman, Australia and China that we did not have in the 1998 water markets book (Easter et al. 1998). Only in the past several decades has data become available to analyze water markets in these three countries. All three provide examples of different types of markets. Two have informal markets, China and Oman, with the ones in Oman centuries old. In Australia the formal markets have developed to reallocate water over a large area in the Murray-Darling River Basin.

17.2 Climate Change and Future Water Supplies

Clearly the demand for clean water has been growing and will continue to grow as population and incomes increase. There will also be a shift in water demands as urbanization continues and diets change with the growth in incomes. On the other side of the equation, water supply will become more variable and uncertain as climate change intensifies (see Chap. 3). Given these pressures on supply and demand there will be increased need to reallocate water and make some critical investments in water storage to help deal with the variability in water supply. For example, Chap. 8 finds that more storage capacity in the water systems in Oman would help them meet critical water demands during dry years. Other countries with extensive irrigation systems will need to determine how they can improve storage capacity and conserve their water supplies.

Water markets can also help rebalance water supply and demand. One market mechanism is to develop water banks or exchange centers where water users can trade water rights, use rights, or allocation rights. Such centers were established in Spain under the 1997 Reform of the Water Law (Chap. 7). They have allowed trading to occur among a few river basins in the southeastern part of Spain. Groundwater trading through water banks are present in Nebraska and California as discussed in Chaps. 5 and 15. They have allowed a limited amount of trading and are still developing. However, if constraints to pumping are not in place water markets can lead to the unintended consequences of increased pumping rates that result in over use of groundwater (see Chaps. 11 and 13). What are needed are tradable permits that specify how much can be pumped such as those in Nebraska (Chap. 15).

Another important market alternative is reliability contracts, which provide the user with the option to buy water during drought periods from a reliable water source for a set price (see Chaps. 3 and 4). Colby et al. in Chap. 4 provide a California example of the use of such contracts to stabilize water supply. They find that new measurement technologies are expanding the opportunity to use reliability contracts for trading water. Rosegrant et al. in Chap. 3 discuss the development of option contracts in Australia to provide more reliable water supplies for urban areas or for farmers with high valued crops.

17.3 Transaction Costs in Market Operations

As discussed in Chap. 2 transaction costs are critical in the success or failure of water markets. Colby et al. in Chap. 4 find that improved measurement technologies are helping to reduce the transaction costs of trading. In Chap. 5 Howitt argues that being willing to address possible third party impacts before water trades are consummated will reduce transaction costs and make option contracts more viable. For example, farmers whose pumping costs might go up because groundwater is being sold to an urban area during a drought period would be guaranteed compensation for any cost increases or crop losses.

In Chap. 7 the authors talk about how the water laws developed in the twentieth century are not adequate for the twenty-first century water demands. These outdated laws have raised the transaction costs such that water trading is primarily restricted to the states in southeastern Spain. In Oman, the village level water markets that have been operating for several centuries have low transaction costs. Yet the development of wells in some areas of Oman have dried up their source of supply and forced many village irrigation systems to close. The informal markets in China also seem to have low transaction costs (Chap. 14) while the more formal markets in Chile have kept their transaction costs low for within basin trades thanks to effective Water User Associations.

17.4 Acceptance and Establishment of Water Markets

In many countries agriculture is the largest user of water. Over 70–80 % of a country's water may be used in agriculture. Thus, for water trading to work farmers need to be convinced that water markets are a good means for allocating water. If farmers don't see how they can benefit from water trading then it is likely to be difficult to set up effective water markets. Consequently a top down approach is not likely to result in much trading. Farmers and other stakeholders need to be involved in building the markets particularly at the local level. Still governments needs to take important steps to make it possible for markets to develop. Governments should change their laws and remove any restrictions on water trading that are outdated or unnecessary. Establishing water rights or use rights or allocation rights that can be registered and traded is a key step (see Chaps. 6 and 9). Water User Associations can serve as important organizers in water markets (Chaps. 6 and 7). They can act as clearing houses for trading and can themselves be a major trader, particularly in interbasin or intersectorial trades.

Water scarcity can be an important force to encourage the use of water markets to reallocate water. Yet concerns about water moving out of the region or that low income and/or small farmers will be disadvantaged have constrained the use of formal water markets in countries such as South Africa and India. In many cases government intervention may be needed to coordinate market participants. This is particularly true in the case of interbasin trades (Chaps. 7 and 10). Chapter 10 talks about early adopters tending to be newer farmers which suggests that long term farmers may need to be a target for education on how water trading can be an important part of a farm management plan and a potential source of investment capital. In many cases stakeholders will find that a well-functioning water market is a better means for allocating water than low paid government officials.

Even though we know that stakeholder participation is important for establishing water markets, we need to know more about these potential participants. We need research such as that in Chap. 12 that studies the characteristics of the potential water market participants and how likely they are to actually participate. The studies should also determine what factors make stakeholders more likely to participate and how this information can be used to increase stakeholder participation. For example, Water User Associations can play an important role in increasing participation.

17.5 Environmental Concerns

It is now pretty clear that environmental impacts need to be taken into account when designing and operating water markets. Environmental demands have increased the need for water markets but they also created new challenges to establish markets for environmental services generated by water. In addition, Chap. 11 illustrates the importance of having good estimates of the economic value of the environmental services provided by water to guide programs such as Australia's water buyback program that aims to increase the flow of the Murray Darling River. It also shows that potential environmental impact must be considered in designing water market institutions and in establishing surface water and groundwater caps on extractions and allocations.

Chapter 15 shows that when the objective is to reduce stream depletion, land retirement schemes (similar to the water buyback program) are more cost-effective when a large reduction is needed. For lower levels of reductions, allowing the trading of pumping permits is a more cost-effective means. This is because a number of wells near the river have very low value of marginal products for water and are the ones most likely to sell their permits.

Chapter 5 highlights the need to develop a more generic definition of the environmental impacts of water transfers out of a watershed or river basin. Once such definitions are established and used at a state or national level transactions costs would drop and water trading would be facilitated. Not having clear definitions of these environmental services make it easier to raise legal barriers to option contracts that delay their execution beyond the point that is beneficial for buyers to exercise their options to transfer water.

Another third party concern that causes both water quantity and quality problems is over drafting of groundwater. Chapter 13 illustrates how informal water markets in India can make these problems worse because groundwater is an open access resource in India and no limits are placed on how much water any given well owner can pump. Different strategies for curbing over pumping need to be tried and evaluated in areas facing such problems. It is not only a problem of setting a limit on pumping but also an enforcement problem. For example, India has tried to meter and charge farmers for the amount of electricity they use. However, something always seems to happen to the meters or the meter doesn't get read. A better option might be a community-based solution with education and village or water user organization monitoring and management.

One question that some economics of groundwater literature asked is whether to manage groundwater at all? A series of studies have found that the gains from management, in many cases, are small in dollar terms and are likely to be outweighed by the costs of regulating groundwater use (e.g. Gisser-Sánchez 1980). However, these studies are of fairly large farms where wells are not close together as they are in India where farms are very small (usually less than a few hectares) and wells quite close together. If the benefit such as gains generated by water market justifies management, one of the research questions that needs to be asked, as we try to develop strategies to limit pumping, is what level of cap should be used. This will require considerable research to determine acceptable levels of pumping in a given area. It will also depend on well locations and their relations to any bodies of surface water, Given that groundwater and surface water in many cases are linked, pumping limits will have to consider the impact of pumping on the surface water (Chap. 15). Chapter 10 also points out that in Australia a number of farmers sold their surface water allocation and used groundwater to replace it. Clearly more hydrologic research is needed to help countries set pumping limits. This will also be needed in nonmarket areas where groundwater is being over used.

Similar research is needed to assist governments in setting caps on surface water allocations (Chaps. 10 and 11). As part of this research we need to develop methods for estimating how much water is likely to be available for allocation over a number of years. To complement this research more work is needed to estimate the likely changes in demand for water from different sectors of the economy. One part of this research could be studies of how opening up of trade (virtual water trade), particularly for intensive water using commodities such as rice, might help reduce the demand for water in water scarce countries. Can production, particularly, of intensive water using commodities shifted to countries or areas within countries that have more abundant water supplies relative to demand?

A different aspect of the environment and water markets needs to be considered. Should countries consider the possibility of pricing water based on its quality and have water market price differences based on water quality? Chapter 7 describes how informal groundwater markets have developed where the price for water depends on the salinity level of the water traded. If water quality pricing is possible then we should estimate the impact high water prices, for high quality water, might have on water pollution. Would high water prices for clean water help reduce water pollution and encourage more water recycling and conservation? High water prices could also encourage firms not to use high quality groundwater for cooling.

17.6 Conclusion

It is not an easy task to set up an effective set of water markets that are well accepted as a means of allocating or reallocating water. Water markets tend to be established in countries where water is scarce and the government organization is fairly effective and operates under a sound legal system. This may help explain why formal water markets have developed in Australia, the U.S. West, Spain and Chile. The big question is can these and other countries facing growing water scarcity use water markets to reallocate water and minimize the negative impacts of water scarcity? The areas that are successful will get an economic growth benefit from the effective reallocation of water. This is likely to be of increasing importance if climate change increases supply variability as many have predicted. As we look at what has worked in countries with water markets we may find as Chap. 11 points out the sequencing of changes may be important especially in preventing third party impacts on the environment or on open access or common property resources (groundwater). We also are likely to find that limiting the number of priority water uses will facilitate the establishment of water markets and improve their effectiveness (Chap. 7). Finally, the Australian experience suggests that separating water and allocation rights can be a key to effective water trading.

References

Easter KW, Rosegrant MW, Dinar A (1998) Markets for water. Potential and performance. Kluwer Academic Publisher, Boston, 298 p

Gisser M, Sánchez DA (1980) Competition versus optimal control in groundwater pumping. Water Resour Res 16(4):638–642

Index

*Only lists first page in chapter where term appears.

K.W. Easter and Q. Huang (eds.), *Water Markets for the 21st Century: What Have We Learned?*, Global Issues in Water Policy 11, DOI 10.1007/978-94-017-9081-9,

The manufacturer's authorised representative in the EU is Springer
Nature Customer Service Centre GmbH, Europaplatz 3, 69115 Heidelberg,
Germany. If you have any concerns regarding our products, please
contact ProductSafety@springernature.com

Printed and bound by CPI Group (UK) Ltd, Croydon, CR0 4YY
30/09/2026
02225445-0010